Concrete Design to EN 1992

Concrete Design to EN 1992

L.H. Martin Bsc, PhD, CEng FICE
J.A. Purkiss Bsc (Eng), PhD

ELSEVIER

AMSTERDAM • BOSTON • HEIDELBERG • LONDON
NEW YORK • OXFORD • PARIS • SAN DIEGO
SAN FRANCISCO • SINGAPORE • SYDNEY • TOKYO

Butterworth-Heinemann is an imprint of Elsevier

Butterworth-Heinemann
An imprint of Elsevier
Linacre House, Jordan Hill, Oxford OX2 8DP, UK
84 Theobald's Road, London WC1X 8RR, UK

First published in Great Britain by Arnold 1996
Reprinted by Butterworth-Heinemann 2000
Second Edition 2006

British Library Cataloguing in Publication Data
A catalogue record for this book is available from the British Library

Library of Congress Cataloging in Publication Data
A catalogue record for this book is available from the Library of Congress

ISBN 13: 978-0-75-065059-5
ISBN 10: 0-75-065059-1

For information on all Butterworth-Heinemann
publications visit our website at http://books.elsevier.com

Typeset by CEPHA Imaging Pvt. Ltd, Bangalore, India
Printed and bound in Great Britain

06 07 08 09 10   9 8 7 6 5 4 3 2 1

Contents

Preface

This book conforms with the latest recommendations for the design of reinforced and prestressed concrete structures as described in Eurocode 2: Design of Concrete Structures – Part 1-1: General rules and rules for buildings. References to relevant clauses of the Code are given where appropriate.

Where necessary the process of design has been aided by the production of design charts.

Whilst it has not been assumed that the reader has a knowledge of structural design, a knowledge of structural mechanics and stress analysis is a prerequisite. The book contains detailed explanations of the principles underlying concrete design and provides references to research where appropriate.

The text should prove useful to students reading for engineering degrees at University especially for design projects. It will also aid designers who require an introduction to the new EuroCode.

For those familiar with current practice, the major changes are:

(1) Calculations may be more extensive and complex.
(2) Design values of steel stresses are increased.
(3) High strength concrete is encompassed by the Eurocode by modifications to the flexural stress block.
(4) There is no component from the concrete when designing shear links for beams. This is not the case for slabs where there is a concrete component.
(5) Bond resistance is more complex.
(6) Calculations for column design are more complicated.
(7) For fire performance, the distance from the exposed face to the centroid of the bar (axis distance) is specified rather than the cover.

NOTE: As this text has been produced before the availability of the National Annex which will amend, if felt appropriate, any Nationally Determined Parameters, the recommended values of such parameters have been used. The one exception is the value of the coefficient α_{cc} allowing for long term effects has been taken as 0,85 (the traditional UK value) rather than the recommended value of 1,0.

Acknowledgements

We would like to thank the British Standards Institute for permission to reproduce extracts from BS EN 1997-1:2004 and BS EN 1991-1-1:2004. British Standards can be obtained from BSI Customer Services, 389 Chiswick High Road, London W4 4AL. Tel: +44(0) 208996 9001. email: cservices@bsi-global.com

	BSI Ref	Book Ref
EN 1997-1	Tables A1/A3	Table 11.1
	Table A4	Table 11.2
	Table A13	Table 11.3
EN 1991-1-1	Table 3.1	Annex A
	Table 7.4N	Table 5.2
	Table 4.1	Table 5.1
	Fig 5.3	Fig 6.10
	Fig 9.13	Fig 11.6
	Fig 9.9	Fig 11.3

Principal Symbols

Listed below are the symbols and suffixes common to Eurocodes

LATIN UPPER AND LOWER CASE

A	accidental action; area
a	distance; shear span
B	width
b	width
D	depth; diameter of mandrel
d	effective depth
E	modulus of elasticity
e	eccentricity
F	action; force
f	strength of a material; stress
G	permanent action; shear modulus
H	total horizontal load or reaction
h	height
I	second moment of area
i	radius of gyration
k	depth factor for shear resistance
L	length; span
l	buckling length; bond length
M	bending moment
N	axial force
n	number
P	prestressing force
Q	variable action
q	uniformly distributed force
R	resistance; reaction; low strength steel
r	radius
s	spacing of links
T	torsional moment; high strength steel
t	thickness
u	perimeter
V	shear force

v	shear stress
x	neutral axis depth
W	load
w	load per unit length
Z	section modulus
z	lever arm

GREEK LOWER CASE

α	coefficient of linear thermal expansion; angle of link; ratio; bond factor
β	angle; ratio; factor
γ	partial safety factor
∂	deflection; deformation
ε	strain
η	strength factor
θ	angle of compression strut; slope
λ	slenderness ratio; ratio
μ	coefficient of friction
ν	strength reduction factor for concrete
ρ	unit mass; reinforcement ratio
σ	normal stress; standard deviation
τ	shear stress
ϕ	rotation; slope; ratio; diameter of a reinforcing bar; creep coefficient
φ	factors defining representative values of variable actions

SUFFIXES

b	bond
c	concrete
d	design value
ef	effective
f	flange
i	initial
k	characteristic
l	longitudinal
lim	limit
m	mean
max	maximum
min	minimum
o	original
p	prestress

R	resistance
req	required
s	steel
sw	self weight
t	time; tensile; transfer
u	ultimate
v	shear
w	web; wires; shear reinforcement
y	yield
x, y, z	axes

1.1 DESCRIPTION OF CONCRETE STRUCTURES

1.1.1 Types of Load Bearing Concrete Structures

The development of reinforced concrete, circa 1900, provided an additional building material to stone, brick, timber, wrought iron and cast iron. The advantages of reinforced concrete are cheapness of aggregates, flexibility of form, durability and low maintenance. A disadvantage is greater self weight as compared with steel, timber or aluminium.

Reinforced concrete structures include low rise and high rise buildings, bridges, towers, floors, foundations etc. The structures are essentially composed of load bearing frames and members which resist the actions imposed on the structure, e.g. self weight, dead loads and external imposed loads (wind, snow, traffic etc.).

Structures with load bearing frames may be classified as:

(a) Miscellaneous isolated simple structural elements, e.g. beams and columns or simple groups of elements, e.g. floors.
(b) Bridgeworks.
(c) Single storey factory units, e.g. portal frames.
(d) Multi-storey units, e.g. tower blocks.
(e) Shear walls, foundations and retaining walls.

Two typical load bearing frames are shown in Fig. 1.1. When subject to lateral loading, some frames (Fig. 1.1(a)) deflect and are called sway, or unbraced, frames. Others (Fig. 1.1(b)) however are stiffened, e.g. by a lift shaft and are called non-sway, or braced, frames. This distinction is important when analysing frames as shown in Chapters 4 and 9. For analysis purposes, frames are idealized and shown as a series of centre lines (Fig. 1.1(c)). Elements of a structure are defined in cl. 5.3.1, EN.

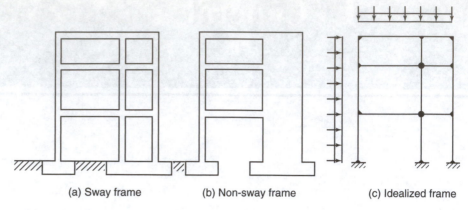

(a) Sway frame   (b) Non-sway frame   (c) Idealized frame

FIGURE 1.1   Typical load bearing frames

1.1.2 Load Bearing Members

A load bearing frame is composed of load bearing members (or elements) e.g. beams and columns. Structural elements are required to resist forces and displacements in a variety of ways, and may act in tension, compression, flexure, shear, torsion, or in any combination of these forces. The structural behaviour of a reinforced concrete element depends on the nature of the forces, the length and shape of the cross section of the member, elastic and plastic properties of the materials, yield strength of the steel, crushing strength of the concrete and crack widths. Modes of behaviour of structural elements are considered in detail in the following chapters.

A particular advantage of reinforced concrete is that a variety of reinforced concrete sections (Fig. 1.2) can be manufactured to resist combinations of forces (actions).

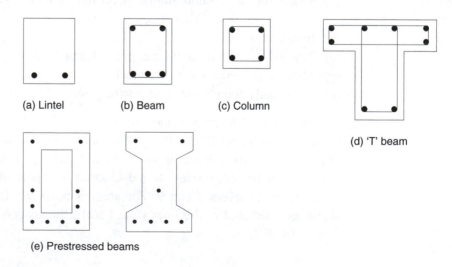

(a) Lintel   (b) Beam   (c) Column   (d) 'T' beam

(e) Prestressed beams

FIGURE 1.2   Typical reinforced concrete sections

The optimization of costs in reinforced construction favours the repeated use of moulds to produce reinforced concrete elements with the same cross sections and lengths. These members are common for use in floors either reinforced or prestressed. The particular advantage of precast concrete units is that the concrete is matured and will take the full loading at erection. The disadvantage, in some situations, is the difficulty of making connections.

1.1.3 Connections

The structural elements are made to act as a frame by connections. For reinforced concrete, these are composed of reinforcing bars bent, lapped or welded which are arranged to resist the forces involved. A connection may be subject to any combination of axial force, shear force and bending moment in relation to three perpendicular axes, but for simplicity, where appropriate, the situation is reduced to forces in one plane. The transfer of forces through the components of a connection is often complex and Chapter 8 contains explanations, research references and typical design examples.

There are other types of joints in structures which are not structural connections. For example, a movement joint is introduced into a structure to take up the free expansion and contraction that may occur on either side of the joint due to temperature, shrinkage, expansion, creep, settlement, etc. These joints may be detailed to be watertight but do not generally transmit forces. Detailed recommendations are given by Alexander and Lawson (1981). A construction joint is introduced because components are manufactured to a convenient size for transportation and need to be connected together on site. In some cases these joints transmit forces but in other situations may only need to be waterproof.

1.2 DEVELOPMENT AND MANUFACTURE OF REINFORCED CONCRETE

1.2.1 Outline of the Development of Reinforced Concrete

Concrete is a mixture of cement, fine aggregate, coarse aggregate and water. It is thus a relative cheap structural material but it took years to develop the binding material which is the cement. Modern cement is a mixture of calcareous (limestone or chalk) and argillaceous (clay) material burnt at a clinkering temperature and ground to a fine powder.

The Romans used lime mortar and added crushed stone or tiles to form a weak concrete. Lime mortar does not harden under water and the roman solution was to grind together lime and volcanic ash. This produced pozzolanic cement named after a village Pozzuoli near Mount Vesuvius.

The impetus to improve cement was the industrial revolution of the eighteenth century. John Smeaton, when building Eddystone lighthouse, made experiments with various mixes of lime, clay and pozzolana. In 1824, a patent was granted to Joseph Aspin for a cement but he did not specify the proportions of limestone and clay, nor did he fire it to such a high temperature as modern cement. However it was used extensively by Brunel in 1828 for the Thames tunnel. Modern cement is based on the work of Johnson who was granted a patent in 1872. He experimented with different mixes of clay and chalk, fired at different temperatures and ground the resulting clinker.

The mass production of cement was made possible by the invention of the rotary kiln by Crampton in 1877. The control, quality and reliability of the cement improved over the years and the proportions of fine and coarse aggregate required for strong durable concrete have been optimized. This lead to the development of prestressed concrete using high strength concrete. Most reinforced concrete construction uses timber as shuttering but where members are repeated, e.g. floor beams, steel is preferred because it is more robust.

Wilkinson in 1854 took out a patent for concrete reinforced with wire rope. Early investigators were concerned with end anchorage but eventually continuous bond between the steel and the concrete was considered adequate. Later, round mild steel bars with end hooks were standardized with a yield strength of approximately 200 MPa which gradually increased until today 500 MPa is common. From 1855 reinforced concrete research, construction and theory was developed, notably in Germany, and results were published by Morsch in 1902.

To overcome the problem of cracking at service load, prestressed concrete was developed by Freyssinet in 1928. Mild steel was not suitable for prestressed concrete because all the prestress was lost but high strength piano wires (700 MPa) were found to be suitable. Initially the post-tensioned system was adopted where the concrete was cast, allowed to mature and the wires, free to move in ducts, were stressed and anchored at the ends of the member. Later the pretensioned system was developed where spaced single wires were prestressed between end frames and the concrete cast round them. When the concrete had matured, the stress was released and the force transferred to the concrete by bond between the steel and concrete.

1.2.2 Modern Method of Concrete Production

The ingredients of concrete, i.e. water, cement, fine aggregate and coarse aggregate are mixed to satisfy the requirements of strength, workability, durability and economy. The water content is the most important factor which influences the workability of the mix, while the water/cement ratio influences the strength and durability. The cement paste (water and cement) should fill the voids in the fine aggregate and the mortar (water, cement and sand) should fill the voids in the coarse aggregate. To optimize this process, the aggregates are graded. Further understanding of the process can be obtained from methods of mix design (Teychenne *et al.*, 1975).

The cement production is carefully controlled and tested to produce a consistent product. The ingredients (lime, silica, alumina and iron) are mixed (puddled) and fed into a rotary kiln in a continuous process to produce a finely ground clinker. The chemistry of cement is complicated and not fully understood. For practical purposes the end product is recognized as either ordinary Portland cement or rapid-hardening Portland cement.

Fine aggregates (sand) are less than 5 mm, while coarse aggregates (crushed stone) range from 5 to 40 mm depending on the dimensions of the member. The aggregates are generally excavated from quarries, washed, mixed to form concrete and, for large projects, transported to the construction site. For small projects, the concrete may be mixed on site from stockpiled materials. Some manufactured aggregates are available, including light-weight aggregates, but they are not often used.

1.3 DEVELOPMENT AND MANUFACTURE OF STEEL

Steel was first produced in 1740 but was not available in large quantities until Bessemer invented the converter in 1856. By 1840, standard shapes in wrought iron were in regular production. Gradually wrought iron was refined to control its composition and remove impurities to produce steel. Further information on the history of steel making can be found in Buchanan (1972); Cossons (1975); Derry (1960); Pannel (1964) and Rolt (1970).

Currently there are two methods of steel making:

(a) Basic Oxygen Steelmaking (BOS) used for large scale production. Iron ore is smelted in a blast furnace to produce pig iron. The pig iron is transferred to a converter where it is blasted with oxygen and impurities are removed to produce steel. Some scrap metal may be used.
(b) Electric Arc Furnace (EAF) used for small scale production. The method, uses scrap metal almost entirely which is fed into the furnace and heated by means of an electrical discharge from carbon electrodes. Little refining is required.

In both processes carbon, manganese and silicon in particular, are controlled. Other elements which may affect coldworking, weldability and bendability may also be limited. BOS steel has lower levels of sulphur, phosphorous and nitrogen while EAF steel has higher levels of copper, nickel and tin.

Traditionally steel was cast into ingots but impurities segregated to the top of the ingot which had to be removed. Ingots have been replaced by a continuous casting process to produce a water cooled billet. The billet is reheated to a temperature of approximately 1150°C and progressively rolled through a mill to reduce its section. This process also removes defects, homogenizes the steel and forms ribs to improve the bond. With the addition of tempering by controlled quenching it is possible to produce a relatively soft ductile core with a hard surface layer (Cares, 2004).

1.4 STRUCTURAL DESIGN

1.4.1 Initiation of a Design

The demand for a structure originates with the client. The client may be a private person, private or public firm, local or national government or a nationalized industry.

In the first stage, preliminary drawings and estimates of costs are produced, followed by consideration of which structural materials to use, i.e. reinforced concrete, steel, timber, brickwork, etc. If the structure is a building, an architect only may be involved at this stage, but if the structure is a bridge or industrial building then a civil or structural engineer prepares the documents.

If the client is satisfied with the layout and estimated costs then detailed design calculations, drawings and costs are prepared and incorporated in a legal contract document. The design documents should be adequate to detail, fabricate and erect the structure.

The contract document is usually prepared by the consultant engineer and work is carried out by a contractor who is supervised by the consultant engineer. However, larger firms, local and national government and nationalized industries, generally employ their own consultant engineer.

The work is generally carried out by a contractor, but alternatively direct labour may be used. A further alternative is for the contractor to produce a design and construct package, where the contractor is responsible for all parts and stages of the work.

1.4.2 The Object of Structural Design (cl. 2.1 EN, Section 2 EN 1990)

The object of structural design is to produce a structure that will not become unserviceable or collapse in its lifetime, and which fulfils the requirements of the client and user at reasonable cost.

The requirements of the client and user may include the following:

(a) The structure should not collapse locally or overall.
(b) It should not be so flexible that deformations under load are unsightly or alarming, or cause damage to the internal partitions and fixtures; neither should any movement due to live loads, such as wind, cause discomfort or alarm to the occupants/users.
(c) It should not require excessive repair or maintenance due to accidental overload or because of the action of weather.
(d) In the case of a building, the structure should be sufficiently fire resistant to, give the occupants time to escape, enable the fire brigade to fight the fire in safety, and to restrict the spread of fire to adjacent structures.
(e) The working life of the structure should be acceptable (generally varies between 10 and 120 years).

The designer should be conscious of the costs involved which include:

(a) Initial cost which includes fees, site preparation, cost of materials and construction.
(b) Maintenance costs, e.g. decoration and structural repair.
(c) Insurance chiefly against fire damage.
(d) Eventual demolition.

It is the responsibility of the structural engineer to design a structure that is safe and which conforms to the requirements of the local bye-laws and building regulations. Information and methods of design are obtained from Standards and Codes of Practice and these are "deemed to satisfy" the local bye-laws and building regulations. In exceptional circumstances, e.g. the use of methods validated by research or testing, an alternative design may be accepted.

A structural engineer is expected to keep up to date with the latest research information. In the event of a collapse or malfunction where it can be shown that the engineer has failed to reasonably anticipate the cause or action leading to collapse or has failed to apply properly the information at his disposal, i.e. Codes of Practice, British Standards, Building Regulations, research or information supplied by the manufactures, then he may be sued for professional negligence. Consultants and contractors carry liability insurance to mitigate the effects of such legal action.

1.4.3 Statistical Basis for Design

When a material such as concrete is manufactured to a specified mean strength, tests on samples show that the actual strength deviates from the mean strength to a varying degree depending on how closely the process is controlled and on the

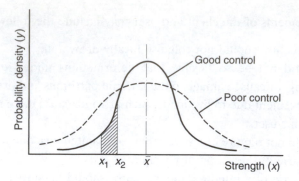

FIGURE 1.3   Normal distribution curve

variation in strength of the component materials. It is found that the spread of results approximates to a normal distribution curve, as shown in Fig. 1.3.

The probability of a test result falling between two values of strength such as x_1 and x_2 is given by the area under the curve between the two values, i.e. the shaded area in the figure. The area under the whole curve is thus equal to unity.

The effect of quality control is as shown, poor control producing a flatter curve. The probability of a test falling below a specified value is clearly greater when the quality of control is reduced.

The equation of the normal distribution curve is

$$y = 1/[\sigma(2\pi)^{0.5}]\exp[-(x - \bar{x})^2/(2\sigma^2)] \tag{1.1}$$

which shows that the curve is fully defined by the mean $\bar{x}$ and the standard deviation σ of the variable x.

For a set of n values of x, the mean is given by

$$\bar{x} = \Sigma x/n \tag{1.2}$$

The standard deviation which is a measure of the dispersion, is the root mean square of the deviations of x from the mean, given by

$$\sigma = [\Sigma(x - \bar{x})^2/n]^{0.5} \tag{1.3}$$

In practice, it is not usually possible to obtain all of the values of x that would theoretically be available. For example, it would not be possible to test all the concrete in a structure and it is therefore necessary to obtain estimates of x and σ by sampling. In this case, the best estimate of the mean is still that given by Eq. (1.2), but the best estimation for the standard deviation is given by

$$\sigma = [\Sigma(x - \bar{x})^2/(n - 1)]^{0.5} \tag{1.4}$$

where n is the number of test results in the sample.

For hand calculations, a more convenient form of Eq. (1.4) is

$$\sigma = [(\Sigma x^2 - n\bar{x}^2)/(n-1)]^{0,5} \qquad (1.5)$$

Statistical distributions can also be obtained to show the variation in strength of other structural materials such as steel reinforcement and prestressing tendons. It is also reasonable to presume that if sufficient statistical data were available, distributions could be defined for the loads carried by a structure. It follows that it is impossible to predict with certainty that the strength of a structural member will always be greater than the load applied to it or that failure will not occur in some other way during the life of the structure. The philosophy of limit state design is to establish limits, based on statistical data, experimental results and engineering experience and judgement, that will ensure an acceptably low probability of failure. At present there is insufficient information to enable distributions of all the structural variables to be defined and it is unlikely that is will ever be possible to formulate general rules for the construction of a statistical model of anything so complicated as an actual structure.

1.4.4 Limit State Design (cl. 2.2 EN)

It is self evident that a structure should be 'safe' during its lifetime, i.e. free from the risk of collapse. There are, however, other risks associated with a structure and the term safe is now replaced by the term 'serviceable'. A structure should not during its lifetime become 'unserviceable', i.e. it should be free from risk of collapse, rapid deterioration, fire, cracking, excessive deflection etc.

Ideally it should be possible to calculate mathematically the risk involved in structural safety based on the variation in strengths of the material and variation in the loads. Reports, such as the CIRIA Report 63, have introduced the designer to elegant and powerful concept of 'structural reliability'. Methods have been devised whereby engineering judgement and experience can be combined with statistical analysis for the rational computation of partial safety factors in codes of practice. However, in the absence of complete understanding and data concerning aspects of structural behaviour, absolute values of reliability cannot be determined.

It is not practical, nor is it economically possible, to design a structure that will never fail. It is always possible that the structure will contain material that is less than the required strength or that it will be subject to loads greater than the design loads.

It is therefore accepted that 5 per cent of the material in a structure is below the design strength, and that 5 per cent of the applied loads are greater than the design loads. This does not mean therefore that collapse is inevitable, because it is extremely unlikely that the weak material and overloading will combine simultaneously to produce collapse.

The philosophy and objectives must be translated into a tangible form using calculations. A structure should be designed to be safe under all conditions of its useful life and to ensure that this is accomplished certain distinct performance requirements, called 'limit states', have been identified. The method of limit state design recognizes the variability of loads, materials construction methods and approximations in the theory and calculations (BS EN 1990).

Limit states may be at any stage of the life of a structure, or at any stage of loading. The limit states which are important for the design of reinforced concrete are at ultimate and serviceability (cl 2.2 EN). Calculations for limit states involve loads and load factors (Chapter 3), and material factors and strengths (Chapter 2).

Stability, an ultimate limit state, is the ability of a structure or part of a structure, to resist overturning, overall failure and sway. Calculations should consider the worst realistic combination of loads at all stages of construction.

All structures, and parts of structures, should be capable of resisting sway forces, e.g. by the use of bracing, 'rigid' joints or shear walls. Sway forces arise from horizontal loads, e.g. winds, and also from practical imperfections, e.g. lack of verticality.

Also involved in limit state design is the concept of structural integrity. Essentially this means that the structure should be tied together as a whole, but if damage occurs, it should be localized. This was illustrated in a tower block (Canning Town Report, 1968) when a gas explosion in one flat caused the progressive collapse of other flats on one corner of the building.

Deflection is a serviceability limit state. Deflections should not impair the efficiency of a structure or its components, nor cause damage to the finishes. Generally the worst realistic combination of unfactored imposed loads is used to calculate elastic deflections. These values are compared with limit states of deformation (cl. 7.4, EN).

Dynamic effects to be considered at the serviceability limit state are vibrations caused by machines, and oscillations caused by harmonic resonance, e.g. wind gusts on buildings. The natural frequency of the building should be different from the exciting source to avoid resonance.

Fortunately there are few structural failures and when they do occur they are often associated with human error involved in design calculations, or construction, or in the use of the structure.

1.4.5 Errors

The consequences of an error in structural design can lead to loss of life and damage to property and it is necessary to appreciate where errors can occur. Small errors in design calculations can occur in the rounding off of figures but these generally do not lead to failures. The common sense advice is that the accuracy of the calculation should match the accuracy of the values given in the European Code.

Errors that can occur in structural design calculations are:

(1) Ignorance of the physical behaviour of the structure under load which introduces errors in the basic theoretical assumptions.
(2) Errors in estimating the loads, especially the erection forces.
(3) Numerical errors in the calculations. These should be eliminated by checking, but when speed is paramount, checks are often ignored.
(4) Ignorance of the significance of certain effects, e.g. creep, fatigue, etc.
(5) Introduction of new materials or methods, which have not been tested.
(6) Insufficient allowance for tolerances or temperature strains.
(7) Insufficient information, e.g. in erection procedures.

Errors that can occur on construction sites are:

(1) Using the wrong number, diameter, cover and spacing of bars.
(2) Using the wrong or poor mix of concrete.
(3) Errors in manufacture, e.g. holes in the wrong position.

Errors that occur in the life of a structure and also affect safety are:

(1) Overloading.
(2) Removal of structural material, e.g. to insert service ducts.
(3) Poor maintenance.

1.5 PRODUCTION OF REINFORCED CONCRETE STRUCTURES

1.5.1 Drawings

Detailed design calculations are essential for any reinforced concrete design but the sizes of the members, dimensions and geometrical arrangement are usually presented as drawings. The drawings are used by the contractor on site, but if there are precast units, the drawing for these may be required for a subcontractor. General arrangement drawings are often drawn to scale of 1:100, while details are drawn to a scale of 1:20 or 1:10. Special details are drawn to larger scales where necessary.

Drawing should be easy to read and should not include superfluous detail. Some important notes are:

(a) Members and components should be identified by logically related mark numbers, e.g. related to the grid system used in the drawings.
(b) The main members should be presented by a bold outline (0,4 mm wide) and dimension lines should be unobtrusive (0,1 mm wide).

(c) Dimensions should be related to centre lines, or from one end; strings of dimensions should be avoided. Dimensions should appear once only so that ambiguity cannot arise when revisions occur. Fabricators should not be put in the position of having to do arithmetic in order to obtain an essential dimension.

(d) Tolerances for erection purposes should be clearly shown.

(e) The grade of steel and concrete to be used should be clearly indicated.

(f) Detailing should take account of possible variations due to fabrication.

(g) Keep the design and construction as simple as possible and avoid changes in section along the length of a member.

(h) Site access, transport and available cranage should be considered.

Reinforcement drawings are prepared primarily for the steel fixer and should conform to the standard method of detailing reinforced concrete (Conc. Soc. and I.S.E report, 1973). The main points to be noted are:

(1) The outline of the concrete should be slightly finer than the reinforcement.

(2) In walls, slabs and columns a series of bars of a particular mark is indicated by the end bars and only one bar in the series is shown in full; the other bar is shown as a short line. In Fig. 1.4(a), for example, the reinforcement consists of two series of bars forming a rectangular grid in plan.

(3) Each series is identified by a code which has the following form: Number, type, diameter-bar mark-spacing, comment. For example in Fig. 1.4(a) 25T20-1-250 indicates that the series contains 25 bars in high yields steel (T), 20 mm diameter, with bar mark 1, spaced at 250 mm between centres. The complete code is used only once for a particular series and wherever possible should appear in the plan or elevation of the member. In sections the bars are identified simply by their bar mark; and again only the end bars need to be shown.

(4) Since the bars are delivered on site bent to the correct radii, bends do not need to be detailed and may be drawn as sharp angles.

(5) Dimensions should be in mm rounded to a multiple of 5 mm. There is no need to write mm on the drawing.

(6) Except with the case of very simple structures the dimensions should refer only to the reinforcement and should be given when the steel fixer could not reasonably be expected to locate the reinforcement properly without them. In many cases, since the bars are supplied to the correct length and shape, no dimensions are necessary, e.g. Fig. 1.4(a). Dimensions should be given from some existing reference point, preferably the face of concrete already been cast. For example, in Fig. 1.4(b) the dimension to the first link is given from the surface of the kicker, which would have been cast with the floor slab and is used to locate the column to be constructed. The formwork for the column is attached to the kicker and the vertical reinforcement starts from the kicker.

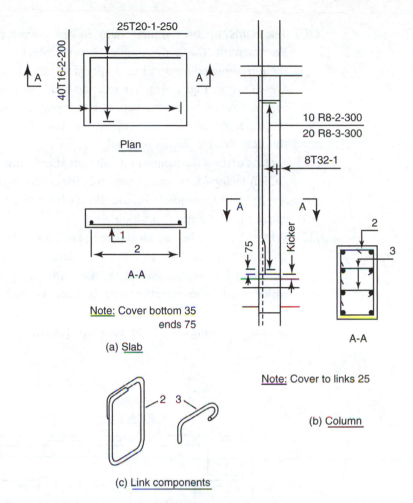

FIGURE 1.4 Reinforcement of slab and column

(7) Cover to the ends and sides of the bars is usually given in the form of notes on the drawing.

(8) Only part of the adjoining members is shown and in sections, only the outlines of the concrete which has been cut through are shown; any concrete beyond the section is generally omitted. The reinforcement in adjoining members is also omitted except where it is necessary to show the relative positions of intersecting or lapping bars. In such cases it is shown as broken lines. In Fig. 1.4(b), for example, it can be seen that the reinforcement of the lower column is to be left protruding from the kicker to form the lap with the reinforcement of the column to be constructed. Bars with diameters less than or equal to 12 mm need not be cranked at the laps.

(9) Sections are drawn to show the relative positions of bars and the shape of links. In the case of beams and columns they are usually drawn to a larger scale than the elevation.

(10) The ends of overlapping bars in the same plane are shown as ticks. For example the conventional representation of the links is shown in the section of the column in Fig. 1.4(b). The actual shapes of the link components are shown in Fig. 1.4(c). There is no really any need to indicate overlapping in links except when the separate components are to be fixed on site, as in this case. Where links are supplied in one piece, as for the beam in Fig. 1.5, the ticks are frequently omitted.

(11) Section arrows for beams and columns should always be in the same direction, i.e. left facing for beams, downwards for columns. The section letters (numbers are not recommended) should always be between the arrows and should be written in the upright position.

(12) When detailing beams all the bars are shown in full. In elevations, the start and finish of the bars in the same place are indicated by ticks. The start is indicated by the full bar code, the end simply by the bar mark (Fig. 1.5) which shows the elevation and sections of part of a beam. Note that the reinforcement in the adjoining column and integral floor slab is not shown; separate drawings would be provided for these. Note also that the dimension

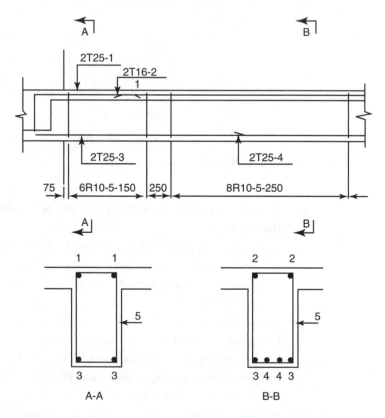

FIGURE 1.5   Reinforcement of a beam

of the start of the first series of links is given from the face of the column already constructed and that a dimension is given between each series of links.

(13) Although the reinforcement in adjoining members is not shown, it is important for the designer to ensure that it will not obstruct the reinforcement of the detailed member. Thus it is especially important in beam–column construction to ensure that the reinforcements of a beam and a column do not intersect in the same plane.

1.5.2 Bar Bending Schedules

Scheduling, dimensioning, bending and cutting of steel reinforcement for concrete is given in BS 8666 (2000). Bars should be designed to have as few bends as possible and should conform to the preferred shapes.

Work on the reinforcement, i.e. cutting, bending and fixing in place before the concrete is placed, is usually dealt with by specialist sub-contractors. Information on reinforcement is conveyed from the engineer to the sub-contractor by use of a bar bending schedule, which is a table which provides the following information about each bar.

Member. A reference identifying a particular structural member, or group of identical members.

Bar mark. An identifying number which is unique to each bar in the schedule.

Type and size. A code letter: 'T' for high yield steel, 'R' for mild steel following by size of bar in mm, e.g. R16 denotes a mild steel bar 16 mm in diameter.

Number of members. The number of identical members in a group.

Number of bars in each.

Total number.

Length of each bar (mm).

Shape code.

Dimensions required for bending. Five columns specifying the standard dimensions corresponding to the particular shape code. These dimensions contain allowances for tolerances. If a bar does not conform to one of the preferred shapes then a dimensioned drawing is supplied.

Bars are delivered to the site in bundles, each of which is labelled with the reference number of the bar schedule and the bar mark. These two numbers uniquely identify every bar in the structure.

1.5.3 Tolerances (cl. 4.4.1.3, EN)

Tolerance are limits placed on unintentional inaccuracies that occur in dimension which must be allowed for in design if structural elements and components are to resist forces and remain durable.

The formwork and falsework should be sufficiently stiff to ensure that the tolerance for the structure, as stipulated by the designer, are satisfied. Tolerance for members are related to controlling the error in the size of section, and vary from $\pm 5\,\text{mm}$ for a 150 mm section to $\pm 30\,\text{mm}$ for a 2500 mm section. Tolerances for position of prestressing tendons are related to the width of depth of a section.

The tolerances associated with concrete cover to reinforcement are important to maintain structural integrity and to resistance corrosion. The concrete cover is the distance between the surface of the reinforcement closest to the nearest concrete surface. The nominal cover is specified in drawings, which is the minimum cover $c_{\min}$ (cl 4.4.1.2, EN) plus an allowance for design deviation Δc_{dev} (cl 4.4.1.3, EN). The usual value of $\Delta c_{\text{dev}} = 10\,\text{mm}$.

1.6 SITE CONDITIONS

1.6.1 General

The drawings produced by the structural designer are used by the contractor on site. Most of the reinforced concrete is cast in situ, on site, and transport and access is generally no problem, except for the basic materials. Most concrete is ready mix obtained from specialist suppliers. This avoids having to provide space for storing cement and aggregates and ensures high quality concrete. Prefabricated products, e.g. floor beams are delivered by road and erected by crane. Most large sites have a crane available. On site, the general contractor is responsible for the assembly, erection, connections, alignment and leveling of the complete structure. During assembly on site it is inevitable that some components will not fit, despite the tolerances that have been allowed. The correction of some faults and the consequent litigation can be expensive.

1.6.2 Construction Rules

To ensure that the structure is constructed as specified by the designers, construction rules are introduced. For concrete, these cover quality of concrete, formwork, surface finish, temporary works, removal of formwork and falsework. For steel-work, the rules cover transport, storage, fabrication, welding, joints and placing. For prestressed concrete there are additional rules for tensioning, grouting and sealing.

1.6.3 Quality Control

Quality control is necessary to ensure that the construction work is carried out to the required standards and rules. Control covers the quality of materials, standard of workmanship and the quality of the components. For materials, it includes the mix of concrete, storage, handling, cutting, welding, prestressing forces and grouting (BS EN 206-1, 2000).

REFERENCES AND FURTHER READING

Alexander, S.J. and Lawson, R.M. (1981) Movement design in buildings, Technical Note 107, CIRIA.

BS EN 206-1 (2000) Concrete Pt.1: Specification, performance, production and conformity, BSI.

BS EN 1990 (2002) Eurocode – Basis of structural design.

BS 8666 (2000) Scheduling dimensioning, bending and cutting of steel reinformcement for concrete, BSI.

Buchanan, R.A. (1972) *Industrial Archeology in Britain*, Penguin.

Canning Town. Min. of Hous. and Local Gov. (1968) *Ronan Point report of the enquiry into the collapse of flats at Ronan Point*.

Cares (2004) Pt.2. *Manufacturing process routes for reinforcing steels*, UK Cares.

CIRIA Report 63. (1977) *Rationalisation of safety and serviceability factors in structural codes*.

Conc. Soc. and I.S.E. (1973) *Standard methods of detailing reinforced concrete*.

Cossons, N. (1975) *The BP Book of Industrial Archeology*.

Derry, T.K. and Williams, T.I. (1960) *A Short History of Technology*, Oxford University Press.

Pannel, J.P.M. (1964) *An illustrated History of Civil Engineering*, Thames and Hudson.

Rolt L.T.C. (1970) *Victorian Engineering*, Penguin.

Teychenne, D.C., Franklin, R.E. and Erntroy, H.C. (1975) Design of normal concrete mixes, Dept. of the Env.

Chapter 2 / Mechanical Properties of Reinforced Concrete

2.1 Variation of Material Properties

The properties of manufactured materials vary because the particles of the material are not uniform and because of inconsistencies in the manufacturing process which are dependent on the degree of control. These variations must be recognized and incorporated into the design process.

For reinforced concrete, the material property that is of most importance is strength. If a number of samples are tested for strength, and the number of specimens with the same strength (frequency) plotted against the strength, then the results approximately fit a normal distribution curve (Fig. 2.1). The equation shown in Fig. 2.1 defines the curve mathematically and can be used to define 'safe' values for design purposes.

2.2 Characteristic Strength

A strength to be used as a basis for design must be selected from the variation in values (Fig. 2.1). This strength, when defined, is called the characteristic strength. If the characteristic strength is defined as the mean strength, then 50 per cent of the material is below this value and this is unsafe. Ideally the characteristic strength should include 100 per cent of the samples, but this is impractical because it is a low value and results in heavy and costly structures. A risk is therefore accepted and it is therefore recognized that 5 per cent of the samples for strength fall below the characteristic strength. The characteristic strength is calculated from the equation

$$f_{ck} = f_{mean} - 1{,}64\,\sigma \tag{2.1a}$$

where for n samples the standard deviation

$$\sigma = [\Sigma(f_{mean} - f)^2/(n-1)]^{0,5} \tag{2.1b}$$

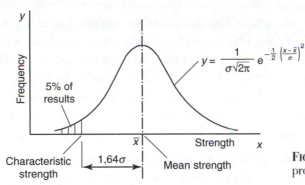

FIGURE 2.1 Variation in material properties

2.2.1 Characteristic Strength of Concrete (cl 3.1.2, EN)

Concrete is a composite material which consists of coarse aggregate, fine aggregate and a binding paste mixture of cement and water. Mix design selects the optimum proportions of these materials, but a simple way to assess the quality of concrete as it matures is to crush a standard cube or cylinder.

The mean strength which must be achieved in the mixing process can be determined if the standard deviation is known from previous experience. For example, to produce a concrete with a characteristic strength of 30 MPa in a plant for which a standard deviation of 5 MPa is expected, required mean strength $= 30 + 5 \times 1{,}64 = 38{,}2$ MPa.

The strength of concrete increases with age and it is necessary to adopt a time after casting as a standard. The characteristic compressive strength for concrete is the 28-day cylinder or cube strength, i.e. the crushing strength of a standard cylinder, or cube, cured under standard conditions for 28 days as described in BS EN 12390-1 (2000). The 28-day strength is approximately 80 per cent of the strength at one year, after which there is very little increase in strength. Strengths higher than the 28-day strength are not used for design unless there is evidence to justify the higher strength for a particular concrete.

The characteristic tensile strength of concrete is the uniaxial tensile strength ($f_{ct,ax}$) but in practice this value is difficult to obtain and the split cylinder strength ($f_{ct,sp}$) is more often used. The relation between the two values is ($f_{ct,ax}) = 0{,}9$ ($f_{ct,sp}$). In the absence of test values of the tensile strength it may be assumed that the mean value of the tensile strength ($f_{ctm}) = 0{,}3(f_{ck})^{2/3}$, where f_{ck} is the characteristic cylinder compressive strength. In situations where the tensile strength of concrete is critical, e.g. shear resistance, the mean value may be unsafe and a 5 per cent fractile ($f_{ctk0,05} = 0{,}7f_{ctm}$) is used. In other situations where the tensile strength is not critical, a 95 per cent fractile ($f_{ctk0,95} = 1{,}3f_{ctm}$) is used. These values are given in Table 3.1, EN (Annex A2).

BS EN 206-1 (2000) specifies the mix, transportation, sampling and testing of concrete. Concrete mixes are either prescribed, i.e. specified by mix proportions, or designed, i.e. specified by characteristic strength. For example, Grade C30P denotes a prescribed mix which would normally give a strength of 30 MPa; Grade C25/30 denotes a designed mix for which a cylinder strength of 25 MPa or cube strength of 30 MPa is guaranteed. The grades recommended by EN are from 12/15 to 50/60 in steps of approximately 5 MPa for normal weight aggregates. The lowest Grades recommended for prestressed concrete are C30 for post-tensioning and C40 for pretensioning.

Although the definition of characteristic strength theoretically allows that a random test result could have any value, however low, a practical specification for concrete in accordance with BS EN 206-1 (2000) would require both of the following conditions for compliance with the characteristic strength, thus excluding very low values:

(a) the average strength determined from any group of four consecutive test results exceeds the specified characteristic strength by 3 MPa for concretes of Grade C16/20 and higher, or 2 MPa for concretes of Grade C7,5–C15;
(b) The strength determined from any test result is not less than the specified characteristic strength minus 3 MPa for concretes of Grade C20 and above, or 2 MPa for concretes of Grade C7,5–C15.

Concrete not complying with the above conditions should be rejected.

2.2.2 Characteristic Strength of Steel (cls 3.2.3 and 3.3.3, EN)

The characteristic axial tensile yield strength of steel reinforcement (f_{yk}) is the yield stress for hot rolled steel and 0,2 per cent proof stress for steel with no pronounced yield stress. The value recommended in the European Code is: High yield steel (hot rolled or cold worked) 500 MPa. For prestressing tendons, the characteristic strength (f_{pk}), at 0.1 per cent proof stress, varies from 1000 to 2000 MPa depending on the size of tendon and type of steel.

2.3 DESIGN STRENGTH

The design strength allows for the reduction in strength between the laboratory and site. Laboratory samples of the material are processed under strictly controlled standard conditions. Conditions on site vary and in the case of concrete: segregation can occur while it is being transported; conditions for casting and compaction differ; there may be contamination by rain; and curing conditions vary especially in hot or cold weather.

The design strength of a material is therefore lower than the characteristic strength, and is obtained by dividing the characteristic strength by a partial safety factor (γ_m). The value chosen for a partial safety factor depends on the susceptibility of the material to variation in strength, e.g. steel reinforcement is less affected by site conditions than concrete.

2.3.1 Design Strength of Steel

The fundamental partial safety factor for steel $\gamma_s = 1,15$. For accidental loading, e.g. exceptional loads, fire or local damage, the value reduces to $\gamma_s = 1,0$. For earthquakes and fatigue conditions values are increased.

2.3.2 Design Strength of Concrete

The fundamental partial safety factor for concrete $\gamma_c = 1,5$. For accidental loading, the value reduces to $\gamma_c = 1,3$. For earthquakes and fatigue conditions, values are increased.

2.4 STRESS–STRAIN RELATIONSHIP FOR STEEL (cl 3.2.7, EN)

A shape of the stress–strain curve for steel depends upon the type of steel and the treatment given to it during manufacture. For reinforcement steel, actual curves for short term tensile loading have the typical shapes shown in Fig. 2.2. Curves for compression are similar. See also BS 4449 (1997).

It can be seen (Fig. 2.2) that hot rolled reinforcing steel yields, or becomes significantly plastic, at stresses well below the failure stress and at strains well below the limiting strain for concrete (0,0035). In a reinforced concrete member, therefore, the steel reinforcement may undergo considerable plastic deformation before the Ultimate Limit State (ULS) is reached, but will not fracture. However large steel strains are accompanied by the formation of cracks in the concrete in the tensile zone, and these may become excessive and result in serviceability failure at loads below the (ULS).

For normal design purposes, the European Code recommends a single, idealized stress–strain curve (Fig. 2.3) for both hot rolled and cold worked reinforcement.

2.4.1 Modulus of Elasticity for Steel (cl 3.2.7(4), EN)

The modulus of elasticity for steel (E_s) is obtained from the linear part of the relationship between stress and strain (Fig. 2.2). This is a material property and values from a set of samples vary between 195 and 205 GPa. For design purposes, this variation is small and the European Code adopts a mean value of $E_s = 200$ GPa.

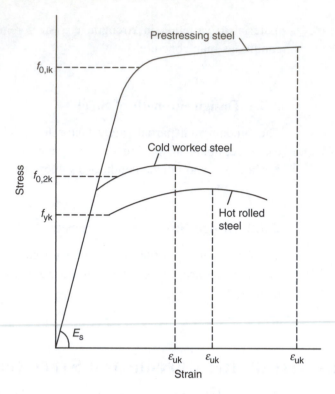

FIGURE 2.2 Typical stress–strain relationships for steel reinforcement

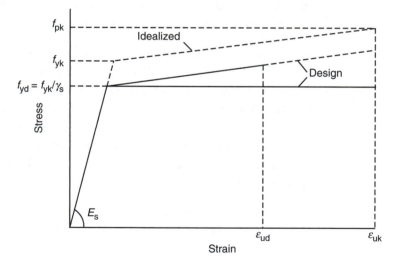

FIGURE 2.3 Design stress–strain relationship for steel reinforcement

The modulus of elasticity for prestressing steel varies with the type of steel from 175 to 195 GPa. Values of the moduli of elasticity are required in calculations involving deflections, loss of prestress and for the analysis of statically indeterminate structures.

2.5 STRESS–STRAIN RELATIONSHIP FOR CONCRETE (cl 3.1.5, EN)

A curve representing the stress–strain relationship for concrete is shown in Fig. 2.4. The important values of peak stress, peak strain and ultimate strain vary with the strength of concrete, rate of loading, age of the concrete, temperature and shrinkage. It is not easy to decide on design values that are safe and realistic.

The maximum stress (Fig. 2.4) is reached at a strain of approximately 0,002, after which the stress starts to fall. Disintegration of the concrete does not commence, however, until the strain reaches 0,0035, which is therefore taken as the limiting strain for concrete at the ultimate limit state. The maximum stress is the characteristic cylinder strength (f_{ck}).

For cross section design, the preferred idealization is shown in Fig. 2.5 and stress and strain values are given in Table 3.1, EN (Annex A2).

2.5.1 Modulus of Elasticity for Concrete (cl 3.1.3, EN)

The value of the modulus of elasticity is related to the type of aggregate and the strength of the concrete. Since the stress–strain curve for concrete is non-linear, a secant or static modulus is used. For normal weight concrete the modulus of elasticity

$$E_{cm} = 22 \, (f_{cm}/10)^{0,3} \, \text{GPa} \qquad (2.2)$$

where the mean cylinder compressive strength $f_{cm} = f_{ck} + 8 \, \text{MPa}$.

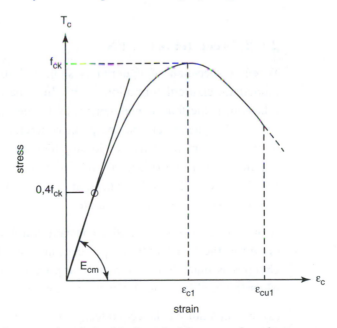

FIGURE 2.4   Stress–strain relationship for uniaxial compression of concrete

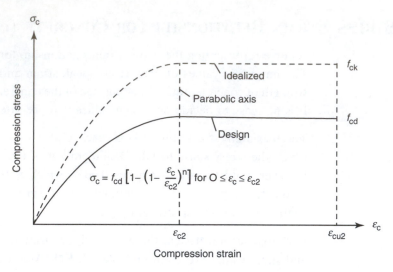

FIGURE 2.5 Parabolic-rectangular stress–strain relationship for concrete in compression

Alternatively the modulus of elasticity can be determined from tests, as described in BS EN 12390 (2000).

2.6 OTHER IMPORTANT MATERIAL PROPERTIES (cl 3.1.3, EN)

2.6.1 Introduction

This section gives a brief description of other properties of concrete which affect design calculations, e.g. deflections and loss of prestress.

2.6.2 Creep (cl 3.1.4, EN)

When a specimen of material is subjected to a stress below the elastic limit, immediate elastic deformation occurs. In some materials, the initial deformation is followed, if the load is maintained, by further deformation over a period of time. This is the phenomena of creep. In concrete, creep is thought to be due to the internal movement of free or loosely bound water and to the change in shape of minute voids in the cement paste. The rate of creep varies with time as shown in Fig. 2.6. Creep affects the deflection of beams at service load and increases the loss of prestress in prestressed concrete.

It may be assumed that under constant conditions 40 per cent of the total creep occurs in the first month, 60 per cent in 6 months and 80 per cent in 30 months. Creep is assumed to be complete in approximately 30 years and is partly recoverable when the load is reduced. Creep in concrete depends on the following:

(a) composition of the concrete,
(b) original water content,

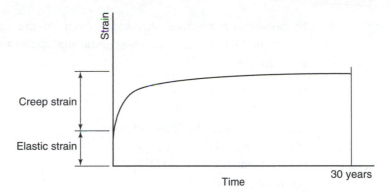

FIGURE 2.6   Rate of creep

(c) effective age at transfer of stress,
(d) dimensions of the member,
(e) ambient humidity,
(f) ambient temperature.

For calculation purposes, creep is expressed as a creep coefficient $\phi_{(\alpha,t_0)}$ which increases the elastic strain to produce the final creep strain, i.e.

$$\varepsilon_c = \text{elastic strain} \times \text{coefficient} = (\text{stress}/E_{cm(t)})\phi_{(\alpha,t_0)} \tag{2.3}$$

where $E_{cm(t)}$ is the elastic modulus of concrete at age of loading.

Values of the creep coefficient ($\phi_{(\alpha,t_0)}$) can be calculated from Fig. 3.1, EN. Further information on creep can be obtained from Annex B, EN and ACI Committee 209 (1972).

2.6.3 Drying Shrinkage (cl 3.1.4, EN)

The total shrinkage strain is composed of;

(a) autogenous shrinkage which occurs during the hardening of concrete,
(b) drying shrinkage which is a function of the migration of water through hardened concrete.

Drying shrinkage, often referred to simply as shrinkage, is caused by the evaporation of water from the concrete. Shrinkage can occur before and after the hydration of the cement is complete. It is most important, however, to minimize it during the early stages of hydration in order to prevent cracking and to improve the durability of the concrete. Shrinkage cracks in reinforced concrete are due to the differential shrinkage between the cement paste, the aggregate and the reinforcement. Its effect can be reduced by the prolonged curing, which allows the tensile strength of the concrete to develop before evaporation occurs.

In prestressed concrete, shrinkage is one of the causes for loss of prestress as shown in Chapter 12. The most important factors which influence shrinkage in concrete are:

(a) aggregate used,
(b) original water content,
(c) effective age at transfer of stress,
(d) effective section thickness,
(e) ambient relative humidity,
(f) reinforcement.

2.6.4 Thermal Strains (cl 2.3.1.2, EN)

Thermal effects should be taken into account when checking Serviceability Limit States (SLS) and at ultimate limit states where significant, e.g. fatigue and second order effects.

Thermal strains in the concrete are given by

$$\varepsilon_{ct} = \alpha_c t \tag{2.4}$$

where

α_c is the coefficient of thermal expansion
t is the rise in temperature

Where thermal change is not of much influence, then $\alpha_c = 10\text{E-}6/°C$. The coefficient of expansion depends on the type of aggregate and the degree of saturation of the concrete. Typical values for dry concrete, corresponding to an ambient relative humidity of 60 per cent, range from 7E-6/°C to 12-6/°C depending on the aggregate. For saturated and dry concrete, these values can be reduced by approximately 2E-6/°C and 1E-6/°C, respectively.

The thermal coefficient of expansion for steel is $\alpha_c = 12\text{E-}6/°C$ and is used in calculations for temperature changes which for the UK are in the range of $-5°C$ to $+35°C$. Notice that the coefficients of expansion for steel and concrete are approximately the same which avoids problems with differential movement.

2.6.5 Fatigue (cls 3.2.6 and 3.3.5 EN)

This is more of a problem with structural steelwork and of lesser importance with reinforced concrete. However problems may be associated with cracks in high strength reinforcement at bends, or where welds have been used. Prestressing wires are also subject to fatigue which should be taken into account.

The term fatigue is generally associated with metals and is the reduction in strength that occurs due to progressive development of existing small pits, grooves or cracks when subject to fluctuating loads. The rate of development of these cracks depends on the size of the crack and on the magnitude of the stress variation in the material and also the metallurgical properties. The number of stress variations, or cycles of stress, that a material will sustain before failure is called fatigue life and there is a linear experimental relationship between the log of the stress range and the log of the number of cycles. Welds are susceptible to a reduction in strength due to fatigue because of the presence of small cracks, local stress concentrations and abrupt changes of geometry. Research into the fatigue strength of welded structures is well documented (Munse (1984); Grundy (1985); and Annex C, EN). Fatigue effects are controlled by using correct welding procedures (BS EN 288-3, 1992)

All structures are subject to varying loads but the variation may not be significant. Stress changes due to fluctuations in wind loading need not be considered, but wind-induced oscillations must not be ignored. The variation in stress depends on the ratio of dead load to imposed load, or whether the load is cyclic in nature, e.g. where machinery is involved. For bridges and cranes, fatigue effects are more likely to occur because of the cyclic nature of the loading which causes reversals of stress. Generally calculations are only required for;

(a) lifting appliances or rolling loads,
(b) vibrating machinery,
(c) wind-induced oscillations,
(d) crowd-induced oscillations.

The design stress range spectrum must be determined, but simplified design calculations for loading may be based on equivalent fatigue loading if more accurate data is not available. The design strength of the steel is then related to the number and range of stress cycles.

2.6.6 Stress Concentrations

This is not a major problem with reinforced concrete. There are stress concentrations in the reinforcing steel where it crosses cracks in the concrete, but if the stresses are high then debonding of the steel occurs and the stress is relieved.

2.6.7 Failure Criteria for Steel and Concrete

2.6.7.1 Steel Failure Criteria

In general, the stresses in the steel reinforcement are axial and at failure become plastic. In special situations, e.g. dowel force resistance (Millard and Johnson, 1984), it is necessary to consider the reduction of axial stress due to shear stress.

The distortion strain energy theory or strain energy theory states that yielding occurs when the shear strain energy reaches the shear strain energy in simple tension. For a material subject to principal stresses σ_1, σ_2 and σ_3, this occurs (Timoshenko, 1946) when

$$(\sigma_1 - \sigma_2)^2 + (\sigma_2 - \sigma_3)^2 + (\sigma_3 - \sigma_1)^2 = 2f_y^2 \qquad (2.5)$$

Alternatively Eq. (2.5) can be expressed in terms of direct stresses σ_b, σ_{bc} and σ_{bt}, and shear stress τ on two mutually perpendicular planes. It can be shown from Mohr's circle of stress that the principal stresses

$$\sigma_1 = (\sigma_b + \sigma_{bc})/2 - [(\sigma_b - \sigma_{bc})^2/4 + \tau^2]^{0,5} \qquad (2.6)$$

and

$$\sigma_2 = (\sigma_b - \sigma_{bc})/2 + [(\sigma_b - \sigma_{bc})^2/4 + \tau^2]^{0,5} \qquad (2.7)$$

If Eqs (2.6) and (2.7) are inserted in Eq. (2.5) with $\sigma_3 = 0$ and f_y is replaced by the design stress f_y/γ_m then

$$(f_y/\gamma_m)^2 = \sigma_{bc}^2 + \sigma_b^2 - \sigma_{bc}\sigma_b + 3\tau^2 \qquad (2.8)$$

If σ_{bc} is replaced by σ_{bt} with a change in sign then

$$(f_y/\gamma_m)^2 = \sigma_{bc}^2 + 3\tau^2 \qquad (2.9)$$

2.6.7.2 Concrete Failure Criteria (cl 12.6.3, EN)

Concrete is composed of coarse aggregate, fine aggregate and a cement paste binder. Since each of these components has variable properties it is not surprising that concrete is also variable. Concrete includes microcracks in the unloaded state and when load is applied, the cracks extend and eventually precipitate failure.

The structural behaviour of concrete in tension is brittle and gives little warning of failure. In compression, behaviour is more plastic with non-linear stress–strain characteristics and warning signs of failure.

For practical convenience, the quality and strength of concrete is based on the crushing strength of a cube or cylinder. The value obtained from this test is sensitive to rate of loading and friction on the patterns. The strength in tension is the uniaxial tensile strength which is difficult to measure and the split cylinder strength is specified as an alternative.

Concrete in practice is not often subject to triaxial stresses and generally it is only necessary to consider biaxial stress conditions. The principal tensile stress failure

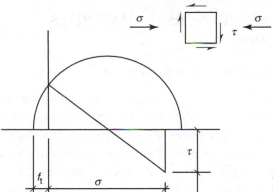

FIGURE 2.7 Principal tensile stress failure criterion for concrete

criterion is simple and has been used in the past. Consider a biaxial state of stress where tensile and shear stresses are combined (Fig. 2.7). From the geometry of Mohr's circle, the direct stress (σ) and the shear stress (τ) are related to the tensile strength of the concrete (f_t)

$$[(\sigma/2)^2 + \tau^2]^{0,5} - \sigma/2 = f_t$$

rearranging to form Eq. 12.5, EN

$$\tau/f_t = (1 + \sigma/f_t)^{0,5} \tag{2.10}$$

This expression becomes unsafe as the direct stress increases (Fig. 2.8) but it is simple and is sometimes useful, e.g. the shear strength of webs of prestressed beams.

When the direct stress exceeds the limit (Fig. 2.8) there is an alternative criterion (Eqs 12.6 and 12.7, EN) as the direct stress approaches compression failure. This criterion could be applied to column joints (Chapter 8). The relationship between the two criteria, for $f_{ck} = 50$ MPa without material factors, is shown in Fig. 2.8.

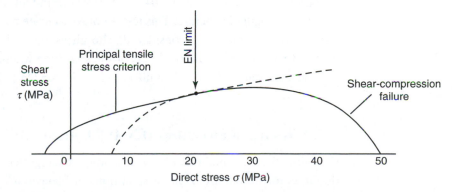

FIGURE 2.8 Shear failure criteria for concrete

2.7 TESTING OF REINFORCED CONCRETE MATERIALS AND STRUCTURES (EN 1990)

2.7.1 Testing of Steel Reinforcement

Reinforcing steel is routinely sampled and tested for tensile strength, during production, to maintain quality. The size, shape, position of the gauges and method of testing of small sample pieces of steel is given in EN 10002 (2001). The tensile test gives values of Young's modulus, limit of proportionality, yield stress or proof stress, percentage elongation and ultimate stress. Occasionally, when failures occur in practice, samples of steel are taken from a structure to assess the strength.

Methods of destructive testing fusion welded joints and weld metal in steel are given in BS EN 288-3 (1992). Welds of vital importance should be subject to non-destructive tests. The defects that can occur in welds are; slag inclusions, porosity, lack of penetration and sidewall fusion, liquation, solidification, hydrogen cracking, lamellar tearing and brittle fracture.

2.7.2 Testing of Hardened Concrete

Concrete supplied, or mixed on a site, is routinely tested by crushing standard cubes or cylinders when they have hardened. The method of manufacture and testing of the specimens is laid down in BS EN 12390-1 (2000). Occasionally some of the cylinders are split to provide the tensile strength of concrete. Rarely, after a structure is complete, doubt may be cast on the strength of the concrete. Cores can be cut from the concrete and tested as cylinders but this is expensive and may not be practical (EN 12504-1, 1999).

Alternatively, non-destructive tests may be used e.g. ultrasonic pulse velocity method (EN 12504-4, 1998). This consists in passing ultrasonic pulses through concrete from a transmitting transducer to a receiving transducer. Velocities range from 3 to 5 km/s and the higher the velocity, higher the strength of concrete. An alternative is the rebound hardness test (EN 12504-2, 1999) where the amount of rebound increases with the strength of concrete. This test is not so accurate but gives a rapid assessment before other methods are considered. The ultrasonic test is the more reliable and gives information over the thickness of the specimen whereas the rebound test give only surface values. Further details are given in Neville (1995).

2.7.3 Testing of Structures (EN 1990, Annex D)

Occasionally, new methods of construction are suggested and there may be some doubt as to the validity of the assumptions of behaviour of the structure. Alternatively if the structure collapses, there may be some dispute as to the strength of a component, or member, of the structure. In such cases, testing of components, or part of the structure, may be necessary. However it is generally expensive because of the accuracy required, cost of material, cost of fabrication, necessity to repeat tests to allow for variations and to report accurately. Tests are described as;

(a) acceptance tests – non-destructive for confirming structural performance,
(b) strength tests – used to confirm the calculated capacity of a component or structure,
(c) tests to failure – to determine the real mode of failure and the true capacity of a specimen,
(d) check tests – where the component assembly is designed on the basis of tests.

Structures which are unconventional, and/or methods of design which are unusual or not fully validated by research, should be subject to acceptance tests. Essentially, these consist of loading the structure to ensure that it has adequate strength to support, e.g.

$$1,0 \text{ (test dead load)} + 1,15 \text{ (remainder of dead load)} + 1,25 \text{ (imposed load)}.$$

REFERENCES AND FURTHER READING

ACI Committee 209 (1972) *Shrinkage and creep in concrete*. ACI Bib. No. 10.

BS EN 288-3 (1992) *Specification and approval of welding procedures for metallic materials. Welding procedure tests for the arc welding of steels*, BSI.

BS 4449 (1997) *Carbon steels for the reinforcement of concrete*, BSI.

BS EN 206-1 (2000) *Concrete Pt 1. Specification, performance, production and conformity*, BSI.

BS EN 12390-1 (2000) *Testing hardened concrete*, BSI.

EN 12504-1 (1999) *Testing concrete in structures- Part 1: Cored specimens – Taking examining and testing in compression*. BSI.

EN 12504-4 (1998) *Testing concrete in structures – Part 4: Determination of ultrasonic pulse velocity*, BSI.

EN 12504-2 (1999) *Testing concrete in structures – Part 2: Non-destructive testing – determination of rebound number*, BSI.

EN 10002 (2001) *Methods of testing of metallic materials*, BSI.

EN 1990 (2002) *Basis of structural design*.

Grundy, P. (1985) Fatigue limit design for steel structures. *Civ. Eng. Trans., ICE*, Australia, CE27, No. 1.

Millard, S.G. and Johnson, R.P. (1984) Shear transfer across cracks in reinforced concrete due to aggregate interlock and dowel action, *Mag. Conc. Res.*, Vol. 36, No 126, March 9–21.

Munse, W.H. (1984) Fatigue of welded structures, *Welding Res. Council*, New York.

Neville, A.M. (1995) *Properties of Concrete*, Longman, London.

Timoshenko, S. (1946) *Strength of Materials Pt II*. Van Nostrand, New York.

Chapter 3 / Actions

3.1 Introduction

Actions are a set of forces (loads) acting on a structure, or/and deformations produced by temperature, settlement or earthquakes.

Ideally, loads applied to a structure during its working life should be analysed statistically and a characteristic load determined. The characteristic load might then be defined as the load above which no more than 5 per cent of the loads exceed, as shown in Fig. 3.1. However data is not available and the characteristic value of an action is given as a mean value, an upper or lower value or a nominal value. Guidance on values is given in EN 1990 (2002), Pts. 1-7.

Actions are classified by:

(1) Variation in time:

    (a) permanent actions (G), e.g. self-weight, fittings, ancillaries and fixed equipment,

    (b) variable actions (Q), e.g. imposed loads, wind actions or snow loads,

    (c) accidental actions (A), e.g. explosions or impact from vehicles.

(2) Spacial variation:

    (a) fixed actions, e.g. self-weight for structures sensitive to self-weight,

    (b) free actions which result in different arrangement of actions, e.g. movable imposed loads, wind actions and snow loads.

These actions are now described in more detail.

3.2 Actions Varying in Time

3.2.1 Permanent Actions (G)

These are due to the weight of the structure, i.e. walls, permanent partitions, floors, roofs, finishes and services. The actual weights of the materials (G_k) should be used in design calculations and values are obtained from EN 1991-1.1.

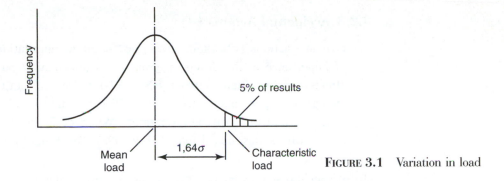

FIGURE 3.1 Variation in load

3.2.2 Variable Actions (Q)

These are due to moveable items such as furniture, occupants, machinery, vehicles, stored materials, snow and wind. Imposed loads generally are expressed as static loads for convenience, although there may be minor dynamic forces involved. Characteristic variable actions (Q_k) are subdivided into:

(a) Imposed floor loads are variable actions, and various dwellings values are given in EN 1991-1.1. These loads include a small allowance for impact and other dynamic effects that may occur in normal occupancy. They do not include forces resulting from the acceleration and braking of vehicles or movement of crowds. The loads are usually given in the form of a distributed load or an alternative concentrated load. The one that gives the most severe effect is used in design calculations.

When designing a floor, it is not necessary to consider the concentrated load if the floor is capable of distributing the load, and for the design of the supporting beams the distributed load is always used. When it is known that mechanical stacking of materials is intended, or other abnormal loads are to be applied to the floor, then actual values of the loads should be used, not those obtained from EN 1991-1.1. In multi-storey buildings, the probability that all the floors will simultaneously be required to support the maximum loads is remote and reductions to column loads are therefore allowed.

(b) Imposed roof loads are variable actions, and are related to snow load and access for maintenance. They are specified in EN 1991-1.3 and, as with floor loads, they are expressed as a uniformly distributed load on plan, or as an alternative concentrated load. The magnitude of the loads decrease as the roof slope increases and in special situations, where roof shapes are likely to result in drifting of snow, then loads are increased.

(c) Wind actions are variable but for convenience they are expressed as static pressures in EN 1991-4. The pressure at any point on a structure is related to the shape of the building, the basic wind speed, topography and ground roughness. The effects of vibration, such as resonance in tall buildings must be considered separately.

3.2.3 Accidental Actions (*A*)

(a) Dynamic actions (vibration, shock, acceleration, retardation and impact) are of importance in the design of cranes and information on loads and impact effects can be obtained from EN 1991-1.7. Vertical static crane loads are increased by percentages from 10 to 25 per cent depending on whether the crane is hand or electrically operated. When designing for earthquakes, the inertial forces must be calculated (EN 1998). This is not of major importance in the UK.

(b) Accidental actions include explosions and fire as described in Chapter 5.

3.3 ACTIONS WITH SPATIAL VARIATION

3.3.1 Pattern Loading (cl 5.1.3, EN)

All possible actions appropriate to a structure should be considered in design calculations. The actions should be considered separately and in realistic combinations to determine which is most critical for strength and stability of the structure.

For continuous structures, connected by rigid joints or continuous over the supports, vertical actions should be arranged in the most unfavourable but realistic pattern for each element. Permanent actions need not be varied when considering such pattern loading, but should be varied when considering stability against overturning. Where horizontal actions are being considered pattern loading of vertical actions need not be considered.

For the design of a simply supported beam, it is obvious that the critical condition for strength is when the beam supports the maximum permanent action and maximum variable action at the ultimate limit state. The size of the beam is then determined from this condition and checked for deflection at the serviceability limit state.

A more complicated structure is a simply supported beam with a cantilever as shown in Fig. 3.2 (a).

Assuming that the beam is of uniform section and that the permanent actions are uniformly applied over the full length of the beam, it is necessary to consider various combinations of the variable actions as shown in Figs 3.2 (b)–(d). Although partial loading of spans is possible, this is not generally considered except in special cases of rolling actions, e.g. a train on a bridge span.

For a particular section, it is not immediately apparent which combination of actions is most critical because it depends on the relative span dimensions and magnitude of the actions. Therefore calculations are necessary to determine the condition and section for maximum bending moment and shear force at the ULS.

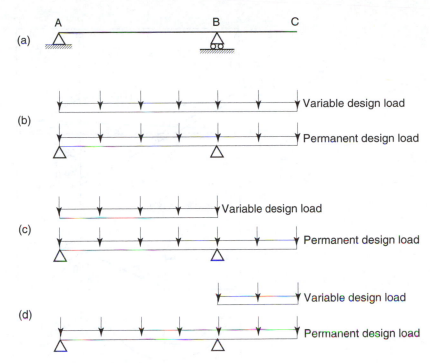

FIGURE 3.2   Pattern loading

In other situations, e.g. when checking the overturning of a structure, the critical combination of actions may be the minimum permanent action, minimum imposed action and maximum wind action (See also Chapter 4).

3.4 DESIGN ENVELOPES

A design envelope is a graph showing, at any point on a structural member, the most critical effect that results from various realistic combinations of actions. Generally, the most useful envelopes are for shear force and bending moment at the ULS. The formation and use of a design envelope is demonstrated by the following example.

EXAMPLE 3.1   Example of a design envelope. The beam ABC in Fig. 3.3 carries the following characteristic loads:

Dead load 10 kN/m on both spans;
Imposed load 15 kN/m on span AB, 12 kN/m on span BC.

Sketch the design envelope for the bending moment and shear force at the ULS. Indicate all the maximum values and positions of zero bending moment (points of contraflexure).

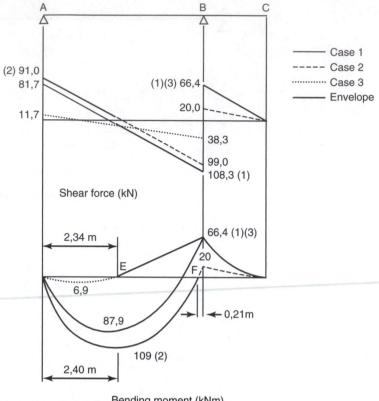

Figure 3.3 Example: design envelope

The maximum and minimum loads on the spans are:

Maximum on AB $= 10 \times 1,4 + 15 \times 1,6 = 38$ kN/m;

Maximum on BC $= 10 \times 1,4 + 12 \times 1,6 = 33,2$ kN/m;

Maximum on AB or BC $= 10 \times 1,0 = 10$ kN/m.

Consider the following load cases:

(1) Maximum on AB and BC.

(2) Maximum on AB, minimum on BC.

(3) Minimum on AB, maximum on BC.

The bending moment and shear force diagrams are shown in Fig. 3.3.

Comments:

(a) Only the numerical value of the shear force is required in design, the sign however may be important in the analysis of the structure.

(b) Positive (sagging) bending moments indicate that the bottom of the beam will be in tension; negative (hogging) moments indicate that the top of the beam will be in tension. Since the beams will need to be reinforced in the tension zone, it is conventional to draw the bending moment on the tension side, as shown.

(c) The envelope, shown as a heavy line, indicates the maximum values produced by any of the load cases. Note that on AB the envelope for shear force changes from case (2) to case (1) at the point where the numerical values of the shear force are equal.

(d) From the bending moment envelope it is evident that tension reinforcement in the top of the beam is governed by case (3) and in the bottom of the beam by case (2). Note that in some regions, tension reinforcement is required in both the top and the bottom of the beam.

(e) The points of contraflexure indicate theoretical points where the reinforcement may be curtailed, i.e. at points E and F on the bending moment envelope.

This process is tedious and an experienced designer often knows the critical action combinations and the positions of the critical values, which avoids some of the work involved. Alternatively the diagrams can be generated from input data using computer graphics. In more complicated structures and loading situations, envelopes are useful in determining where a change of member size could occur and where splices could be inserted. In other situations, wind action is a further alternative to combinations of permanent and variable actions.

3.5 OTHER ACTIONS

(a) Temperature forces and moments are induced into reinforced concrete structures which are restrained or highly redundant. Where it is necessary to allow for temperature change in the UK, the range is usually $-5°C$ to $+35°C$. Thermal effects should be taken into account for serviceability limit states but at ultimate limit states only when significant (cl 2.3.1.2, EN). See also EN 1991-1.5.

(b) Erection forces may be of importance in reinforced concrete construction where prefabrication is used. Suspension points for members or parts of structures may have to be specified to avoid damage to components. It is extremely difficult to anticipate all possible erection forces and the contractor is responsible for erection which should be carried out with due care and attention. Nevertheless a designer should have knowledge of the most likely method of erection and design accordingly. If necessary temporary stiffening or supports should be specified, and/or instructions given. See also EN 1991-1.6.

(c) Earth loads occur in situations such as a retaining walls and foundations. Advice on earth pressures and allowable bearing pressures for different types of soil at service load conditions is given in EN 1997-1.

(d) Differential settlement is likely to occur in mining areas and values that are likely to occur in the vicinity of the structure should be determined. Methods of accommodating the settlement include use of adjustable bearings, pin joints and adjustable stiffeners.

REFERENCES AND FURTHER READING

EN - 1991 (2003) Parts 1–7. *Actions on structures*.

EN - 1990 (2002) *Basis of structural design*.

EN - 1998 (2006) *Designs of structures for earthquake resistance*.

Chapter 4 / Analysis of the Structure

Chapter 3 has considered the imposition of structural actions be they loads, deformations or thermal movements together with the concepts of limit state design, whether at ultimate (ULS) or serviceability (SLS). This chapter considers the methods that may be used to determine the internal forces in the structure due to the actions applied to it.

4.1 GENERAL PHILOSOPHY FOR ANALYSIS OF THE STRUCTURE

The engineer is free to choose the idealization (or model) to represent the structure. This model must however be realistic, and not be chosen for, say, sheer simplicity. If in doubt, any model should give a lower bound solution for the internal forces in all critical elements within the structure. Generally analysis methods will provide the designer with the internal forces or stress resultants, namely bending moments, shear forces and axial forces and may provide displacements. Methods which give internal stresses and strains, such as finite element methods, are not precluded although care is needed in the interpretation of results. The common mathematical idealizations of material (and hence element) behaviour are elastic, elastic with redistribution, plastic (with or without second order effects due to additional moments induced by large deformations).

4.1.1 Ultimate Limit State

Whatever the method used to determine the internal forces, the actions applied to the structure must be factored using those values pertaining to ULS, whether for single or multiple variable loads. Full plastic analysis may not be used for frames unless the plastic hinges, other than the last one to form, occur in the beams. This is because columns with high axial loads will not have any ductile capacity (and will not therefore satisfy the maximum limits for neutral axis depth to effective depth ratios (x/d) (Section 6.1.1)). This therefore means that frames are normally analysed elastically and the moment field then adjusted using redistribution to simulate plastic behaviour.

For slabs, plastic analysis using the Johansen Yield Line methods may be used, but there are severe restrictions on the x/d ratio to ensure ductility. In practice, however, due to the relatively low levels of reinforcement, the x/d ratio is rarely critical. It must be noted the maximum amount of redistribution is reduced and plastic analysis prohibited if Class A reinforcement is used.

Care needs taking where high membrane forces can be generated, in such as multi-bay flat slabs, as yield line analysis cannot then be used as again ductility will be limited. The method may however give valuable insights into the behaviour of flat slabs around column heads (Regan, 1981 and Oliviera, Regan and Melo, 2004). In all cases a lower bound approach such as the Hillerborg Strip method can be used (See Chapter 10).

In elastic analyses, the section properties are generally calculated on the gross concrete section (with no allowance for the reinforcement), and Young's modulus taken as the secant modulus, E_{cm} (in GPa),

$$E_{cm} = 22\left(\frac{f_{ck} + 8}{10}\right)^{0,3} \tag{4.1}$$

where f_{ck} (MPa) is the concrete grade strength (Table 3.1 of EN 1991-1-2).

4.1.2 Serviceability Analysis

This must be performed using an elastic analysis and contain all actions applied to the structure including any deformations or settlement effects. It is not generally necessary to consider time-dependant effects such as creep, except where they will cause substantial changes in internal forces. This last will occur in hyperstatic prestressed concrete structures such as continuous bridge decks. Although normal reinforced concrete undergoes creep, the effect is less as the stress levels are potentially lower and thus is effectively ignored.

Generally a full serviceability analysis (and design) will be unnecessary. For most structures, it is sufficient to use 'deemed to satisfy' approaches for, say, deflection or cracking. Exceptions to this approach are where there is an explicit requirement to determine crack widths in liquid-retaining structures or bridgeworks, or a design which fails the 'deemed to satisfy' clauses, but needs to be shown to be adequate. It is accepted that 'deemed to satisfy' clauses are conservative.

4.2 BEHAVIOUR UNDER ACCIDENTAL EFFECTS

Accidental actions should be considered in a number of possible circumstances:

- Explosions, whether due to gas or terrorism
- Impact due to vehicles or aircraft

Should the risk of such an incident be high and the effects be catastrophic, or in certain circumstances the need to check be mandatory (e.g. vehicular impact on bridge piers), then the designer must ensure that the structure is designed and detailed to ensure that should an accidental situation occur, the structure does not suffer complete or partial collapse from either the accidental situation itself or subsequent events such as spread of fire.

Four now classic cases where collapse occurred due to accidental or terrorist action are:

- Ronan Point block of flats, where a gas explosion blew out a wall panel causing progressive collapse of a corner of the structure (Wearne, 1999).
- The Alfred P. Murrah Building in Oklahoma City where collapse was caused by a terrorist bomb blast (Wearne, 1999).
- World Trade Centre Towers in New York on 11 September 2001 due to impact from deliberate low-flying aircraft and subsequent fire spread. Although the towers were steel-framed structures, there is still a lesson to be learnt for any form of construction (ISE, 2002).
- Pentagon Building on 11 September 2001 into which a commercial airline was crashed. The structure suffered partial but not extensive collapse due to the impact and subsequent fire (Mlakar *et al.*, 2003).

4.2.1 Progressive Collapse

EN 1990 identifies the need to consider that in the event of an accident such as explosion or fire, the structure should not exhibit disproportionate damage. This is reinforced by the requirements of the relevant Building Regulations within the UK, e.g. the recently revised Approved Document A of the England and Wales Building Regulations.

Such damage can be mitigated by:

(a) Attempting to reduce or limit the hazard
    In the case of fire, this could be done by full consideration given to the use of non-flammable materials within the structure and by the provision of relevant active fire protection measures such as sprinklers. In the case of industrial processes where explosions are a risk then such potential risks need to be taken into account by, say, enclosing the process in blast-proof enclosures.

(b) Maintenance
    The recent Pipers Row Car Park collapse (Wood, 2003) indicates the need for adequate design, especially where punching shear may be critical as this is a quasi-brittle failure and the need for adequate inspection and maintenance where it is known that environmental effects will be severe. Although the failure

was due to poor maintenance, it was exacerbated by uneven reaction distribution at the slab–column interface and no tying through the lower face of the slab. Admittedly the construction technique used would have made the latter extremely difficult although it contributed to the former.

(c) Consideration of the structural form

A case where this is relevant is when the structure could be subject to externally provoked explosions. One reason why the damage was extensive following the Oklahoma City explosion was that the blast was amplified by an overhanging portion of the structure above ground floor level. A solution for structures known to be at potential risk is to provide an external curtain which is not structural and is easily blown out whilst ensuring the structure is stabilized by a core which is in the centre of the building. Additionally the risk can be mitigated by ensuring vehicles cannot come within certain limiting distances of the structure. More information on this including assessment of blast forces is given in a SCI publication (Yandzio and Gough, 1999) most of which is relevant to concrete construction.

(d) Removal of key elements to establish stability

In this approach a risk analysis is carried out to identify the key elements in a structure which are vital for its stability. A structural analysis is carried out with any one these elements removed to establish the stability of the remaining structure and its ability to carry the loading so induced. As this is an accidental limit state, lower partial safety factors are used on both the loading and the material strengths determining the member strengths. It is not necessary to check any serviceability limit states.

(e) Ensuring the structure is adequately tied to resist collapse

This needs considering on two levels. The first is to determine the magnitudes of the likely levels of tying force required. The second is to provide properly detailed reinforcement to the structure. Such ties are generally peripheral and vertical. It is permitted to use the reinforcement provided for structural strength for these and not normally necessary to supply additional reinforcement unless that needed for normal structural strength is inadequate (cl 9.10.1). The following system of ties is recommended by EN 1992-1-1:

- Peripheral Ties

  Peripheral ties round the external structural envelope (or internal voids) should be designed to take a force of $F_{\text{tie,per}}$ given by

  $$F_{\text{tie,per}} = L_i q_1 \leq q_2 \tag{4.2}$$

  where L_i is the length of the appropriate span, q_1 has a recommended value of 10 kN/m and q_2 of 70 kN.

- Internal Ties

  These should be at roof and each floor level, be continuous and effectively anchored to the peripheral ties. The recommended design force $F_{tie,int}$ is 20 kN. In floors without screeds (i.e. where ties cannot be placed in the span direction), the minimum force on the beam line F_{tie} should be taken as

$$F_{tie} = \frac{L_1 + L_2}{2} q_3 \leq q_4 \qquad (4.3)$$

  where L_1 and L_2 are the span lengths of the floor slab on either side of the beam, q_3 has a recommended value of 20 kN/m and q_4 of 70 kN. It is clear that where punching shear is a critical failure mode, bottom steel needs to be continuous through the column to avoid disproportionate collapse (Reagan, 1981; Wood, 2003).

- Horizontal ties to columns and walls

  Edge columns and walls should be tied horizontally at the roof and each floor level as appropriate. Corner columns should be tied in each direction, but the reinforcement provided for peripheral ties may also be used for the horizontal ties. For walls, the recommended value of the tying force $F_{tie,fac}$ is 20 kN, and for columns $F_{tie,col} = 150$ kN.

- Vertical Ties

  In panel buildings of 5 storeys or more, vertical ties must be provided to limit the damage to a floor should the wall or column immediately below be lost. There is no recommendation made on the actual values of force to be resisted as each case needs evaluating on an individual basis.

4.2.2 Structure Stability

Any structure with high lateral loading or where vertical loading can be applied outside the frame envelope must be checked for overturning. Also any continuous member with a cantilever must be checked for the possibility of uplift on any support.

In all these cases the loading which causes overturning is deemed to be unfavourable, and the loading tending to restore the situation is known as favourable. This is defined in EN 1990 as verification of static equilibrium EQU.

For this situation the partial safety factors given in Table 11.1 (corresponding to EQU) must be used.

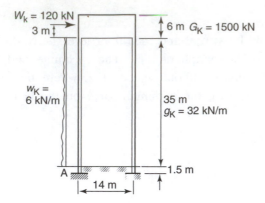

FIGURE 4.1   Design data for Example 4.1

EXAMPLE 4.1   High wind load. Consider the possibility of overturning for the water tower shown in Fig. 4.1.

Taking Moments about Point A:
Unfavourable effects due to wind:

$$1{,}5 \times 120 \times (3 + 35 + 1{,}5) + 1{,}5 \times 6 \times 35(35/2 + 1{,}5) = 13\,095\,\text{kNm}$$

Favourable effects due to self weight:

$$0{,}9 \times 1500 \times 14/2 + 0{,}9 \times 32 \times 35 \times 14/2 = 16\,506\,\text{kNm}$$

The moment due to the favourable effects exceeds that due to the unfavourable effect, thus the structure will not overturn.

EXAMPLE 4.2.   External loading. Consider the cantilevering frame shown in Fig. 4.2.

Here the check is more complex, as the permanent load of 220 kN per storey is both favourable and unfavourable, in that the load acting on the cantilever is

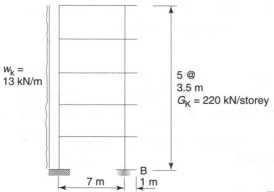

FIGURE 4.2   Design data for Example 4.2

unfavourable as it contributes to the overturning effect. The permanent load per storey may be expressed as 27,5 kN/m run.

The favourable part of the permanent loading on the internal span takes a factor of 0,9, and that on the cantilever of 1,1.

Taking moments about B:
Effect due to the wind loading:

$$1,5 \times 13 \times 17,5^2/2 = 2986 \text{ kNm}$$

Effect due to permanent action (per storey):

$$0,9 \times 27,5 \times 7^2/2 - 1,1 \times 27,5 \times 1^2/2 = 591 \text{ kNm per storey}$$
$$\text{Total} = 5 \times 591 = 2955 \text{ kNm}$$

The structure will just overturn as the disturbing moment is marginally greater than the restoring moment. Thus either the structure needs tying down using, say, tension piles or ground anchors, or the permanent loading could be marginally increased. The latter is the cheaper option owing to the small margin involved.

4.2.3 Member Stability

Individual members may also have requirements placed on them to ensure that in the case of beams lateral torsional buckling will not occur, or in the case of columns their slenderness is kept below that giving severe buckling problems.

4.2.3.1 Beams

Limits are placed on the section dimensions (cl 5.9(3)). These restrictions are seldom onerous as conventional member sizing usually provides stockier members. The limits are that the ratio of the distance between lateral restraints l_{0t} to the width of the compression flange b is given by

- For persistent situations

$$\frac{l_{0t}}{b} \leq \frac{50}{(h/b)^{1/3}} \tag{4.4}$$

  where h is the overall depth of the beam, and the ratio is subject to the limit $h/b \leq 2,5$.

- For transient situations

$$\frac{l_{0t}}{b} \leq \frac{70}{(h/b)^{1/3}} \tag{4.5}$$

  where h is the overall depth of the beam, and the ratio is subject to the limit $h/b \leq 3,5$.

4.2.3.2 Columns

For columns there is a minimum slenderness set for the condition where second order effects need not be considered. This is equivalent to fixing a transition slenderness between short and slender columns where second order effects will need considering. This is dealt with in Section 4.3.2.

4.3 FRAME IMPERFECTIONS (cl 5.2)

Cl 5.2(1)P states that **the unfavourable effects of possible deviations in the geometry of the structure and the position of loads shall be taken into account in the analysis of members and structures. These imperfections must be considered at ULS whether persistent or accidental limit states.** They need not be taken into account at serviceability. The reason for imposition of such imperfections is that no construction is perfect. This is not a licence for ignoring the need to work to limits of verticality usually imposed on construction. The frame is considered to be out of vertical by a rotation θ_1 given by

$$\theta_1 = \alpha_h \alpha_m \theta_0 \tag{4.6}$$

where the basic value θ_0 is taken as 1/200 radians. The parameter α_h is a reduction factor to allow for the building height l (in m) and is given by

$$\alpha_h = \frac{2}{\sqrt{l}} \tag{4.7}$$

subject to the limit $2/3 \le \alpha_h \le 1,0$. The parameter α_m allows for the number of members m, and is given by

$$\alpha_m = \sqrt{0,5\left(1 + \frac{1}{m}\right)} \tag{4.8}$$

Conservatively the effects of α_h and α_m can be ignored and θ_1 set equal to θ_0.

4.3.1 Application of Notional Imperfections

- For isolated members, l is the actual length and $m = 1$;
- For a bracing system l is the building height, and m is the number of vertical members contributing to the horizontal force in the bracing system.

A similar style of definition is used for floor or roof diaphragms.

Rather than apply a geometric imperfection, the analysis may be replaced by applying notion horizontal forces:

- for isolated members the vertical force will act at an eccentricity of e_1 given by

$$e_1 = \theta_1 \frac{l_0}{2} \tag{4.9}$$

Alternatively a transverse force equal to $\theta_1 N$ for braced members or $2\theta_1 N$ for unbraced members may be applied at the point of maximum bending moment, where N is the axial load.

- for complete structures, the horizontal H_i force is given by

$$H_i = \theta_1 (N_b - N_a) \tag{4.10}$$

for bracing systems,

$$H_i = \theta_1 \frac{N_b + N_a}{2} \tag{4.11}$$

for floor diaphragms, and

$$H_i = \theta_1 N_a \tag{4.12}$$

for roof diaphragms.

In all cases H_a and H_b are the column loads above and below the levels being considered.

Cl 5.2 (1)P implies that the imperfections (or their resultant horizontal loadings) must be taken into account when considering analysis under vertical forces due to permanent and variable loads. This, therefore, also implies that the structure, if a sway frame, must be analysed as a whole since most simple subframes are restrained against lateral deflection. Subframes which extend across the whole structure could possibly be used. For non-sway frames the effect of the horizontal loads will not be significant on the columns or floor units.

4.3.2 Second Order Effects (cl 5.8.3.1)

For single members, second order effects may be ignored if the slenderness ratio λ is less than the limiting value λ_{lim} given by

$$\lambda_{\text{lim}} = \frac{20ABC}{\sqrt{n}} \tag{4.13}$$

where λ is defined as

$$\lambda = \frac{l_0}{i} \tag{4.14}$$

where l_0 is the effective length and i the lesser radius of gyration of the section based on the uncracked concrete section.

The parameter n is given by

$$n = \frac{N_{Ed}}{A_c f_{cd}} \quad (4.15)$$

where N_{Ed} is the applied axial force in the column, A_c is the gross area and f_{cd} is the design concrete strength. The parameter A is dependant on the effective creep ratio ϕ_{ef},

$$A = \frac{1}{1 + 0{,}2\phi_{ef}} \quad (4.16)$$

The parameter B is dependant upon the mechanical steel ratio ω ($= A_s f_{yd}/A_c f_{cd} = (A_s f_{yk}/\gamma_s)/(A_c f_{ck})/\gamma_c)$,

$$B = \sqrt{1 + 2\omega} \quad (4.17)$$

where A_s is the reinforcement area and f_{yd} its design strength.

The parameter C is dependant upon the ratio of the end moments r_m,

$$C = 1{,}7 - r_m \quad (4.18)$$

where r_m is defined as M_{01}/M_{02}, where M_{01} and M_{02} are the absolute values of the end moments such that $r_m \leq 1{,}0$.

Where the values of ϕ_{ef}, ω and r_m are not known, then A may be taken as 0,7, B as 1,1 and C as 0,7. In this case Eq. (4.13) reduces to

$$\lambda_{lim} = \frac{10{,}78}{\sqrt{n}} \quad (4.19)$$

The value given by Eq. (4.19) will be conservative. Thus if the value of λ marginally exceeds λ_{lim}, it will be worth iterating the procedure when more accurate values of A, B, and C may be determined.

4.4 FRAME CLASSIFICATION

A frame may be classified as sway or non-sway. A non-sway frame is one where horizontal deformations are limited by the provision of substantial bracing members often in the form of lift shafts or stairwells. Where there are substantial shear walls limiting sway deformation or cores containing lifts shafts, etc. the structure may be considered braced. Note, however, a long thin structure (in plan) is often braced in the transverse direction but not braced in the longitudinal direction as the design horizontal forces are lower and can be resisted by a much greater number of columns. A frame can also be considered as non-sway if the additional moments determined using the applied horizontal actions and the deformations calculated from a first order elastic analysis are less than 10 per cent of the primary moments (cl 5.8.2 (6)).

Annex H of EN 1992-1-2 provides criteria for neglecting global second order effects.

For bracing systems without significant shear deformations, e.g. shear walls, then second order effects may be ignored if

$$F_{V,Ed} \leq 0,1F_{V,BB} \tag{4.20}$$

where $F_{V,Ed}$ is the total vertical load on the braced members and the bracing and $F_{V,BB}$ is the nominal global buckling load for global bending which may be determined from

$$F_{V,BB} = \xi \sum \frac{EI}{L^2} \tag{4.21}$$

where EI is the sum of the flexural stiffnesses of the bracing members (including the effects of cracking) and L the height of the building above the level of moment restraint. The bending stiffnesses may be taken as an approximation as

$$EI = 0,4\frac{E_{cm}}{\gamma_{cE}}I_c \tag{4.22}$$

The value of γ_{cE} is recommended to be taken as 1,2. If the cross section can be shown to be uncracked, then the coefficient of 0,4 in Eq. (4.22) may be replaced by 0,8.

The value of ξ reflects the effects of the number of storeys, stiffness variation, base rigidity and load distribution, and may be taken as

$$\xi = 7,8\frac{n_s}{n_s + 1,6}\frac{1}{1 + 0,7k} \tag{4.23}$$

where n_s is the number of stories and k is the relative flexibility at the base defined as

$$k = \frac{\theta}{M}\frac{EI}{L} \tag{4.24}$$

where θ/M is the rotation per unit moment, EI is the stiffness from Eq. (4.22) and L the height of the bracing unit.

For bracing systems with significant global shear deformations, second order effects may be ignored if

$$F_{V,Ed} \le 0,1 F_{V,B} = 0,1 \frac{F_{V,BB}}{1 + (F_{V,BB}/F_{V,BS})} \qquad (4.25)$$

where $F_{V,B}$ is the global buckling load including shear, and $F_{V,BS}$ the global buckling load for pure shear. This may be taken as the total shear stiffness of the bracing ΣS.

If global second order effects are important then the structure should be analysed under magnified horizontal design forces $F_{H,Ed}$ whether forces due to wind, geometrical imperfections, etc. where $F_{H,Ed}$ is given by

$$F_{H,Ed} = \frac{F_{H,0,Ed}}{1 - (F_{V,Ed}/F_{V,B})} \qquad (4.26)$$

where $F_{H,0,Ed}$ are the first order horizontal forces. Equation (4.26) is the classic sway magnification factor.

4.5 FRAME ANALYSIS

Before considering any possible analysis simplifications, the loading cases for vertical loading must be considered.

4.5.1 Load Cases

For continuous beams and slabs without cantilevers subject to loading which can be considered as uniformly distributed, the following load combinations need considering (cl 5.1.3):

(a) Alternate spans carrying the full design load $\gamma_Q Q_k + \gamma_G G_k$ with the intermediate spans carrying $\gamma_G G_k$, and

(b) Any two adjacent spans carrying $\gamma_Q Q_k + \gamma_G G_k$ with the remainder carrying $\gamma_G G_k$.

Note that for the spans carrying loading of $\gamma_G G_k$, this loading is usually favourable and thus γ_G takes a value of 1,0.

For continuous beams with a large number of spans, load combination (a) can be replaced by all spans loaded with the maximum load. For the three span continuous beam in Example 4.3 this would underestimate the maximum support (hogging) moment by about 7 per cent.

4.5.2 Frame Analysis

Most structures are three dimensional, but it is generally possible to consider the structure as a series of interlinked two dimensional frames which can be analysed separately. This is possible due to the principle of elastic superposition and that in most structures the predominant mode of resisting the applied loading is flexure. Where the structure is such that torsion and shear become important modes then the system can only be decoupled with extreme care.

In design where the three-dimensional structure has been decoupled into a series of two-dimensional frames, it is important that ALL the forces are included especially in the column and foundation design, but the self-weight of such members must not be included twice.

With the advent of small, fast, powerful desktop computing facilities available in even the smallest design practices, there is now less need to consider the analysis of parts of the structure rather than the whole, thus the use of subframes for analysis is now of less importance and is here relegated Annex C in the text (Fig. 4.3).

However, to obtain quick estimates of column moments, Subframes 1 to 4 are sometimes useful. Where only a single beam is taken on one or both sides of a column, the stiffness of such a beam can be halved to allow for the remainder of the structure. Alternatively the end(s) of the beam remote from the column can be considered pinned and the full stiffness taken.

Continuous beam subframes can be useful for a rapid assessment of beam or slab moments, though small dedicated analysis programs for continuous beams are also readily available. Also the need to consider notional horizontal forces on the whole structure will further mitigate against hand analysis techniques unless the frame is braced (non-sway).

4.5.2.1 Analysis under Horizontal Loading

Since as mentioned above, an analysis is required to consider the effects of geometry imperfections using either notional vertical out-of-plumb displacements or notional horizontal forces, the same analysis can be used to determine the effects of design horizontal loading.

4.5.2.2 Vertical Loading

As a full analysis is required for horizontal loading, then there is little point in using smaller subframes for the vertical loading for the final design of the structure. They may however be useful if a preliminary analysis is required very quickly for one or two potentially critical members. Figure 4.3 indicates possible subframes that may be used in a given structure. The results for various subframe analyses are given in Annex C.

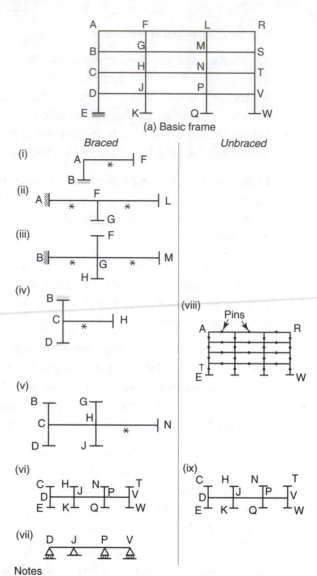

(a) Basic frame

FIGURE 4.3 Basic frame with typical subframes for both the braced and unbraced case

4.5.2.3 Bending Moment and Shear Force Envelopes

Where multiple load cases become necessary due to patterned loading the individual BMD's and SFD's are combined, or superimposed, to produce a set of BM and SF envelopes. Most frame analysis packages allow the complete bending moment diagrams or envelopes to be output. Where the analysis only gives the BM (and SF) at the ends of the beam or column segment, it is necessary to determine the maximum midspan moment (usually sagging). It may be sufficient to determine the moment at midspan rather than the true maximum at the point of zero shear.

The maximum moment for the case where the variable loading comprises point loads may be easily calculated as such maxima usually lie under one (or more) of the point loads. For uniformly distributed loads, the following results may be found useful,

- Moment at midspan

$$M_{mid} = \frac{qL_{AB}^2}{8} + \frac{M_{AB} - M_{BA}}{2} \qquad (4.27)$$

- Maximum moment

$$M_{max} = \frac{qL_{AB}^2}{8} + \frac{M_{AB} - M_{BA}}{2} + \frac{1}{2q}\left(\frac{M_{AB} + M_{BA}}{L_{AB}}\right)^2 \qquad (4.28)$$

This value occurs at a position x_{max} measured from point A (in Fig. 4.4) given by

$$x_{max} = \frac{L_{AB}}{2} - \frac{M_{AB} - M_{BA}}{qL_{AB}} \qquad (4.29)$$

The notation is given in Fig. 4.4.

The sign convention for both M_{mid} and M_{max} is sagging positive. For beams with approximately numerically equal values of M_{AB} and M_{BA}, the values of M_{mid} and M_{max} are little different. The case when either M_{BA} or M_{AB} is zero (i.e. either end of a continuous beam) will require the use of M_{max} rather than M_{mid}.

The sign of the shear force V is given by

$$V = \frac{dM}{dx} \qquad (4.30)$$

4.5.3 Member Stiffness

In general, it is sufficient to use a member stiffness calculated using the second moment of area determined on the basis of the concrete section alone.

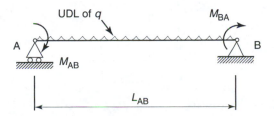

FIGURE 4.4 Comparison between maximum and midspan moments

4.5.3.1 Rectangular Beams (or Columns)

The second moment of area is calculated as $bh^3/12$ where b and h are the width and depth of the section, respectively.

4.5.3.2 Flanged Beams

Owing to the effect of shear lag (See Section 6.4), the full width of the flange cannot be used, but a reduced width has to be taken. This reduced width however strictly should vary along the section, especially where there are hogging and sagging moments.

This means that of course the stiffness should vary along the length. This is possible, but it is generally accepted that the stiffness should be calculated on the effective flange width used for design under sagging moments (cl 5.3.2.1 (4)). The calculation of effective widths are covered in Section 6.4.1.

The stiffness of a 'T' beam (Fig. 4.5) is given by

$$\frac{3I}{b_w h^3} = \frac{b_{\text{eff}}}{b_w}\left(1 - \frac{x}{h}\right)^3 + \left(\frac{x}{h}\right)^3 - \left(\frac{b_{\text{eff}}}{b_w} - 1\right)\left(1 - \frac{h_t}{h} - \frac{x}{h}\right)^3 \tag{4.31}$$

where x/h is given by

$$\frac{x}{h} = \frac{1}{2}\frac{(b_{\text{eff}}/b_w) + ((b_{\text{eff}}/b_w) - 1)(1 - (h_t/h))}{(b_{\text{eff}}/b_w) + (1 - (h_t/h))((b_{\text{eff}}/b_w) - 1)} \tag{4.32}$$

4.5.3.3 Flat Slabs

Unless substantial drop heads are detailed, it will be difficult to fix the shear reinforcement satisfactorily, even if proprietary shear heads are used (Chana and Clapson, 1992). It is not a normal practice to use flat slabs in sway frames partly for this reason and partly the excessive amounts of flexural steel required due to sway effects from lateral loading.

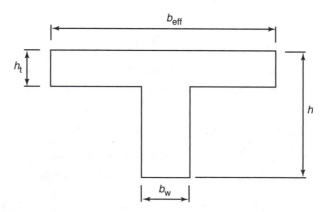

FIGURE 4.5   'T' beam dimensions

4.5.3.4 Ribbed or Waffle Slabs

These may be treated as solid slabs, provided the rib spacing is less than 1500 mm, the depth of the rib below the slab does not exceed 4 times the rib width, the depth of the flange is maximum of 50 mm or 1/10 the clear distance between the ribs, and transverse ribs are provided at a clear spacing not exceeding 10 times the overall depth of the slab. In other cases it is probable that a grillage-type analysis will need to be employed (See Section 10.8).

4.6 COLUMN LOADS

In a multi-storey structure, the frame analysis will provide the reactions in the columns. However such analyses assume that all floors are fully loaded at all times. Statistically this is conservative, thus the axial loads due to variable loading may be reduced in the lower columns of a multi-storey structure. Note, there is clearly no reduction in loads due to permanent (or quasi-permanent) loading, and each individual floor must be designed under full variable and permanent actions. The variable loading may also be adjusted for loaded area, as the loading will be more concentrated over smaller floor areas (and less concentrated over large areas). This means that column reactions may be reduced for this reason also. However reductions may be made either for the number of floors OR loaded area. These reductions are only for loading categories A to D of Table 6.1 of EN 1991-1-1. This prohibits any reduction on situations where the loads are predominantly storage or industrial.

The reduction factor α_n for the number of stories n ($n > 2$) is given by

$$\alpha_n = \frac{2 + (n-2)\psi_0}{n} \tag{4.33}$$

The reduction factor α_A for the loaded area is given by

$$\alpha_A = \frac{5}{7}\psi_0 + \frac{A_0}{A} \leq 1{,}0 \tag{4.34}$$

where A is the loaded area, A_0 the reference value of $10\,\text{m}^2$. The reduction factor is also subject to the restriction that for load categories C and D it may not be less than 0,6.

4.7 REDISTRIBUTION

This may be employed on the results of an elastic analysis in order to simulate a 'plastic' distribution of internal forces, provided sufficient ductility exists to allow the formation and action of (hypothetical) plastic hinges. An elastic analysis will always provide a lower bound solution to structural problems, but will generate higher support moments than a plastic analysis. As it is generally the supports which cause the most problem in detailing, then a method of reducing such moments may be beneficial. It should also be noted that if the 'deemed to satisfy' clauses specifying axis distance and member sizes are used for the fire limit state, then if the redistribution is greater than 15 per cent (i.e. $\delta = 0,85$), then any beams or slabs must be treated as simply supported and NOT continuous.

The existence of ductility is dependant upon the amount of reinforcement (tension or compression) (Fig. 6.4), and will always be present in an 'under-reinforced' section (See Section 6.1). The existence of such ductility will depend upon the position of the neutral axis within a section, thus limits are placed upon the neutral axis depth dependant upon the amount of redistribution allowed. This restriction effectively means that redistribution cannot be permitted in columns as the neutral axis depth can be outside the section owing to high axial forces.

Allowable plastic rotation need not be checked in reinforced concrete, provided the x/d limits given in Eq. (6.6) or (6.7) are satisfied where redistribution is used or the plastic analysis limits given in Section 6.1.1 are satisfied.

For reinforced concrete whose x/d values do not satisfy the requisite limits (e.g. columns) or for prestressed concrete the cl 5.6.3 of EN 1991-1-2 must be satisfied. This clause indicates that if the rotational capacity $\theta_{pl,d}$ is less than the calculated rotation θ_s determined over a length $0,6h$ on either side of a support (h is the overall depth of the member). The rotations θ_s are determined from the design values of both actions and materials and on the basis of the mean values of any prestressing. The neutral axis depth ratios in such cases are limited to 0,45 for concrete of Grade C50 or less, and 0,35 for higher grades. The allowable plastic rotation $\theta_{pl,d}$ may be determined by multiplying a basic value (for which recommended values are given in Table 5.6N for a shear slenderness k_λ of 3) by a correction factor k_λ given by

$$k_\lambda = \sqrt{\frac{\lambda}{3}} \qquad (4.35)$$

where λ is determined from

$$\lambda = \frac{M_{Sd}}{V_{Sd}d} \qquad (4.36)$$

where M_{Sd} and V_{Sd} are co-existent values of moment and shear.

Cl 5.6.3(3) implies this check is only required for prestressed concrete or members with a high axial load as reinforced concrete predominantly in flexure can be made to satisfy the x/d condition.

4.7.1 Redistribution Process

The redistributed moments are in equilibrium with the applied loads. This is common sense, but is frequently a condition that can be overlooked, even by the most experienced of designers!

The amount of redistribution δ is limited as follows, to 70 per cent for Class B or C reinforcement and 80 per cent for Class A.

The redistribution ratio δ is defined as the redistributed moment to the moment before redistribution. There appears no lower limit on the ratio δ except that as δ is reduced, the value of x/d is reduced and thus means that substantial amounts of compression steel will be needed in order to satisfy the moment-carrying capacity of the section. Owing to the x/d restriction, no redistribution will be possible on column moments, and the raw elastic moments should be used (cl 5.5(6)).

There are a number of drawbacks to (excessive) redistribution:

(1) It may cause increased requirements for shear reinforcement as it is the shear at a support which is generally the critical design case and the amount of tension steel at the support is reduced by redistribution.
(2) Deflections and cracking may be increased as there is less effective continuity at the support with the moment being transferred to midspan.
(3) Fire performance will be reduced as the moment reduction at the support (where the flexural tensile steel is in the top face) will control the carrying capacity as the moment at midspan rapidly decreases owing to the relatively high steel temperatures. This is the reason why data for simply supported beams is used when the redistribution is high.

Redistribution does not reduce the amount of flexural steel (it might even increase it owing to the neutral axis depth restriction requiring extra compression steel). Redistribution should only be carried out if under elastic analysis there is an excessive requirement for the top (tensile) steel at the support. This may give a bar spacing, even after checking the effect of multi-layering or bundling, such that the concrete cannot be properly compacted.

4.7.2 Reduction of Support Moments Due to Continuity

Where a beam or slab spans continuously over a support such as a wall or bearings which supply no restraint to rotation, the design moment at the support based on spans equal to the centre to centre distance may be reduced by an amount ΔM_{Ed} given by

$$\Delta M_{Ed} = F_{Ed,sup} \frac{t}{8} \tag{4.37}$$

where $F_{Ed,sup}$ is the design support reaction and t the breadth of the support. It is not clear whether this reduction applies to moments after redistribution.

Two examples demonstrating the principles will be given; the first is a basic one on an encastré beam and the second is on a continuous beam.

EXAMPLE 4.3   Encastré beam. Consider an encastré beam of 6 m span carrying a total load at ULS of 30 kN/m (Fig. 4.6(a)). Draw the resultant BM and SF diagram after 25 per cent redistribution, (i.e. $\delta = 0,75$)

The support moment from an elastic analysis is

$$qL^2/12 = 30 \times 6^2/12 = 90\,\text{kNm}$$

and the midspan moment

$$qL^2/24 = 45\,\text{kNm}$$

The point of contraflexure is at 1,27 m from the support.

The elastic (unredistributed) BMD is plotted in Fig. 4.6(b). After redistribution the support moment is $0,75 \times 90 = 67,5$ kNm. The free $BM = qL^2/8 = 135$ kNm, thus the midspan moment after redistribution is $135 - 67,5 = 67,5$ kNm.

The point of contraflexure is now at 0,88 m from the support. The final BMD is plotted in Fig. 4.6(c).

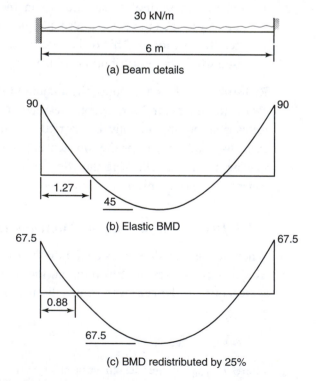

(a) Beam details

(b) Elastic BMD

(c) BMD redistributed by 25%

FIGURE 4.6   Redistribution for an encastré beam

EXAMPLE 4.4. Redistribution on a three span continuous beam (Fig. 4.7). A three span continuous beam of constant section has end spans of 8 m and a central span of 10 m and carries a characteristic permanent load of 50 kN/m and a characteristic variable load of 35 kN/m. Determine the Bending Moment Distribution (BMD) after 30 per cent redistribution (Class B or C reinforcement must be used).

A number of load cases need considering:

(i) All spans fully loaded: $q = 1{,}35 \times 50 + 1{,}5 \times 35 = 120$ kN/m.
(ii) End spans loaded with permanent (50 kN/m) and the central 10 m span with full load.
(iii) One 8 m span and the centre 10 m span with full load, and the remaining 8 m span with permanent load.

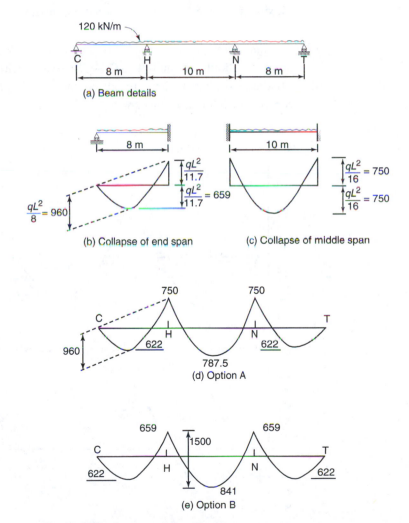

FIGURE 4.7   Plastic design for a continuous beam

(iv) One 8 m span loaded with full load, and the 10 m span and the other 8 m span with permanent load only.

(v) End 8 m spans loaded with full load and centre 10 m span with permanent (50 kN/m).

The analysis was carried out using (C61) to (C63), and the resultant superimposed BM diagrams are plotted in Fig. 4.8(a).

Strictly load case (i) is not part of the requirement of EN 1991-1-1, but is included to demonstrate that for a three span beam it produces moments around 7 per cent lower than load case (iv). As the number of spans increases the error will decrease. Load case (i) will not be considered in the redistribution. Redistribution is carried out in Table 4.1, and the redistributed BMD plotted in Fig. 4.8(b).

Design notes:

(1) Maximum numerical design moments; hogging 743 kNm, sagging 685 kNm (load case (v)).
(2) The x/d ratio for the supports is limited to 0,208. That for internal midspan moments should be limited to 0,448 (the value with no redistribution).
(3) Load case (v) has the lowest midspan moment in the central span BC. This will control curtailment in this span.

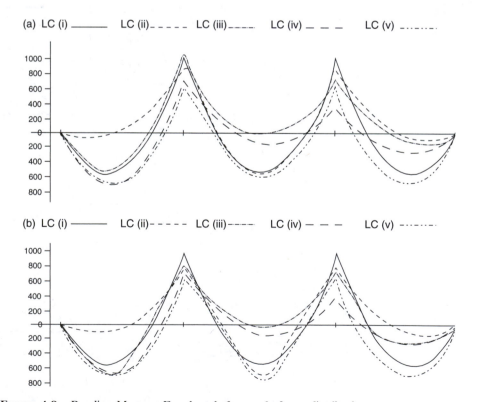

FIGURE 4.8 Bending Moment Envelope before and after redistribution

TABLE 4.1   Redistribution in Example 4.4.

Load case	A (kNm)		B (kNm)		C (kNm)		D (kNm)
(a) Before redistribution							
(i)	0	357	−987	513	−987	357	0
(ii)	0	90	−842	658	−842	90	0
(iii)	0	502	−1062	616	−717	122	0
(iv)	0	650	−681	128	−336	300	0
(v)	0	681	−606	19	−606	681	0
(b) 30 per cent redistribution on (iii) at B							
(iii)	0	**624**	**−743**	**770**	−717	122	0
(c) 12 per cent redistribution on (ii) to give the same hogging moments as (iii)							
(ii)	0	**115**	**−743**	**757**	**−743**	**115**	0

Notes
a) Sign convention: Sagging positive, hogging negative;
b) Redistributed moments (and resultant changes to 'midspan' moments) shown in **bold**;
c) The midspan moments are maximum moments.

It is prudent to carry out shear design based on the worst of the cases before and after redistribution. The results for the end shears are given in Table 4.2.

The Shear Force diagrams are plotted in Figs 4.9(a) and (b), respectively for the unredistributed and redistributed load cases.

Use the following numerical values of Shear Force (kN) for design:

Span	End			
AB	A	404	B	613
BC	B	634	C	634 (symmetry)

TABLE 4.2   End shears before and after redistribution.

	A	B		C		D
Load case	AB (kN)	BA (kN)	BC (kN)	CB (kN)	CD (kN)	DC (kN)
(a) Before redistribution						
(i)	357	−603	600	−600	603	−357
(ii)	95	−305	600	−600	305	−95
(iii)	347	−613	634	−566	290	−110
(iv)	395	−565	284	−216	242	−158
(v)	404	−556	250	−250	556	−404
(b) After redistribution						
(ii)	107	−293	600	−600	293	−107
(iii)	387	−573	597	−603	290	−110

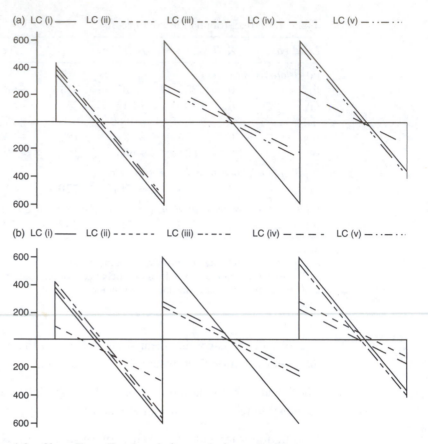

Figure 4.9 Shear Force Envelope before and after redistribution

In all cases, the contribution from the tension reinforcement will be based on the design AFTER redistribution, and for the end support after any curtailment.

4.8 Plastic Analysis

Whilst plastic analysis is common for slabs when yield line analysis is used, it is less common for beams. If plastic collapse is used, it must be remembered that the x/d value for flexural design is limited to 0,25 for Grade C50 or less and 0,15 for higher Grades. This will be more onerous than the limits imposed following redistribution after elastic analysis at midspan hinges. It may not be more onerous at the support as Example 4.4 demonstrates with a limit after redistribution of 0,204. The principles of plastic analysis are covered in any structural analysis text book (e.g. Moy, 1996). The method is best illustrated by an example.

Example 4.5. Plastic analysis of a continuous beam. Consider the same beam as Example 4.4.

It is generally sufficient to consider each span under full loading, so using standard solutions,

Central 10 m span (Fig. 4.7(c)):

$$M_{Sd} = q_{Sd}L^2/16 = 120 \times 10^2 2/16 = 750\,\text{kNm}$$

End 8 m span (Fig. 4.8(b)2):

$$M_{Sd} = (1{,}5 - \sqrt{2})q_{Sd}l^2 = (1{,}5 - \sqrt{2}) \times 120 \times 8^2 = 659\,\text{kNm}$$

The individual spans do not collapse simultaneously. The centre span is critical as the required design moment is higher. There are two alternative solutions that may now be adopted.

(a) Accept that the centre of the 8 m span remains elastic with a design value determined from Eq. (4.28)

$$M_{max} = \frac{qL_{AB}^2}{8} + \frac{M_{AB} - M_{BA}}{2} + \frac{1}{2q}\left(\frac{M_{AB} + M_{BA}}{L_{AB}}\right)^2$$

$$= \frac{120 \times 8^2}{8} + \frac{-750}{2} + \frac{1}{2 \times 120}\left(\frac{-750}{8}\right)^2$$

$$= 622\,\text{kNm}$$

at a position x_{max} from Eq. (4.29)

$$x_{max} = \frac{L_{AB}}{2} + \frac{M_{AB} + M_{BA}}{qL_{AB}} + \frac{8}{2} - \frac{750}{120 \times 8} = 3.22\,\text{m}$$

(b) The alternative is to keep the sagging moment in the end spans and the hogging support moment as 659 kNm and to increase the midspan design moment in the 10 m span. The increased moment is given from Eq. (4.27) as

$$M_{mid} = \frac{qL_{AB}^2}{8} + \frac{M_{AB} - M_{BA}}{2}$$

$$= \frac{120 \times 10^2}{8} + \frac{-659 - 659}{2} = 841\,\text{kNm}$$

For each solution, the resultant BMD's and SFD's are illustrated in Figs 4.7(d) and (e), respectively.

From a practical point of view the second solution is probably preferable since the hogging moment and hence the quantity of tension reinforcement in the top face is reduced. The requirements for shear steel are unlikely to be much different as, although the end reaction is slightly higher for (b), the tension steel requirement at midspan is also higher. The second solution is slightly less satisfactory in fire as the sagging moments will reduce faster than the hogging moments. If 'deemed to satisfy' clauses are used for the fire limit state, it would be prudent, although possibly conservative, to use the data for simply supported beams.

REFERENCES AND FURTHER READING

Chana, P. and Clapson, J. (1992) Innovative shearhoop system for flat slab construction. *Concrete*, **26(1)**, 21–24.

EN 1990: Eurocode – Basis of structural design.

ISE (2002) *Safety in Tall Buildings and Other Buildings with Large Occupancy*. Institution of Structural Engineers.

Mlakar, P.F., Dusenberry, D.O., Harris, J.R., Haynes, G., Phan, L.T. and Sozen, M.A. (2003) Pentagon Building Performance Report, *Civil Engineering*, **73(2)**, 43–55.

Moy, S.S.J. (1996) *Plastic Methods for Steel and Concrete Structures* (2nd Edn), MacMillan, Hampshire, UK.

Oliviera, D.R.C., Regan, P.E. and Melo, G.S.S.A. (2004) Punching resistance of RC slabs with rectangular columns. *Magazine of Concrete Research*, **56(3)**, 123–138.

Regan, P.E. (1981) *Behaviour of Reinforced Concrete Flat slabs*. CIRIA Report 89.

Wearne, P. (1999) *Collapse – Why Buildings Fall Down*. Channel 4 Books, London, UK.

Wood, J.G.M. (2003) Pipers Row Car Park Collapse: Identifying risk. *Concrete*, **37(9)**, 29–31

Yandzio, E. and Gough, M. (1999) *Protection of Buildings against Explosions*. SCI Publication 244, SCI.

Chapter 5 / Durability, Serviceability and Fire

Traditionally engineers have been content to consider the control of durability using 'deemed to satisfy' approaches with respect to phenomena such as cracking or deflection. This traditional approach, although still valid, needs to be supplemented by more fundamental approaches. For a large number of cases, control of cracking or deflection by bar spacing rules or span to effective depth ratios is satisfactory, but where control of cracking is vital to the use or appearance of a structure, calculations may become necessary. Further in order to allow the formulation of simple rules, assumptions need to be made which are conservative in nature, thus it may be more economic in the use of materials to use more rational methods to assess, say, deflections.

Further, one of the major problems besetting concrete structures are those due to loss of durability through reinforcement corrosion, alkali aggregate reaction or sulfate attack at an age below the design life of the structure.

5.1 MECHANISMS CAUSING LOSS OF DURABILITY

The ideal situation would be to be able to determine when loss of durability would occur such that acceptable limits were breached, i.e. durability would become a true limit state phenomenon. In practice, the requirements for detailing such as cover are set to ensure that the loss of durability should not occur within the service life of the structure, provided there is good quality assurance on the concrete and on site procedures to ensure that specified cover is met (Clark *et al.*, 1997). The major concerns with loss of durability are corrosion of the reinforcement, alkali-aggregate reaction and sulfate attack. The Concrete Society have issued a Technical Report (Bamforth, 2004) which provides an excellent background to measures which may be taken to enhance durability.

5.1.1 Reinforcement Corrosion

At its extreme, corrosion of the reinforcement will lead to spalling of the concrete cover with the resultant exposure of such reinforcement, together with loss in strength of the structural member as all or part of the cross-sectional area of the rebar is lost. In a less extreme form, unsightly surface rust staining will occur. When the first signs of spalling are observed it will often be too late to combat corrosion and recourse will be needed to expensive remedial repairs or mitigation. In the ultimate case, expensive and often very difficult replacement of the structural member will be necessary. Spalling occurs as the chemical products of the corrosion reaction occupy a larger volume than the original rebar.

Corrosion occurs when the pH value of the concrete reduces from its original high value of around 13 during placing to below 9 at a later stage (Page and Treadaway, 1982). This change in pH is known as a loss in passivity and is due to two possible causes:

(a) Carbonation, and
(b) Chloride attack.

5.1.1.1 Carbonation

This is a gradually occurring attack caused by the penetration of acidic gasses (mostly carbon dioxide) and reaction with any free alkali present. This reaction lowers the pH value. Since the mechanism is diffusion controlled, the depth of carbonation is proportional to the square root of time. The covers to reinforcement specified in EN 1992-1-1 are such that for good quality concrete the depth of carbonation during the design service life of the structure should not depassify the concrete between the concrete surface and any rebar. The process of carbonation is illustrated in Fig. 5.1.

The minimum cover $c_{min,dur}$ required by EN 1992-1-1 together with the environmental conditions and structure classification applicable to each are summarized in Table 5.1. The standard structure classification is Class 4 (design life of 50 years). However, the structure classification is adjusted for parameters such as increased design life, higher strength concretes than those noted in Table 5.1, slab type elements, and special quality control. Increasing the design life to 100 years increases the structure classification to two levels higher and the remaining factors reduce the structure classification by one level lower. The actual cover required c_{min} is given by

$$c_{min} = \max[c_{min,b}; c_{min,dur} + \Delta c_{dur,\gamma} - \Delta c_{dur,st} - \Delta c_{dur,add}: 10mm] \qquad (5.1)$$

where $c_{min,b}$ is the minimum cover due to bond requirement, $\Delta c_{dur,\gamma}$ additive safety element (recommended value of zero), $\Delta c_{dur,st}$ reduction in cover due to use of stainless steel (recommended value of zero), and $\Delta c_{dur,add}$ reduction due to use of

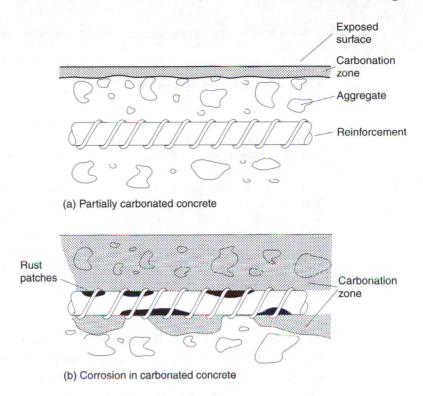

Exposed
surface

Carbonation
zone

Aggregate

Reinforcement

(a) Partially carbonated concrete

Rust
patches

Carbonation
zone

(b) Corrosion in carbonated concrete

FIGURE 5.1   Corrosion due to carbonation

additional protection (recommended value is zero). The minimum cover due to bond $c_{min,b}$ for reinforced concrete is bar diameter for separated bars and equivalent diameter given by $\phi_n = \phi/\sqrt{n_b}$ ($\leq 55\,mm$) where n is the number of bars (from Table 4.2 of EN 1992-1-1). For prestressed concrete, it is the duct diameter for circular post tensioning ducts, twice the diameter of pretensioning strand or 3 times the diameter of indented wire.

The nominal cover c_{nom} used in design is given by

$$c_{nom} = c_{min} + \Delta c_{dev} \tag{5.2}$$

The recommended value for Δc_{dev} is $10\,mm$, although the code does allow reductions where there is a quality control system (more likely in precast concrete) or where a very accurate device is used for monitoring and any resulting non-conforming items rejected.

The covers specified in Table 5.1 are only adequate when the concrete has the requisite low coefficient of permeability. This can only be ensured if the concrete has a sufficiently high workability to enable the concrete to be properly placed and vibrated. Workability should be controlled through the use of admixtures such as plasticisers or super-plasticisers and NOT by the addition of extra water which may give problems with shrinkage cracking (also causing potential loss in durability

TABLE 5.1 Minimum Covers $c_{dur,min}$ (mm) for reinforced concrete.

Exposure Class	Environment and type	Structure class						Min strength class
		S1	S2	S3	S4	S5	S6	
X0	Unreinforced; reinforced in buildings	10	10	10	10	15	20	C30/37
XC1	Dry; permanently wet low humidity	10	10	10	15	20	25	C30/37
XC2/XC3	Wet, rarely dry; foundations; external sheltered concrete	10	15	20	25	30	35	C35/45
XC4	Cyclic wetting and drying	15	20	25	30	35	40	C40/50
XD1	Airborne chlorides	20	25	30	35	40	45	C40/50
XS1	Airborne salt — near to coast	20	25	30	35	40	45	C40/50
XD2	Swimming pools; industrial chlorides	25	30	35	40	45	50	C40/50
XS2	Permanently submerged in sea	25	30	35	40	45	50	C45/55
XD3	Cyclic due to chloride spray	30	35	40	45	50	55	C45/55
XS3	Marine tidal and splash zones	30	35	40	45	50	55	C45/55

unless designed for). It is therefore essential when detailing the reinforcement, especially at laps including those between starter bars and main bars to ensure the concrete can flow. Construction joints which are often at points of high internal forces are also areas where the concrete quality needs considering as honeycombing will give a high coefficient of diffusion and thus allow easy ingress of carbon dioxide or chloride-bearing fluids.

Problems with construction joints can also be eased by kickerless construction methods and the use of retarding agents to delay the initial set of the concrete.

A further influence on achieving acceptable levels of permeability is adequate curing of the concrete. Concrete which is not properly cured is a potential source for concern. The economics of modern construction has forced reductions in times before the formwork is struck. EN 206-1 provides some data on minimum curing times related to ambient conditions, concrete temperature during curing and rate of strength development. The earlier the formwork is struck, the more the surface-free water will evaporate rather than combine chemically with the cement. It is this surface layer which is important to ensure the concrete member is relatively impermeable. The reduction of striking times has been made possible by grinding Ordinary Portland Cements (OPC) very fine, thereby increasing the specific surface area and increasing hydration rates, and turning OPC's into effectively Rapid Hardening Cements (RHC). This increase in hydration is also made possible by increasing the proportion of C_3A in the cement. Drawbacks to this trend are the resultant increase in heat of hydration and early thermal or shrinkage cracks and that at normal ages the concrete strength becomes potentially high enough to cause spalling problems in fire (Bailey, 2002).

Beeby (1978) has indicated that cracking parallel to the rebar is more harmful in corrosive conditions than that of the normal to the bar, as such cracking allows an

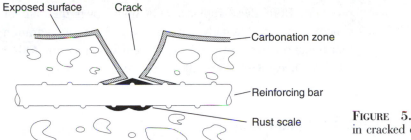

FIGURE 5.2 Corrosion in cracked concrete

easier path to the surface of the bar. Such parallel cracks are generally due to shrinkage or early thermal movement and can be reduced by the provision of adequate distribution reinforcement. Such reinforcement is more effective when it is of small diameter at relatively low spacing, and may need to exceed the minimum laid down in any design code.

5.1.1.2 Chloride Attack

Some free chloride ions exist in the hydrating cement but in low proportions. The once historically popular use of admixtures containing $CaCl_2$ is now expressly forbidden, but contaminants containing chlorides can still be introduced when sea-dredged aggregates or sea water is used in the production of concrete. Both these should be avoided wherever possible. However, two possible sources of chloride ingress cannot be avoided. These are where the structure is exposed to a marine environment or to de-icing salts on highways. The major problem with both these is chloride ingress through cracks parallel to rebar (Fig. 5.2), and it is thus necessary to control the width of such cracks by explicit calculations.

5.1.2 Alkali-Aggregate Reaction

In this case a chemical reaction occurs between the alkalinity of the pore water (usually Na_2O or K_2O), due either to alkaline elements within the cement or an external source such as sea water or de-icing salts, and certain types of silica in the aggregate (this is known as Alkali–Silica Reaction, or ASR). The reaction causes a positive volume change with associated microcracking, which will eventually result in loss of cover to the rebar. Most, but not all, British aggregates are not susceptible to this type of reaction. Unfortunately, there is no simple test to determine any such susceptibility, and thus care needs taking to limit the alkali content of the mix, the effective alkali content of the cement itself and where the structural element is exposed to an alkaline environment, consideration should be given to provision of an impermeable surface membrane. The alkali content of a cement can be reduced by using a cement replacement such as Pulverized Fuel Ash (PFA) or Ground Granulated Blast Furnace Slag (GGBFS), both of which are industrial waste products. Both these replacements slow down the rate of strength gain in the concrete and additionally have a lower heat of hydration, thus reducing early thermal shrinkage and cracking. Recent guidance on ASR has been published by BRE in Digest 330 (BRE, 2004).

5.1.3 Sulfate Attack

This occurs due to attack from free sulfates contained in groundwater or soil, and is therefore generally limited either to buried structures including foundations or structures in an aggressive environment where sulfates are naturally present, e.g. certain types of sewage, industrial effluent or chemical process plants. It is generally sufficient to combat sulfate attack by using Sulfate-Resisting Portland Cements (SRPC). In these cases, the requirement to ensure the existence of low permeability concrete in the structure is paramount as any ingress of sulfates should be prevented (Dunster and Crammond, 2003).

5.1.4 Taumasite Attack

This phenomenon is caused by a chemical reaction between cement and certain types of clay in the presence of groundwater. It will therefore affect foundations or buried concrete. Fortunately the types of clay involved only occur in limited areas and thus the phenomenon is not widespread (Taumasite Experts Group, 1999). A useful overview of concrete deterioration, methods of overcoming it and specialist techniques for concrete below ground is given in a CIRIA Report (Henderson *et al.*, 2002).

5.2 Serviceability Limit States

As mentioned in Chapter 1, these are simply a series of criteria which are applied to the performance of the structure during its normal working life under the action of service (or working) loads or actions.

For the majority of structures, the two most important criteria are deflections and crack widths. These both help ensure that durability is not unduly impaired and that the structure has no visible damage (or detrimental effects). In some cases, vibration response needs to be considered where machinery is involved or where wind loading can cause vortex shedding and subsequent oscillatory motion (Tacoma Narrows Suspension Bridge). Concrete structures other than some footbridges are seldom slender enough to cause problems under normal variable loading. A recent, but, classic case of such problems occurring was the Millennium Bridge in London described in Dallard *et al.* (2001). In all cases, it is necessary for the engineer to assess the risks of any phenomenon likely to affect any structure and then carry out any requisite design checks.

The design for fire resistance has been included in this chapter partly because the subject can be handled using simple clauses specifying member thicknesses or rebar cover (as in durability) and partly because, although fire is an ultimate accidental limit state in that collapse is being considered, loading lower than that at ULS is used to determine the internal stress resultants.

5.2.1 Deflections

In general structural deflections are traditionally limited to span/250, although where partitions or finishes may be damaged, this limit is reduced to span/500 (cl 7.4.1). The deflections may be controlled either by the use of span/effective depth ratios or by explicit calculations.

5.2.1.1 Span/Effective Depth Ratios

This approach can be justified in outline as follows:
For any beam the midspan deflection Δ for a given loading pattern is obtained from

$$\Delta = k_1 \frac{W L^3}{E_s I_s} \tag{5.3}$$

where k_1 is a coefficient dependant upon the loading pattern and the support conditions, W the total service load, L the span, and $E_s I_s$ is flexural rigidity of the section in 'steel' units (assumed constant at the value obtaining at the maximum moment).

The maximum moment M_s is given by

$$M_s = k_2 W L \tag{5.4}$$

where k_2 is a factor dependant upon the load pattern.
The stress in the tensile reinforcement σ_s is given by

$$\sigma_s = \frac{M_s(d-x)}{I_s} \tag{5.5}$$

where d is the effective depth and x the depth to the neutral axis.

Substitute Eqs (5.4) and (5.5) into Eq. (5.3) to give

$$\frac{\Delta}{L} = \frac{k_1}{k_2} \frac{\sigma_s}{d-x} \frac{L}{d} \tag{5.6}$$

TABLE 5.2 Recommended Basic K values for deflection (from Table 7.4N of EN 1992-1-1).

		l/d	
	K	$\rho=1,5\ \%$	$\rho=0,5\ \%$
Simply supported beam			
One or two way simply supported slab	1,0	14	20
End span of a continuous beam or	1,3	18	26
one way spanning slab			
Two way spanning slab continuous			
over the long edge			
Interior span of continuous beam or slab	1,5	20	30
Flat slab (longer span)	1,2	17	24
Cantilever	0,4	6	8

Note: the reinforcement $\rho=0,5\%$ represents lightly stressed concrete members and $\rho=1,5\%$ represents highly stressed concrete members.

Equation (5.6) indicates that if the deflection requirement is expressed as a fraction of the span, then it is directly related to the span/effective depth ratio and is also a function of the stress level in the reinforcement.

The method used in EN 1992-1-1 is to take basic span/effective depth ratios and modify them appropriately to allow for factors such as type of reinforcement (cl 4.4.3.2).

The basic K factors for deflection are given in Table 5.2 (from Table 7.4N of EN 1992-1-1).

The Code indicates that the span effective depth ratios have been obtained from a parametric study and correspond to a service stress in the reinforcement of 310 MPa (with a yield stress of 500 MPa).

The basic span effective depth ratio l/d is given by,
for $\rho \le \rho_0$

$$l/d = K\left[11 + 1,5\sqrt{f_{ck}}\left(\frac{\rho_0}{\rho}\right) + 3,2\sqrt{f_{ck}}\left(\frac{\rho_0}{\rho} - 1\right)^{1,5}\right] \qquad (5.7)$$

and for $\rho > \rho_0$

$$l/d = K\left[11 + 1,5\sqrt{f_{ck}}\left(\frac{\rho_0}{\rho - \rho'}\right) + \frac{1}{12}\sqrt{f_{ck}}\sqrt{\frac{\rho'}{\rho_0}}\right] \qquad (5.8)$$

where ρ_0 is a reference steel ratio defined as $0,001\sqrt{f_{ck}}$, f_{ck} the characteristic strength of the concrete, ρ the tensile steel ratio $(=A_s/bd)$ and ρ' the compressive steel ratio $(=A_s'/bd)$. The values of ρ and ρ' are based the required values of A_s and A_s', respectively.

For two way spanning slabs, the shorter span is used to determine the critical span/effective depth ratio, but for flat slabs the larger span should be taken.

Additional modifications may be made for

- Geometry

  The values from Eqs (5.7) or (5.8) are multiplied by a factor when flanged sections are used. The factor takes a value of 0,8 for $b_{eff}/b_w \geq 3$, and 1 for $b_{eff}/b_w = 1,0$, where b_{eff} is the effective width of the flange (Section 6.4) and b_w the web width. Linear interpolation can be used between these values such that the factor is given by

$$0,8 + 0,1\left(3 - \frac{b_{eff}}{b_w}\right) \leq 0,8 \qquad (5.9)$$

- Reinforcement stress levels

  As noted above, the tabular values have been based on a service stress σ_s in the reinforcement of 310 MPa, corresponding to the service stress taken as approximately $(5/8)f_{yk}$ with a reinforcement strength of 500 MPa.

For reinforcement with other characteristic strengths, the tabular values should be multiplied by a factor $310/\sigma_1$ given by

$$\frac{310}{\sigma_s} = \frac{500}{f_{yk}(A_{s,req}/A_{s,prov})} = \frac{500}{f_{yk}}\frac{A_{s,prov}}{A_{s,req}} \qquad (5.10)$$

where f_{yk} is the characteristic strength in the reinforcement, $A_{s,req}$ is the area of tension reinforcement required to resist the applied moment and $A_{s,prov}$ is the actual amount of tensile steel. This latter correction factor takes account of the fact that any extra reinforcement over that actually needed will reduce the service stress in the steel by increasing the second moment of area.

Additionally, where supporting partitions (or finishes) may be damaged by excessive deflections the values from Eqs (5.7) or (5.8) should be multiplied by $7/l_{eff}$ for beams and slabs (other than flat slabs) with spans exceeding 7 m, and for flat slabs where the greater span l_{eff} exceeds 8,5 m, by $8,5/l_{eff}$, where l_{eff} is the effective span.

EXAMPLE 5.1 Deflection control using span/effective depth ratios. Check the design of the cantilever shown in Fig. 5.3 is suitable.

$$A_{s,prov}/bd = 5026/635 \times 370 = 0,0214$$
$$\rho = A_{s,req}/bd = 3738/635 \times 370 = 0,0159$$
$$A'_s/bd = 1608/635 \times 370 = 0,00684$$
$$f_{ck} = 25 \text{ MPa}, \rho_0 = 0,001\sqrt{f_{ck}} = 0,001 \times \sqrt{25} = 0,005$$

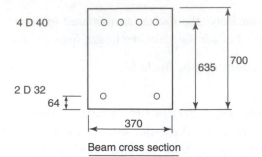

4 D 40

635 | 700

2 D 32
64

370

Beam cross section

Beam data : Transient load 40 kN/m
Permanent load 30 kN/m
Span (Cantilever) 4 m
$f_{ck}1<$ 25 MPa
f_{yk} 500 MPa

FIGURE 5.3   Beam data for Examples 5.1 and 5.2

As $\rho > \rho_0$, so use Eq. (5.8)

$$l/d = K\left[11 + 1{,}5\sqrt{f_{ck}}\left(\frac{\rho_0}{\rho - \rho'}\right) + \frac{1}{12}\sqrt{f_{ck}}\sqrt{\frac{\rho'}{\rho_0}}\right]$$

$$= 0{,}4\left[11 + 1{,}5\sqrt{25}\left(\frac{0{,}005}{0{,}0159 - 0{,}00684}\right) + \frac{1}{12}\sqrt{25}\sqrt{\frac{0{,}00684}{0{,}005}}\right] = 6{,}25$$

Correction for service stress and reinforcement.

$$310/\sigma_s = 500/(f_{yk}A_{s,req}/A_{s,prov}) = 500/(500 \times 3738/5026) = 1{,}34$$

Actual allowable span/effective depth ratio $= 6{,}25 \times 1{,}34 = 8{,}38$.
Actual span/effective depth ratio $= 4/0{,}635 = 6{,}3$ (or minimum effective depth $= 4000/8{,}38 = 477$ mm). The beam is therefore satisfactory.

5.2.1.2 Direct Calculation Method

This requires knowledge of the curvatures in the section in both a cracked and uncracked state. In a cracked section, the contribution of the concrete in tension is ignored, whereas in an uncracked section the concrete is assumed to act elastically in both the tension and compression zones with the maximum tensile capacity f_{ctm} of the concrete. For concrete grades not greater than C50, this is taken as

$$f_{ctm} = 0{,}3(f_{ck})^{2/3} \tag{5.11}$$

where f_{ck} is the characteristic strength of the concrete. The effective Young's modulus for the concrete $E_{c,eff}$ is given by

$$E_{c,eff} = \frac{E_{cm}}{1 + \phi(\infty, t_0)} = \frac{22\left(\frac{f_{ck}+8}{10}\right)^{0,3}}{1 + \phi(\infty, t_0)} \tag{5.12}$$

where $\phi(\infty, t_0)$ is the creep coefficient.

The curvature (or other appropriate deformation parameter) α under any condition is given by

$$\alpha = \zeta\alpha_{II} + (1 - \zeta)\alpha_I \tag{5.13}$$

where α_I is calculated on the uncracked section, α_{II} is calculated on the cracked section, and ζ is a distribution parameter which allows for tension stiffening and is given by

$$\zeta = 1 - \beta\left(\frac{\sigma_{sr}}{\sigma_s}\right)^2 \tag{5.14}$$

where β allows for the duration of the loading and is 1,0 for a single short term load and 0,5 for sustained or repetitive loading, σ_s is the stress in the steel calculated on a cracked section and σ_{sr} is the stress in the steel as the section is just about to crack. Beeby, Scott and Jones (2005) suggest that as the decay of tensile stress is relatively fast (i.e. a matter of hours), then β should always be taken as 0,5. In flexure, the ratio σ_s/σ_{sr} can be replaced by M/M_{cr} (where M_{cr} is the moment which just causes cracking) or in tension by N/N_{cr} (where N_{cr} is the force which just causes cracking). For an uncracked section ζ is zero.

The curvature due to shrinkage $(1/r_{cs})$ is given by

$$\frac{1}{r_{cs}} = \varepsilon_{cs}\alpha_e\frac{S}{I} \tag{5.15}$$

where ε_{cs} is the shrinkage strain, α_1 the modular ratio (defined as $E_s/E_{c,eff}$), S the first moment of area of the reinforcement about the centroidal axis and I the second moment of area. Both S and I should be calculated for the cracked and uncracked section and combined using Eq. (5.13) with α replaced by $1/r_{cs}$.

The section properties required are determined as follows:

(a) Uncracked section (Fig. 5.4):
    In an uncracked section the concrete is taken as linear elastic with equal Young's moduli for both compression and tension. The reinforcement will be linear elastic.

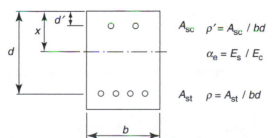

FIGURE 5.4   Cross section data

The second moment of area I_{gross} is given by

$$\frac{I_{\text{gross}}}{bd^3} = \frac{1}{3}\left(\frac{x}{d}\right)^3 + \frac{1}{3}\left(\frac{h}{d} - \frac{x}{d}\right)^3 \tag{5.16}$$

where x/d is given by

$$\frac{x}{d} = \frac{1/2(h/d)^2 + \alpha_e(\rho'(d'/d) + \rho)}{(h/d) + \alpha_e(\rho' + \rho)} \tag{5.17}$$

The uncracked moment M_{uncr} that can be taken by the section is given by

$$M_{\text{uncr}} = \frac{f_{\text{ct}}I_{\text{gross}}}{h - x} \tag{5.18}$$

where f_{ct} is the relevant tensile stress in the concrete, and the curvature $1/r_{\text{M,uncr}}$ is given by

$$\frac{1}{r_{\text{M,uncr}}} = \frac{1}{h - x}\frac{f_{\text{ct}}}{E_{\text{c,eff}}} \tag{5.19}$$

The moment M_{cr} and curvature $1/r_{\text{cr}}$ at the point of cracking is obtained by setting $f_{\text{ct}} = f_{\text{ctm}}$ in Eq. (5.19).

(b) Cracked section:

In a cracked section, the concrete is taken as linear elastic in compression and cracked in tension. The reinforcement will be linear elastic.

The second moment of area I_{cr} is given by

$$\frac{I_{\text{cr}}}{bd^3} = \frac{1}{3}\left(\frac{x}{d}\right)^3 + \rho'\alpha_e\left(\frac{x}{d} - \frac{d'}{d}\right)^2 + \rho\alpha_e\left(1 - \frac{x}{d}\right)^2 \tag{5.20}$$

where x/d is given by,

$$\frac{x}{d} = -\alpha_e(\rho' + \rho) + \left[\alpha_e^2(\rho' + \rho)^2 + 2\alpha_e\left(\rho'\frac{d'}{d} + \rho\right)\right]^{0,5} \tag{5.21}$$

The moment M_{cr} taken by the section is given by

$$M_{\text{cr}} = \frac{f_{\text{cs}}I_{\text{cr}}}{x} \tag{5.22}$$

where f_{cs} is the compressive stress in the concrete, and the curvature $1/r_{\text{cr}}$ is given by

$$\frac{1}{r_{\text{M,cr}}} = \frac{1}{x}\frac{f_{\text{cs}}}{E_{\text{c,eff}}} \tag{5.23}$$

Where there is no compression steel Eqs (5.20) and (5.21), reduce to

$$\frac{I_{cr}}{bd^3} = \frac{1}{3}\left(\frac{x}{d}\right)^3 + \rho\alpha_e\left(1 - \frac{x}{d}\right)^2 \tag{5.24}$$

where x/d is given by,

$$\frac{x}{d} = -\alpha_e\rho + \left[\alpha_e^2\rho^2 + 2\alpha_e\rho\right]^{0.5} = \alpha_e\rho\left[\sqrt{1 + \frac{2}{\alpha_e\rho}} - 1\right] \tag{5.25}$$

The resultant deflections can be determined using two methods of deceasing accuracy:

- Integration of the curvatures calculated at discrete increments along the length of the member,
- direct use of the curvature at the point of maximum moment interpolating between the cracked and uncracked values section using Eq. (5.13).

In the exact method, the total curvature $1/r_{tot}$ at any point is related to the deflection w by the following expression

$$\frac{1}{r_{tot}} = \frac{d^2w}{dx^2} \tag{5.26}$$

The second order differential can be replaced by its finite difference form to give

$$\frac{1}{r_{tot}} = \frac{w_{x+1} + w_{x-1} - 2w_x}{(\Delta x)^2} \tag{5.27}$$

where the deflections are determined at the point x and either side of it, and Δx is the incremental length along the beam between points at which the curvature is calculated. For a simply supported (or continuous) beam the deflections are zero at the supports and for a cantilever the deflection and slope are zero at the (encastré) support.

In the approximate method, the deflections are calculated assuming the whole member is either cracked or uncracked and the resultant deflection determined using the maximum curvature derived from substituting Eq. (5.4) into Eq. (5.3),

$$\Delta = \frac{k_1}{k_2}\frac{M}{EI}L^2 = k_3\frac{1}{r}L^2 \tag{5.28}$$

where k_3 depends on the loading pattern and L is the span. For common load cases, values of k_3 are given in Fig. 5.5.

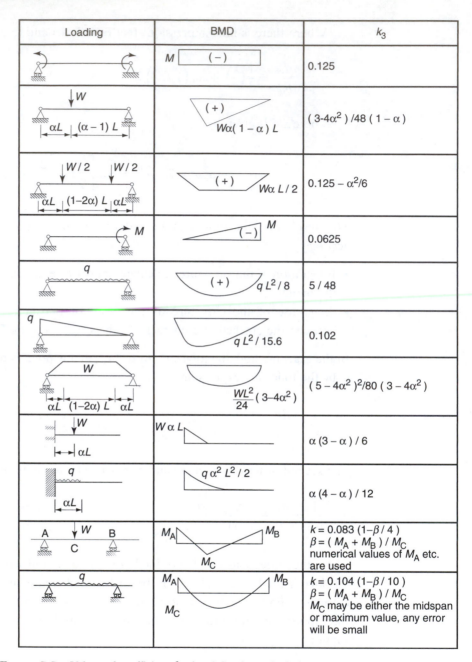

Loading	BMD	k_3
	M $\boxed{(-)}$	0.125
$\downarrow W$ / αL / $(\alpha - 1) L$	$(+)$ $W\alpha(1-\alpha)L$	$(3-4\alpha^2)/48(1-\alpha)$
$\downarrow W/2$ $\downarrow W/2$ / αL $(1-2\alpha)L$ αL	$(+)$ $W\alpha L/2$	$0.125 - \alpha^2/6$
M	$(-)$ M	0.0625
q	$(+)$ $qL^2/8$	$5/48$
q	$qL^2/15.6$	0.102
W / αL $(1-2\alpha)L$ αL	$\dfrac{WL^2}{24}(3-4\alpha^2)$	$(5-4\alpha^2)^2/80(3-4\alpha^2)$
$\downarrow W$ / αL	$W\alpha L$	$\alpha(3-\alpha)/6$
q / αL	$q\alpha^2 L^2/2$	$\alpha(4-\alpha)/12$
A $\downarrow W$ B / C	M_A M_B / M_C	$k = 0.083(1-\beta/4)$ $\beta = (M_A + M_B)/M_C$ numerical values of M_A etc. are used
q	M_A M_B / M_C	$k = 0.104(1-\beta/10)$ $\beta = (M_A + M_B)/M_C$ M_C may be either the midspan or maximum value, any error will be small

FIGURE 5.5 Values of coefficient k_3 for deflection calculations

EXAMPLE 5.2 Deflection calculations. Determine the actual deflection for the cantilever shown in Fig. 5.3.

Basic data:

$$f_{ctm} = 0{,}30f_{ck}^{2/3} = 0{,}30 \times 25^{2/3} = 2{,}56\,\text{MPa}.$$
$$E_{cm} = 22((f_{ck}+8)/10)^{0,3} = 22((25+8)/10)^{0,3} = 31{,}5\,\text{GPa}.$$

Shrinkage strain (Table 3.2 EN 1992-1-1):
The shrinkage strain ε_{cs} comprises two components: drying shrinkage (ε_{cd}) and autogenous shrinkage (ε_{ca}).

Drying shrinkage ε_{cd}:
Assuming 40 per cent relative humidity, the nominal shrinkage strain $\varepsilon_{cd,0}$ is $0{,}60 \times 10^{-3}$. Assume loading is at 28 days (i.e. $t = 28$) and that curing finished at a time of 3 days (i.e. $t_s = 3$), then the shrinkage strain at time of loading is

$$\varepsilon_{cd}(t) = \beta_{ds}(t)k_h\varepsilon_{cd,0}$$

where k_h is a parameter dependant upon the notional size h_0 given by $2A_c/u$, and $\beta_{ds}(t,t_s)$ is given by

$$\beta_{ds}(t,t_s) = \frac{t - t_s}{t - t_s + 0{,}04\sqrt{\left(\dfrac{2A_c}{u}\right)^3}}$$

where A_c is the concrete gross sectional area and u is the perimeter exposed to drying.

$$h_0 = 2 \times (700 \times 370)/(2(700 + 370)) = 242\,\text{mm}.$$

From Table 3.3 (EN 1992-1-1), $k_h = 0{,}808$ (by linear interpolation)

$$\beta_{ds}(t,t_s) = \frac{28 - 3}{28 - 3 + 0{,}04\sqrt{242^3}} = 0{,}142$$

So, $\varepsilon_{cd}(t) = 0{,}142 \times 0{,}808 \times 0{,}60 \times 10^{-3} = 69 \times 10^{-6}$

Autogenous shrinkage strain, ε_{ca}:

$$\varepsilon_{ca}(t) = \beta_{as}(t)\varepsilon_{ca}(\infty)$$

where the limiting autogenous shrinkage strain $\varepsilon_{ca}(\infty)$ is given by

$$\varepsilon_{ca}(\infty) = 2{,}5(f_{ck} - 10) \times 10^{-6}$$
$$= 2{,}5(25 - 10) \times 10^{-6} = 37{,}5 \times 10^{-6}$$

the loading correction factor $\beta_{as}(t)$ is given by

$$\beta_{as}(t) = 1 - e^{-0{,}2t^{0,5}}$$
$$= 1 - e^{-0{,}2 \times 28^{0,5}} = 0{,}653$$

Thus, the autogenous shrinkage strain ε_{ca},

$$\varepsilon_{ca}(t) = \beta_{as}(t)\varepsilon_{ca}(\infty)$$
$$= 0{,}653 \times 37{,}9 \times 10^{-6} = 24 \times 10^{-6}$$

The total shrinkage strain ε_{cs}:

$$\varepsilon_{cs} = \varepsilon_{cd} + \varepsilon_{ca} = 69 \times 10^{-6} + 24 \times 10^{-6}$$
$$= 93 \times 10^{-6}$$

Creep coefficient $\phi(t,t_0)$:

Use Annex B of EN 1992-1-1:

$$\phi(t,t_0) = \phi_0 \beta_c(t,t_0)$$

where the basic creep coefficient ϕ_0 is given by

$$\phi_0 = \phi_{RH} \beta(f_{cm}) \beta(t_0)$$

For concrete grade C25, ϕ_{RH} is given by

$$\phi_{RH} = 1 + \frac{1 - RH/100}{0,1\sqrt[3]{h_0}}$$
$$= 1 + \frac{1 - 40/100}{0,1\sqrt[3]{242}} = 1,963$$

The modification factor for mean concrete strength, $\beta(f_{cm})$:

$$\beta(f_{cm}) = \frac{16,8}{\sqrt{f_{cm}}} = \frac{16,8}{\sqrt{25 + 8}} = 2,92$$

The correction factor for the time of loading of the element $\beta(t_0)$:

$$\beta(t_0) = \frac{1}{0,1 + t_0^{0,20}} = \frac{1}{0,1 + 3^{0,20}} = 0,445$$

or ϕ_0 is given as

$$\phi_0 = \phi_{RH} \beta(f_{cm}) \beta(t_0)$$
$$= 1,963 \times 2,92 \times 0,445 = 2,55$$

The factor $\beta(t,t_0)$ to allow for the time relation between time at loading t_0 and time considered, t, is given by

$$\beta_c(t,t_0) = \left[\frac{t - t_0}{\beta_H + t - t_0} \right]^{0,3}$$

where for concrete grade less than C35, β_H is given by

$$\beta_H = 1,5 \frac{1 + (0,012 \text{RH})^{18}}{h_0} + 250$$

$$= 1,5 \frac{1 + (0,012 \times 40)^{18}}{242} + 250 = 613$$

Thus $\beta(t,t_0)$ is

$$\beta_c(t,t_0) = \left[\frac{t - t_0}{\beta_H + t - t_0} \right]^{0,3}$$

$$= \left[\frac{28 - 3}{613 + 28 - 3} \right]^{0,3} = 0,378$$

so $\phi(t,t_0)$ becomes

$$\phi(t,t_0) = \phi_0 \beta_c(t,t_0) = 2,55 \times 0,378 = 0,964$$
$$E_{c,eff} = E_{cm}/(1 + \phi(t,t_0)) = 31,5/(1 + 0,964) = 16,04 \, \text{GPa.}$$
$$\alpha_e = 200/16,04 = 12,5$$

Determine the cracked and uncracked second moments of area:

$$\rho' = A'_s/bd = 1608/635 \times 370 = 0,0068;$$
$$\rho = A_s/bd = 5026/635 \times 370 = 0,0214;$$
$$d'/d = 64/635 = 0,101; \; h/d = 700/635 = 1,102$$
$$\beta = 0,5 \; \text{(sustained loads)}$$

Uncracked section:
Use Eq. (5.17) to determine x/d:

$$x/d = (0,5 \times 1,102^2 + 12,5(0,0214 + 0,0068 \times 0,101))/ \ldots$$
$$(1,102 + 12,5(0,0068 + 0,0214)) = 0,607$$
$$x = 0,607 \times 635 = 385 \, \text{mm}$$

and Eq. (5.16) for I_{gross},

$$I_{gross}/bd^3 = 0,607^3/3 + (1,102 - 0,607)^3/3 + 12,5 \times 0,0068$$
$$(0,607 - 0,101)^2 + 12,5 \times 0,0214(1 - 0,607)^2 = 0,178$$
$$I_{gross} = 0,178 \times 370 \times 635^3 = 16,86 \times 10^9 \, \text{mm}^4$$

Use Eq. (5.18) with f_{ct} replaced by f_{ctm} to determine M_{cr}

$$M_{cr} = f_{ctm} I_{gross}/(h - x) = 2,56 \times 16,86 \times 10^9/(700 - 385) \, \text{Nmm} = 137 \, \text{kNm.}$$

Cracked section:
Use Eq. (5.21) to determine x/d

$$x/d = -12{,}5(0{,}0068 + 0{,}0214) + \ldots$$
$$\sqrt{(12{,}5^2(0{,}0068 + 0{,}0214)^2 + 2 \times 12{,}5(0{,}0068 \times 0{,}101 + 0{,}0214))} = 0{,}470$$
$$x = 0{,}470 \times 635 = 298 \text{ mm}$$

and Eq. (5.20) to determine I_{cr}

$$I_{cr}/bd^3 = 0{,}47^3/3 + 12{,}5 \times 0{,}0068(0{,}101 - 0{,}47)^2 + 12{,}5 \times 0{,}0214(1 - 0{,}47)^2 = 0{,}121$$
$$I_{cr} = 0{,}121 \times 370 \times 635^3 = 11{,}46 \times 10^9 \text{ mm}^4$$

Shrinkage parameters:

$$\varepsilon_{cs}\alpha_e = 93 \times 10^{-6} \times 12{,}5 = 1163 \times 10^{-6}$$

Calculate S/bd^2:
Uncracked section:

$$S/bd^2 = 0{,}0214(1 - 0{,}607) - 0{,}0068(0{,}101 - 0{,}607) = 0{,}0119$$
$$S = 0{,}0119 \times 370 \times 635^2 = 1{,}775 \times 10^6 \text{ mm}^3$$
$$S/I_{gross} = 1{,}775 \times 10^6/16{,}86 \times 10^9 = 0{,}1053 \times 10^{-3}/\text{mm}$$

Cracked section:

$$S/bd^2 = 0{,}0214(1 - 0{,}47) - 0{,}0068(0{,}101 - 0{,}47) = 0{,}0139$$
$$S = 0{,}0139 \times 370 \times 635^2 = 2{,}074 \times 10^6 \text{ mm}^3$$
$$S/I_{cr} = 2{,}074 \times 10^6/11{,}46 \times 10^9 = 0{,}181 \times 10^{-3}/\text{mm}$$

Integration method:
Divide the beam into four segments each 1 m long ($x=0$ at free end)

Total UDL (assuming the whole of the transient load acts on a quasi-permanent basis) $= 40 + 30 = 70$ kN/m
$M_{Sd} = 70x^2/2 = 35x^2$ kNm.
As the curvature $1/r_{cs}$ on the cracked and uncracked sections is constant along the beam these may be determined first.

Shrinkage:
Uncracked section:

$$1/r_{cs,unc} = (\varepsilon_{cs}\alpha_e)(S/I)_{unc} = 0{,}001163 \times 0{,}1053 \times 10^{-3}$$
$$= 0{,}1225 \times 10^{-6}/\text{mm}$$

Cracked:

$$1/r_{cs,cr} = (\varepsilon_{cs}\alpha_e)(S/I)_{cr} = 0,001163 \times 0,181 \times 10^{-3}$$
$$= 0,211 \times 10^{-6}/mm$$

Point 1 ($x=0$): curvature is zero (zero moment)
Point 2 ($x=1$)
$M_{Sd} = 35\,kNm < M_{cr}$, thus $\zeta = 0$:

Bending curvature:

$$f_{ct} = M_{Sd}(h-x)/I_{gross} = 35 \times 10^6 \times (700-385)/16,89 \times 10^9 = 0,653\,MPa$$
$$1/r_{M,unc} = (f_{ct}/E_{c,eff})/(h-x) = (0,653/16,04 \times 10^3)/(700-385)$$
$$= 0,129 \times 10^{-6}/mm$$

Total uncracked curvature, $1/r_I$:

$$1/r_I = 1/r_{cs,unc} + 1/r_{M,unc} = (0,129 + 0,1225) \times 10^{-6} = 0,252 \times 10^{-6}/mm$$

As the concrete is uncracked, $1/r_{tot} = 1/r_I$
Point 3 ($x=2$)

$$M_{Sd} = 140\,kNm > M_{cr}, \text{ thus } \zeta \text{ needs calculating:}$$
$$\sigma_s/\sigma_{sr} = M_{Sd}/M_{cr} = 140/137 = 1,022$$

Use Eq. (5.14) to calculate ζ:

$$\zeta = 1 - 0,5/1,022^2 = 0,478$$

Uncracked curvatures:
Bending:

$$f_{ct} = M_{Sd}(h-x)/I_{gross} = 140 \times 10^6 \times (700-385)/16,89 \times 10^9 = 2,61\,MPa$$
$$1/r_{M,unc} = (f_{ct}/E_{c,eff})/(h-x) = (2,61/16,04 \times 10^3)/(700-385)$$
$$= 0,517 \times 10^{-6}/mm$$

Total uncracked curvature $1/r_I$:

$$1/r_I = 1/r_{cs,unc} + 1/r_{M,unc} = (0,517 + 0,1225) \times 10^{-6} = 0,640 \times 10^{-6}/mm$$

Cracked curvatures:
Bending:

$$f_{cs} = M_{Sd}\,x/I_{cr} = 140 \times 10^6 \times 298/11,46 \times 10^9 = 3,64\,MPa$$
$$1/r_{M,cr} = (f_{cs}/E_{c,eff})/x = (3,64/16,04 \times 10^3)/298 = 0,762 \times 10^{-6}/mm$$

Total cracked curvature, $1/r_{II}$

$$1/r_{II} = 1/r_{cs,cr} + 1/r_{M,cr} = 0{,}762 \times 10^{-6} + 0{,}211 \times 10^{-6} = 0{,}973 \times 10^{-6}/\text{mm}$$

Effective total curvature $1/r_{tot}$ is given by Eq. (5.13),

$$1/r_{tot} = 0{,}478 \times 0{,}973 \times 10^{-6} + (1 - 0{,}478) \times 0{,}64 \times 10^{-6} = 0{,}799 \times 10^{-6}/\text{mm}$$

Point 4 ($x = 3$)

$$M_{Sd} = 315\,\text{kNm} > M_{cr}, \text{ thus } \zeta \text{ needs calculating:}$$
$$\sigma_s/\sigma_{sr} = M_{Sd}/M_{cr} = 315/137 = 2{,}3$$

Use Eq. (5.14) to calculate ζ:

$$\zeta = 1 - 0{,}5/2{,}3^2 = 0{,}906$$

Uncracked section:
Bending curvature:

$$f_{ct} = M_{Sd}(h - x)/I_{gross} = 315 \times 10^6 \times (700 - 385)/16{,}89 \times 10^9 = 5{,}87\,\text{MPa}.$$
$$1/r_{M,unc} = (f_{ct}/E_{c,eff})/(h - x) = (5{,}87/16{,}04 \times 10^3)/(700 - 385)$$
$$= 1{,}162 \times 10^{-6}/\text{mm}$$

Total uncracked curvature $1/r_I$:

$$1/r_I = 1/r_{cs,unc} + 1/r_{M,unc} = 1{,}162 \times 10^{-6} + 0{,}1225 \times 10^{-6}$$
$$= 1{,}285 \times 10^{-6}/\text{mm}$$

Cracked section:
Bending:

$$f_{cs} = M_{Sd}\,x/I_{cr} = 315 \times 10^6 \times 298/11{,}46 \times 10^9 = 8{,}411\,\text{MPa}$$
$$1/r_{M,cr} = (f_{cs}/E_{c,eff})/x = (8{,}411/16{,}04 \times 10^3)/298 = 1{,}760 \times 10^{-6}/\text{mm}$$

Total cracked curvature, $1/r_{II}$

$$1/r_{II} = 1/r_{cs,cr} + 1/r_{M,cr} = 1{,}760 \times 10^{-6} + 0{,}211 \times 10^{-6} = 1{,}971 \times 10^{-6}/\text{mm}$$

Effective curvature $1/r_{tot}$ is given by Eq. (5.13),

$$1/r_{tot} = 0{,}906 \times 1{,}971 \times 10^{-6} + (1 - 0{,}906) \times 1{,}285 \times 10^{-6} = 1{,}907 \times 10^{-6}/\text{mm}$$

Point 5 $(x=4)$

$M_{Sd} = 560\,\text{kNm} > M_{cr}$, thus ζ needs calculating:

$$\sigma_s/\sigma_{sr} = M_{Sd}/M_{cr} = 560/137 = 4{,}088$$

Use Eq. (5.14) to calculate ζ:

$$\zeta = 1 - 0{,}5/4{,}088^2 = 0{,}970$$

Uncracked section:
Bending curvature:

$$f_{ct} = M_{Sd}(h-x)/I_{gross} = 560 \times 10^6 \times (700-385)/16{,}89 \times 10^9 = 10{,}44\,\text{MPa}$$
$$1/r_{M,cr} = (f_{ct}/E_{c,eff})/(h-x) = (10{,}44/16{,}04 \times 10^3)/(700-385)$$
$$= 2{,}066 \times 10^{-6}/\text{mm}$$

Total uncracked curvature, $1/r_I$:

$$1/r_I = 1/r_{cs,unc} + 1/r_{M,unc} = 2{,}066 \times 10^{-6} + 0{,}1225 \times 10^{-6}$$
$$= 2{,}189 \times 10^{-6}/\text{mm}$$

Cracked section:

$$f_{cs} = M_{Sd}\,x/I_{cr} = 560 \times 10^6 \times 298/11{,}46 \times 10^9 = 14{,}56\,\text{MPa}$$
$$1/r_{II} = (f_{cs}/E_{c,eff})/x = (14{,}56/16{,}04 \times 10^3)/298 = 3{,}046 \times 10^6/\text{mm}$$

Total cracked curvature $1/r_{II}$

$$1/r_{II} = 1/r_{cs,cr} + 1/r_{M,cr} = 3{,}046 \times 10^6 + 0{,}211 \times 10^{-6}$$
$$= 3{,}257 \times 10^6/\text{mm}$$

Effective curvature $1/r_{tot}$ is given by Eq. (5.13),

$$1/r_{tot} = 0{,}970 \times 3{,}257 \times 10^6 + (1 - 0{,}970)2{,}189 \times 10^{-6} = 3{,}225 \times 10^{-6}/\text{mm}$$

The results from determining $(\Delta x)^2 1/r_{tot}$ at each node point are given in Table 5.3. To allow the boundary conditions to be evaluated at point 5 in the beam, a fictitious point 6 needs to be introduced (Fig. 5.6). The boundary conditions at node 5 are:

(a) deflection is zero, or $w_5 = 0$
(b) slope is zero, or $(w_6 - w_4)/(2\Delta x) = 0$; or, $w_6 = w_4$.

Thus applying Eq. (5.27) at each node point, together the boundary conditions, gives in matrix form

$$AW = \Delta x^2 \frac{1}{r} \tag{5.29}$$

TABLE 5.3 Deflection parameters for Example 5.2.

Node Point	$(\Delta x)^2 1/r_{tot}$ (mm)
1	0
2	0,252
3	0,799
4	1,907
5	3,225

where $(\Delta x)^2 = 1000^2 = 10^6\,mm^2$.

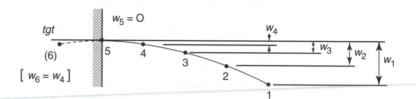

FIGURE 5.6 Deflection profile for Example 5.2

or, solving

$$W = A^{-1}\Delta x^2 \frac{1}{r}$$ (5.30)

Substituting in numerical values gives,

$$\begin{pmatrix} & & & 2 \\ & & 1 & -2 \\ & 1 & -2 & 1 \\ 1 & -2 & 1 & \end{pmatrix} \begin{pmatrix} w_1 \\ w_2 \\ w_3 \\ w_4 \end{pmatrix} = \begin{pmatrix} 3,225 \\ 1,907 \\ 0,799 \\ 0,225 \end{pmatrix}$$

or

$$w = (14,00 \quad 9,451 \quad 5,132 \quad 1,612)^{-1}$$

Thus the deflection at the free end, $w_1 = 14{,}0\,mm$.
This is equivalent to span/285 which is acceptable.

Approximate method:
The deflection w is given by Eq. (5.28). The calculation must be determined for loading and shrinkage separately as each has different k_3 factors.

For a UDL on a cantilever $k_3 = 0{,}25$ (Fig. 5.5).

At the support, the loading curvature $1/r_I$ based on an uncracked section is $2{,}066 \times 10^{-6}$/mm, and $1/r_{II}$ for a cracked section is $3{,}046 \times 10^{-6}$/mm. Equation (5.13) may be used to determine the resultant deflection, so with $\zeta = 0{,}97$ (see earlier), the resultant curvature is given by

$$1/r_{res} = 0{,}97 \times 3{,}046 \times 10^{-6} + (1 - 0{,}97)2{,}066 \times 10^{-6} = 3{,}02 \times 10^{-6}/\text{mm}$$

thus, the deflection due to the loading is given by

$$w = 0{,}25 \times 4000^2 \times 3{,}02 \times 10^{-6} = 12{,}1 \, \text{mm}$$

For shrinkage, $1/r_{cs,uncr} = 0{,}1225 \times 10^{-6}$/mm and $1/r_{cs,cr} = 0{,}211 \times 10^{-6}$/mm, resultant shrinkage curvature $1/r_{cs,res}$ is given by

$$1/r_{cs,res} = 0{,}97 \times 0{,}211 \times 10^{-6} + (1 - 0{,}97)0{,}1225 \times 10^{-6} = 0{,}208 \times 10^{-6}/\text{mm}$$

$k_3 = 0{,}5$ for this case, so

$$w = 0{,}5 \times 4000^2 \times 0{,}208 \times 10^{-6} = 1{,}7 \, \text{mm}$$

Total estimated deflection $= 12{,}1 + 1{,}7 = 13{,}8 \, \text{mm}$.

The exact calculations and the approximate calculations give virtually identical results.

NOTE: Under normal circumstances, deflections due to shrinkage are not calculated, but in this example it should be noted that the shrinkage deflection is around 14 per cent of the total. Johnson and Anderson (2004) indicate that European Design Codes appear to give higher shrinkage strains than those traditionally from current British Codes, and thus neglect of shrinkage strains may no longer be justified.

5.2.2 Crack Widths (cl 4.4.2)

Cracking should be limited to an extent which will not impair the proper functioning or durability or cause its appearance to be unacceptable (cl 7.3.1 (1)(P)). Clause 7.3.1 (2)(P) accepts cracking in reinforced concrete is a normal occurrence.

In reinforced concrete, for exposure Classes XC2 to XC4 it is generally acceptable that the maximum design crack width should not exceed 0,3 mm. This also applies to Classes XD1, XD2 and XS1 to XS3. For Class X0 and XC1 exposure (internal) the 0,3 mm width is relaxed to 0,4 mm with the Code accepting that this limit is only required for aesthetic purposes and not durability.

For prestressed concrete, these limits are reduced to 0,2 mm but with the need to check the section does not go into tension under severe exposure.

The average crack width that would be observed in a structure will be less than the design crack width, owing to there being a statistical variation in crack width distributions which will depend on the size and type of member.

Cracking is due to two essential mechanisms; the first is due to dimensional changes in the member due to, for example, shrinkage or early thermal cracking; the second is due to imposed deformations by, say, settlement of a support. Further, consideration needs to be given to the state of stress in the member whether complete tension or tensile strains induced by flexure (in which part of the cross section is in compression).

5.2.2.1 Explicit Calculation of Crack Widths

The background to the ideas behind this were given by Beeby (1979) and are still valid. Indeed the principles laid down by Beeby are now presented in a much more transparent manner in EN 1992-1-1 than in earlier British Codes. This is especially true for determination of crack widths for closely spaced bars.

The design crack width w_k is given by

$$w_k = s_{r,max}(\varepsilon_{sm} - \varepsilon_{cm}) \qquad (5.31)$$

where $s_{r,max}$ is maximum crack spacing, ε_{sm} is the mean strain in the reinforcement and ε_{cm} is the mean strain in the concrete between the cracks.

The effective strain $\varepsilon_{sm} - \varepsilon_{cm}$ is calculated from

$$\varepsilon_{sm} - \varepsilon_{cm} = \frac{\sigma_s - k_t(f_{ct,eff}/\rho_{p,eff})(1 + \alpha_e \rho_{p,eff})}{E_s} \geq 0.6\frac{\sigma_s}{E_s} \qquad (5.32)$$

where E_s is the Young's modulus for reinforcement, σ_s is the steel stress based on a cracked section, $\alpha_e = E_s/E_{cm}$, k_t takes values of 0,6 for short term loading or 0,4 for sustained loading.

The effective reinforcement ratio $\rho_{p,eff}$ is defined by

$$\rho_{p,eff} = \frac{A_s + \xi_1^2 A_{\rho'}}{A_{c,eff}} \qquad (5.33)$$

where A_s is the area of tension reinforcement, $A_{\rho'}$ is the area of prestressing tendons within the effective concrete area $A_{c,eff}$ and ξ_1 is a ratio between the bond stresses for reinforcement and prestressing tendons taking account of the difference in bar diameters. $A_{c,eff}$ is the area of concrete surrounding the reinforcement to a depth of 2,5 times the axis distance of the reinforcement to the tension face (Fig. 5.7), subject

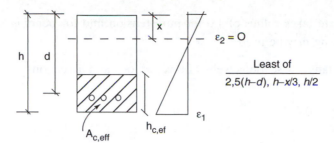

FIGURE 5.7 Definition of $A_{c,eff}$ and $h_{c,ef}$

to the limit for slabs in flexure of $(h-x)/3$. For members in tension an additional limit of $h/2$ is imposed.

For closely spaced reinforcement (i.e. spacing less than $5(c+\phi/2)$, where c is the cover and ϕ the bar diameter) the maximum crack spacing $s_{r,max}$ (in mm) is calculated from

$$s_{r,max} = 3,4c + 0,425k_1k_2\frac{\phi}{\rho_{p,eff}} \tag{5.34}$$

where ϕ is the bar diameter (or for a mixture of bars sizes, the equivalent bar diameter), k_2 is taken as 1,0 for pure tension and 0,5 for flexure (intermediate values can be used and are determined from $(\varepsilon_1+\varepsilon_2)/2\varepsilon_1$, where ε_1 and ε_2 are the greater and lesser tensile strains on the surface). The reinforcement ratio $\rho_{p,eff}$ is defined as $A_s/A_{c,eff}$. The value of k_1 is taken as 0,8 for ribbed bars and 1,6 for plain bars.

For members with wide bar spacing $s_{r,max}$ should be taken as $1,3(h-x)$, and for walls where the base is restrained and is subject to early thermal cracking only $s_{r,max}$ should be taken as the 1,3 times the height of the wall.

5.2.2.2 Normal Rules

The incidence of cracking is generally controlled by imposing a minimum area of steel criterion. The minimum reinforcement area $A_{s,min}$ is given by

$$A_{s,min} = k_c\,k f_{ct,eff}\frac{A_{ct}}{\sigma_s} \tag{5.35}$$

where A_{ct} is the area of the concrete in tension just before the concrete cracks, σ_s is the maximum permitted stress in the reinforcement which may be taken as f_{yk}, $f_{ct,eff}$ is the effective tensile strength in the concrete based on the strength of the concrete at which cracks are expected and can generally be taken as f_{ctm} unless cracking is expected before 28 days, k_c is a coefficient allowing for the stress distribution due to loading and imposed deformations immediately prior to cracking and for reinforced concrete takes values of 1,0 for pure tension and for bending and axial forces, the following may be used

• Rectangular sections webs of box sections and T sections

$$k_{\mathrm{c}} = 0,4 \left[1 - \frac{\sigma_c}{k_1 f_{\mathrm{ct,eff}} \frac{h}{h^*}} \right] \le 1,0 \tag{5.36}$$

• Flanges of box and T sections

$$k_{\mathrm{c}} = 0,9 \frac{F_{\mathrm{cr}}}{A_{\mathrm{ct}} f_{\mathrm{ct,eff}}} \ge 0,5 \tag{5.37}$$

where σ_c is the mean stress in the concrete $(=N_{\mathrm{Ed}}/bh)$, $h^*=h$ for $h \le 1,0\,\mathrm{m}$ or $1,0\,\mathrm{m}$ for $h > 1,0\,\mathrm{m}$, $k_1 = 1,5$ if N_{Ed} is compressive or $2h^*/3h$ if N_{Ed} is tensile, and F_{ct} is the absolute value of the force in tension flange at incipient cracking.

The parameter k takes a value of 1,0 for members less than 300 mm thick and 0,65 for greater than 800 mm (with linear interpolation). The minimum reinforcement level determined from Eq. (5.35) is further subject to the limit of the greater of $0,26\,(f_{\mathrm{ctm}}/f_{\mathrm{yk}})b_{\mathrm{t}}d$ or $0,0013b_{\mathrm{t}}d$, where b_{t} is the mean width of the tension zone (cl 9.2.1.1).

For additional design rules for slabs, see Section 10.5.3.

For beams, cracking will be kept below the design values for the minimum steel areas given by Eq. (5.35) if EITHER the bar diameter or spacing given in Table 5.4 (from Tables 4.11 and 4.12 from EN 1992-1-1) are complied with. The required spacings or diameters are controlled by the stress in the reinforcement under loads which can be considered quasi-permanent. Thus the variable actions should be multiplied by the factor ψ_2 (from Table A1.1 of EN 1990)

TABLE 5.4   Bar detailing data (from Tables 4.11 and 4.12 EN 1992-1-1).

Steel stress (MPa)	Maximum bar size (mm)			Maximum bar spacing (mm)		
	0,4	0,3	0,2	0,4	0,3	0,2
160	40	32	25	300	300	200
200	32	25	16	300	250	150
240	20	16	12	250	200	100
280	16	12	8	150	75	50
320	16	12	8	150	100	--
360	10	8	5	100	50	--
400	8	6	4	--	--	--
450	6	5	--	--	--	--

In reinforced concrete the maximum bar diameter may be modified as follows:
Bending (at least part of the section in compression)

$$\phi_s = \phi_s^* \frac{f_{ct,eff}}{2,9} \frac{k_c h_{cr}}{2(h-d)} \quad\quad (5.38)$$

tension,

$$\phi_s = \phi_s^* \frac{f_{ct,eff}}{2,9} \frac{h_{cr}}{24(h-d)} \quad\quad (5.39)$$

where ϕ_s^* is the maximum bar diameter from Table 5.4, ϕ_s is the adjusted value, h is the depth of the section, h_{cr} is the depth of the tensile zone immediately prior to cracking under quasi-permanent load combinations and k_c is given by Eq. (5.36) or Eq. (5.37). For beams of depth greater than 1 m, additional reinforcement must be provided in the tensile zone. It should have a minimum value obtained from Eq. (5.35) with $k = 0,5$ and $\sigma_s = f_{yk}$. The spacing and size of such bars may be obtained from Table 5.4 assuming pure tension and a steel stress half of that used in the assessment of the main tensile reinforcement.

EXAMPLE 5.3   Crack width calculations. Check design crack width for the beam of Examples 5.1 and 5.2

The critical zone is at the support.
Determine the clear spacing between bars:

$$s = (370 - 2 \times 45 - 4 \times 40)/3 = 40 \, mm$$

Limiting spacing for closely spaced bars is $5(c + \phi/2) = 5(45 + 40/2) = 325$ mm. Bars are thus closely spaced, thus $s_{r,max}$ is given by Eq. (5.31),

$$s_{r,max} = 3,4c + 0,425 k_1 k_2 \phi / \rho_{p,eff}$$

$k_1 = 0,8$ (high bond ribbed bars)
$k_2 = 0,5$ (flexure)

Effective loading:
Permanent: 30 kN/m
Quasi-permanent: $0,3 \times 40 = 12$ kN/m (assuming $\psi_2 = 0,3$)

$$M_{Sd} = 42 \times 4^2/2 = 336 \, kNm,$$

From Example 5.2:

$I_{cr} = 11,46 \times 10^9$ mm4; $x = 298$ mm

$\sigma_s = \alpha_e M_{sd}(d-x)/I_{cr} = 12,5 \times 366 \times 10^6 \times (635 - 298)/11,46 \times 10^9 = 135$ MPa.

$A_{c,eff} = 2,5 (h-d)b = 2,5(700 - 635)370 = 60125$ mm^2.

subject to the limits:

$$b(h-x)/3 = 370(700 - 298)/3 = 49580 \, \text{mm}^2$$
$$bh/2 = 370 \times 700/2 = 129500 \, \text{mm}^2$$

Thus $A_{c,eff}$ should be taken as 49580 mm²

$$A_s = 5026 \, \text{mm}^2;$$
$$\rho_r = A_s/A_{c,eff} = 5026/49580 = 0{,}101$$
$$\phi = 40 \, \text{mm}, \, c = 45 \, \text{mm}$$

Thus from Eq. (5.31),

$$s_{r,max} = 3{,}4 \times 45 + 0{,}425 \times 0{,}8 \times 0{,}5 \times 40/0{,}101 = 220 \, \text{mm}$$

Determination of $\varepsilon_{sm} - \varepsilon_{cm}$ (from Eq. (5.32)):
$f_{ct,eff}$ may be taken as $f_{ctm} = 0{,}3 \times f_{ck}^{(2/3)} = 2{,}56 \, \text{MPa}$

Duration factor k_1:
Conservatively assume long term loading, i.e. $k_t = 0{,}4$, so from Eq. (5.32)

$$
\begin{aligned}
\varepsilon_{sm} - \varepsilon_{cm} &= \frac{\sigma_s - k_t(f_{ct,eff}/\rho_{p,eff})(1 + \alpha_e \, \rho_{p,eff})}{E_s} \\
&= \frac{135 - 0{,}4(2{,}56/0{,}101)(1 + 12{,}5 \times 0{,}101)}{200 \times 10^3} = 653 \times 10^{-6}
\end{aligned}
$$

Limiting value is given by $0{,}6\sigma_s/E_s = 0{,}6 \times 135/200 \times 10^3 = 405 \times 10^{-6}$
Thus, the value of $\varepsilon_{sm} - \varepsilon_{cm}$ to be used is 653×10^{-6}.
Determine w_k from Eq. (5.31)

$$
\begin{aligned}
w_k &= s_{r,max}(\varepsilon_{sm} - \varepsilon_{cm}) = 220 \times 653 \times 10^{-6} \\
&= 0{,}144 \, \text{mm}
\end{aligned}
$$

This is below the limit of 0,3 mm, and therefore acceptable.

If the whole of the transient load is considered as permanent, then $M_{Sd} = 560 \, \text{kNm}$; $\sigma_s = 206 \, \text{MPa}$; $\varepsilon_{sm} - \varepsilon_{cm} = 1008 \, \mu\text{strain}$; $w_k = 0{,}22 \, \text{mm}$ (this is still satisfactory).

EXAMPLE 5.4   Bar spacing rules. The data are as before.

Determination of reinforcement stress σ_s:
Effective loading:
Permanent: 30 kN/m
Quasi-permanent: $0{,}3 \times 40 = 12 \, \text{kN/m}$ (assuming $\psi_2 = 0{,}3$)

$$M_{Sd} = 42 \times 42/2 = 336 \, \text{kNm},$$

From Example 5.2:

$$I_{cr} = 11{,}46 \times 10^9 \, \text{mm}^4; x = 298 \, \text{mm}$$

$$\sigma_s = \alpha_e M_{sd}(d-x)/I_{cr} = 12{,}5 \times 366 \times 10^6 \times (635 - 298)/11{,}46 \times 10^9 = 135 \, \text{MPa}.$$

Using the lowest value in Table 5.4 of $\sigma_s = 160$ MPa, the maximum bar spacing for $w_k = 0{,}3$ is 300 mm. This is clearly satisfied as there are four bars in a total width of 370 mm.

The maximum bar size allowed (before modification) is 32 mm.

Use Eq. (5.38) for sections in flexure to determine the modified bar diameter:

$\phi_s^* = 32$ (from Table 5.4)

$f_{ct,eff}$:

This may be taken as $f_{ctm} = 0{,}3 \times f_{ck}^{2/3} = 2{,}56$ MPa

The depth of the tensile zone at incipient cracking, h_{cr}:

From Example 5.2, The depth of the uncracked neutral axis is 385 mm, so, $h_{cr} = h - x = 700 - 385 = 315$ mm.

Use Eq. (5.36) to determine k_c:

As there is no applied axial force, $\sigma_c = 0$, thus k_c reduces to a value of 0,4.

So using (5.38)

$$\phi_s = \phi_s^* \frac{f_{ct,eff}}{2{,}9} \frac{k_c h_{cr}}{2(h-d)}$$
$$= 32 \frac{2{,}56}{2{,}9} \frac{0{,}4 \times 315}{2(700 - 635)} = 27{,}4 \, \text{mm}$$

Note in this case the modification factor produces a reduced bar diameter.

Interpolating between values of σ_s of 200 and 240 from Table 5.4 gives a bar spacing of around 240 mm. The actual is well below this. However, as EITHER the bar diameter rules OR bar spacing rules need be satisfied, the design is satisfactory.

Note, a further example on crack width calculations is given in Example 11.5.

5.3 FIRE

Design for fire is required, where applicable, in order to ensure any structure is capable of withstanding the accidental effects of exposure from high temperatures caused by the occurrence of fire. Although all multi-storey structures are required to be checked for such effects, single storey structures may only need evaluating to ensure the structure does not collapse outside its effective boundary. All building structures need to be designed to ensure that evacuation of persons can be carried out safely and that the fire can be fought with no ill-effects to the fire fighters

(Purkiss, 1996). It should also be noted that for other structures such as bridges the risk of fire exposure is generally very low and therefore can be ignored.

The England and Wales Building Regulations, Approved Document B (Department of the Environment, 2000) allows either the use of a full fire engineering approach to the evaluation of structures exposed to the accidental effects of fire, or an approach using tabulated data where the structure is considered to be exposed to the standard furnace test temperature–time curve. It should be noted that the fire endurance periods in the current version of Approved Document B are being revised (Kirby *et al.*, 2004).

5.3.1 Tabular Data

These are used when the fire resistance period is either specified according to standard requirements (e.g. Approved Document B) or determined from the time equivalence approach. This latter allows the effects of a real or parametric fire to EN 1991-1-2 to be related to the curve in the standard furnace test. It should be considered where the combustible fire load or the available ventilation is low as it may demonstrate that the equivalent time is such that no additional measures need be taken to give adequate fire performance. Time equivalence was originally conceived for protected steelwork, care therefore should be taken when applying it to concrete structures, especially columns as the maximum average concrete tempera-ture and the maximum reinforcement temperatures will not be coincident. Also time equivalence will give no indication of rate of heating as it only assesses the total heat input. It should also only be considered where the structure is of fixed usage throughout its design life.

Tabular data are for members within a structure, and it should not be assumed that if members are detailed for, say, 60 minutes standard exposure that the complete structure would last for 60 minutes in that exposure – if such a situation could be evaluated. Whole building behaviour whether observed in actual fires or the Cardington test (Bailey, 2002) would seem to suggest that whole building behaviour is better than that of individual elements as a structure can allow redistribution of internal thermal or mechanical forces generated during the fire. The exceptions to this are that failure of compartment walls caused by excessive deflection of the floors above would not be picked up in a standard furnace test, nor would lateral movement of the complete structure. Equally the efficiency of fire stopping between the edges of slabs and external curtain walls cannot be evaluated in furnace tests. The importance of such fire stopping was shown in the recent Madrid fire (Redfern, 2005) where fire spread between floors of a 32 storey concrete frame structure to such an extent that the structure needed demolition.

Unless otherwise indicated, tabular data imply that the load level is approximately 0,70 times that at ULS and thus are likely to be conservative at not all levels, but the variable load may be present during a fire other than in the extremely early period during evacuation when any temperature rise within the concrete will be very low. This is recognized in EN 1990 where a ψ factor is applied to variable loading in the calculation approach. EN 1992-1-2 does give a procedure whereby the critical temperature and resultant axis distance can be modified for lower load levels (cl 5.2). The critical temperatures θ_{cr} for a load level of 0,7 are taken as 500°C for reinforcing steel, 400°C for prestressing in the form of bars, and 350°C for strand or wire. This means that for bar prestressing, the tabulated values of the axis distance, a, should be increased by 10 mm, and for strand or wires by 15 mm.

The major departure from current practice is that EN 1992-1-2 uses axis distance not covered to reinforcement. This is actually more scientifically correct as it is observed in computer-based heat transfer calculations that the temperatures at the centre of a reinforcing bar are identical to that at the same position in plain concrete (Ehm, 1967, quoted in Hertz, 1981). Becker *et al.* (1974) indicate this is reasonable up to a reinforcement area of 4 per cent of the gross section. Where the reinforcement is in more than one layer, the weighted mean axis distance is used. The axis distance used for any bar is the least of that to any fire exposed face. Rather than cover each type of structural element, it is only proposed to cover those elements where interpretation of the Code is required. Note in this section, all Table and clause references are to EN 1992-1-2.

5.3.1.1 Columns

The Code provides two methods denoted A and B.

- Method A:

  This was developed from an empirical procedure originally proposed by Dortreppe and Franssen (undated). It can only be used under the following recommended limits:

  - Effective length $l_{0,fi} \leq 3$ m, where $l_{0,fi}$ is taken as l_0 (the effective length at ambient) subject to the condition that if the standard exposure is greater than 30 minutes, then $l_{0,fi}$ is taken as $0,5l$ for intermediate floors and $0,5l \leq l_{0,fi} \leq 0,7l$ for the upper floor (l is the height of the column between centre lines).
  - Eccentricity of load, e, defined as $M_{0Ed,fi}/N_{0Ed,fi}$, should be less than e_{max} which has a recommended value of the lesser of $0,15h$ or b (although the code does suggest an upper limit of $04h$ or b). $M_{0Ed,fi}$ and $N_{0Ed,fi}$ are the first order moment and axial force in the fire limit state.
  - Maximum reinforcement A_s should not exceed $0,4A_c$.

  NOTE: The design values of section size and cover given in Table 5.2a of EN 1992-1-2 are for a value of α_{cc} of 1,00. Thus if a value of 0,85 is adopted (as is likely in the UK), the formula attached to the Table MUST be used. This gives the value of the fire resistance R as

$$R = 120 \left(\frac{R_{\eta fi} + R_a + R_l + R_b + R_n}{120} \right)^{1,8} \tag{5.40}$$

where the correction factor $R_{\eta fi}$ for load level μ_{fi} and α_{cc} is given by

$$R_{\eta fi} = 83 \left(1,00 - \mu_{fi} \frac{1 + \omega}{(0,85/\alpha_{cc}) + \omega} \right) \tag{5.41}$$

where μ_{fi} is defined as $N_{Ed.fi}/N_{Rd}$ (i.e. the ratio of the axial load in the fire limit state to the design resistance at ambient) and ω is the mechanical reinforcement ratio $(= A_s f_{yd}/A_c f_{cd} = (A_s f_{yk}/\gamma_s)/(A_c f_{ck}/\gamma_c)$,
the correction factor R_a for axis distance a ($25 \leq a \leq 80$ mm),

$$R_a = 1,6(a - 30) \tag{5.42}$$

the correction factor R_l for effective length $l_{0,fi}$,

$$R_l = 9,60 \left(5 - l_{0,fi} \right) \tag{5.43}$$

the correction factor R_b for a parametric width b' ($200 \leq b' \leq 450$ mm),

$$R_b = 0,9b' \tag{5.44}$$

where for a circular column b' is the diameter, and for a rectangular column b' is given by

$$b' = \frac{2A_c}{b+h} \tag{5.45}$$

subject to the limit that $h \leq 1,5b$.

- Method B:
  The background to Method B is given in Dortreppe, Franssen and Vanderzeypen (1999).
  This is covered at a basic level in Table 5.2b with more detailed tables in Annex C of EN 1992-1-2.

  The tables operate in terms of a load level n defined as

$$n = \frac{N_{oEd,fi}}{0,7(A_c f_{cd} + A_s f_{yd})} \tag{5.46}$$

  and the mechanical reinforcement ratio ω.
  Table 5.2b also has the restrictions that $e/b \leq 0,25$ (subject to a maximum value of e of 100 mm) and the slenderness ratio in the fire limit state $\lambda_{fi} < 30$.

5.3.1.2 Beams and Slabs

For beams, it should be noted there are requirements on section geometry for I beams (cl 5.6.1(5)) and that for other beams the width b is determined at the centroid of the reinforcing.

Where continuous beams (or slabs) have had elastic moments redistributed, then where the redistribution exceeds 15 per cent, the beam (or slab) must be treated as simply supported.

For situations where the standard fire resistance period is R90 or above, then for a distance of $0,3l_{eff}$ from any internal support, the reinforcement $A_{s,req}(x)$ at a distance x from the support should be not less than,

$$A_{s,req}(x) = A_{s,req}(0)\left(1 - 2,5\frac{x}{l_{eff}}\right) \qquad (5.47)$$

where l_{eff} is the effective span and $A_{s,req}(0)$ is the area of reinforcement required at the support.

Where the minimum dimension for b is adopted and the reinforcement in a beam is in a single layer, then the axis distance to a corner bar must be increased by 10 mm. This also applies to ribbed slabs (without the restriction on single layered reinforcement).

5.3.1.3 High Strength Concrete

Tabulated data may be used but with an increase on cross-sectional dimensions of $(k-1)a$ for walls and slabs with single sided exposure and $2(k-1)a$ for other structural elements where a is the axis distance determined from the tables. Additionally the axis distance actually required should be factored by k, where k is 1,1 for concrete C55/67 or 60/75 and 1,3 C70/85 or 80/95 given in cl 6.4.3.1(3). The potential problems due to spalling are dealt with in Section 5.3.3.1.

5.3.2 Calculation Methods

EN 1992-1-2 allows a series of methods for calculation:

(a) End point analysis.
This may be carried out using either the 500° isotherm method or the method of slices. The 500° isotherm method was originally proposed by Anderberg (1978) and involves taking the concrete within the 500° isotherm at full strength, concrete outside at zero strength with a temperature reinforcement strength. The method of slices developed by Hertz (1981, 1985, 1988) replaces a variable strength concrete stress block by an equivalent reduced width stress block using the temperature determined at the centre of the section. In both cases, temperature-reduced steel strengths are used.

(b) Full analysis.

An analysis using acceptable materials models incorporating transient strain for the concrete can be used. This is a specialist approach and involves the use of Finite Element techniques for both the determination of the temperature field within the structure and the resultant stresses, strains and deformations.

Design methods by calculation will not be covered in this text and reference should be made to Purkiss (1996).

5.3.3 Additional Comments

5.3.3.1. Spalling

Spalling occurs in one of two main forms in a fire. The first is explosive spalling which occurs very early in a fire and is likely to lead to loss of cover to the main reinforcing and hence to more rapid rises in temperature and resultant strength loss leading to reduced fire performance. The second form is known as sloughing whereby the concrete gradually comes away due to loss of effective bond and strength loss. This mode tends to occur toward the end of a fire and is rarely critical. It should be noted that although there can be severe spalling in a fire, the structure may still be capable of withstanding the effects of the fire without collapse. This was demonstrated in the fire test on the large frame structure at Cardington (Bailey, 2002).

The exact mechanism of explosive spalling is still not completely understood, although Connolly (1995, 1997) has indicated that there is unlikely to be one single mechanism. High Strength Concrete (HSC) and Self-Compacting Concrete (SCC) may give potentially worse problems in fire due to spalling than normal strength concrete due to their extremely low porosity. These concretes should both have polypropylene fibres with a dosage rate of 2 per cent included in the mix (EN 1992-1-2; Persson, 2003). The following factors affecting spalling have been identified (Malhotra, 1984; Connolly, 1995 and 1997):

- **Moisture content**
  A concrete with a high moisture content is more likely to spall, since one of the possible mechanisms of spalling is due to the build up of high vapour pressures by moisture clog in the element. The blanket limit of a moisture content of 3 per cent, below which EN 1991-1-2 indicates spalling will not occur, should be questioned. The original proposal based on work by Meyer-Ottens (1975) also suggested stress limits. If the permeability is high then a high moisture content may not cause problems.

- **Concrete porosity**

  A more porous concrete will allow the dissipation of vapour pressure, and thus relieve any build up within the section. A reminder needs to be made of the relationship between porosity and strength, in that the lower the porosity, the higher the strength. However, it must be pointed out that a porous concrete will give a poor performance with respect to durability.

- **Stress conditions**

  From evidence of fire tests and observations in fires, it has been noted that spalling is likely to be more severe in areas where the concrete cross section is in compression, i.e. areas of hogging moments in beams or slabs, or in columns. This can partly be explained by the fact that in areas of compressive stresses, cracks cannot open up to relieve internal pressures. This does not mean that spalling cannot occur in areas of sagging moments where tensile cracks exist, since it is possible that pressure build up will still occur as in general tension cracks are discrete and not part of a continuum.

- **Aggregate type**

  Historically, the evidence available appeared to suggest that the aggregate most likely to give spalling is siliceous aggregate, with limestone producing less spalling and lightweight concrete the least. This is likely to be linked to the basic porosity of the aggregate in that siliceous aggregate is impermeable compared to the others and that moisture transport has to occur through the mortar matrix. However, there is now some evidence that limestone and lightweight aggregates may give problems, especially in younger concretes as the pore structure provides convenient reservoir storage for free water (Connolly, 1995).

- **Section profile and cover**

  There is some evidence to suggest that sharp profiles will produce more spalling than rounded or chamfered edges. Spalling is also exacerbated in thin sections, partly since the depth of spalling is a greater proportion of the section dimension and hence proportionally worse, and partly due to the fact that there is less of a cool reservoir for any moisture to migrate towards.

  High covers are also likely to produce greater amounts of spalling as the concrete is not restrained in a triaxial state of stress in the surface layers. Thus design codes frequently place restrictions when high covers are needed at high fire resistance periods in order to maintain low temperatures in the reinforcing. EN 1992-1-2 indicates that for axis distances of greater than 70 mm a mesh with wires of at least 4 mm and spacing of 100 mm should be provided at the surface (cl 4.5.2(2)). It should, however, be noted that tests have shown that even with high covers such additional reinforcement is not absolutely necessary to give fire resistance periods of up to 4 hours (Lawson, 1985).

- **Heating Rate**

  The higher the heating rate, the less chance the pore water has migrated from the outside hot layers to the relatively cooler central layers of a member (Shorter and Harmathy, 1965). Thus the rate of heat given off by the fire is critical for an assessment of the likelihood of spalling. A hydrocarbon-type fire will therefore be worse than a cellulosic-type fire as the rate of heat release is higher.

5.3.3.2 Shear and Bond

For simply supported or continuous reinforced concrete, construction shear has rarely been observed as a problem. For large prestressed concrete beams, Bobrowski and Bardhan-Roy (1969) indicated that a potentially critical section for shear was between $0,15L$ and $0,2L$ from the support where L is the span. This is due to the moments induced in the section by the prestress itself which will not be constant along the length of the beam owing to the varying profile of the tendons. Shear was thought to be potentially critical in conventional precast prestressed concrete floor units as shear resistance is provided entirely by the tendons which are however placed close to the fire exposed face of the units. However work reported by Acker (2003/2004) and Lennon (2003) indicates that this appears not to be the case, provided the precast units are constrained to act as a diaphragm by being tied together by reinforcement in the plane of the floor. The tests reported by Lennon which were under a compartment fire régime rather than the standard furnace curve also indicated there was no spalling and that the load carrying capacity was satisfactory up to a time-equivalent of around 60 minutes. The test results for load carrying capacity should not be extrapolated beyond this point. Bond is generally not a problem even though bond strengths are severely reduced in a fire.

5.3.3.3 Detailing

It should be self-evident that the prime factor controlling fire performance is axis distance to the reinforcement which in practical terms means any specified cover must be guaranteed. Clark et al. (1997) indicated that from surveys carried out, 15 per cent of the measured covers failed to achieve their minimum values. Although no indication was given on the magnitude of such failure, it is clear that there could be problems with fire performance. A parametric study by Purkiss and Du (2000) indicates that for columns, variations in cover have little effect on fire performance. For the top reinforcement in slabs or beams (other than that adjacent to the fire exposed surface), there will be little effect on the fire performance. However, for the bottom reinforcement in beams or slabs to have insufficient cover is potentially very dangerous as the fire endurance will be severely compromised. Purkiss, Claridge and Durkin (1989) noted in a simple simulation on slab behaviour in fire that variations in cover was the most significant variable for changes in fire performance. If there is adequate top reinforcement correctly detailed, then some moment transfer could occur. It should be admitted that the tabular data are likely to be conservative, provided the points noted below are considered. Lennon (2004a,b) has indicated that the data in BS 8110 were set below test results. This situation is also likely for EN 1992-1-2 as the data are not dissimilar. It should be noted that the only column test data available then were for columns heated on four sides under axial loads. Thus the margin of safety for edge or corner columns heated on two or three sides with a moment gradient is unquantifiable. However, the level of safety will be reduced, should the correct covers not be achieved on site. There are, however, two additional issues which should be raised:

(1) It was demonstrated in the Cardington tests, both on the concrete and steel frame, that damage to the structure also occurred during the cooling phase owing to the effects of contraction on a tied structure.

(2) The relevance of the BS 476 Part 20 (or similar) curve must be questioned for modern buildings. It has often been suggested with no hard evidence that the heating phase of the standard curve is equivalent to the heating curve for a moderate fire in an office-type environment. This might just have been feasible where the contents were totally, or almost totally, cellulosic.

It is not true for modern offices where the contents are plastic (or hydrocarbon based). Hydrocarbon fires are of shorter duration but have a higher rate of heat release. This means that whilst it is possible that the critical temperatures reached in concrete elements (or structures) are lower, the effect on spalling will be to potentially increase its occurrence and severity owing to the higher thermal gradients close to the surface.

The Cardington Test also showed the importance of ensuring the concrete design strength is adhered to during construction, as concrete strengths increased for whatever reason can lead to an increased probability of spalling. Thus any consideration of enhanced strengths over design values must be cleared by the designer in order to assess any possible detrimental effects. It is possible therefore that the designer may need to specify both minimum and maximum strengths for any concrete that could potentially be exposed to fire.

5.3.3.4 Whole Building Behaviour

The one fire test on the concrete frame at the large test facility at Cardington reported by Bailey (2002) showed up some interesting features. The structure was tested at an age of around three and a half years. The first was that the columns cast from high strength concrete did not spall in spite of a moisture content of per cent, although the dosage of polypropylene fibres of $2,7 \, \text{kg/m}^3$ was higher than the recommended value of $2 \, \text{kg/m}^3$ in EN 1992-1-2. There was some splitting on the arises of the column although this was likely to have been caused late in the test.

The slab, however, underwent substantial spalling, the reasons for this were the use of a flint-type aggregate, the moisture content of per cent, a concrete cube strength of 61 MPa at 28 days (cylinder strength of approximately 50 MPa) compared with the design mix strength class C37.

However before extrapolating the results from Cardington to other structures, a number of points need to be noted (Chana and Price, 2003). The first is that the original emphasis on the project was to investigate optimization of the construction process. The second was that the high moisture contents were due to the structure being in a relatively cool unventilated enclosure where drying would have been at the very least slow. The third is that the structure was of flat slab construction with steel cross bracing. This means that the lateral movements noted during the test would almost certainly have been less had the structure had in situ concrete lift shafts or stair wells, but the increased restraint may have exacerbated the spalling of the slab. The movements noted were enough to have buckled the steel flats in the bracing.

REFERENCES AND FURTHER READING

Acker, A. Van, (2003/2004) Shear Resistance of prestressed hollow core floors exposed to fire, *Structural Concrete, Journal of the fib*, **2**, June 2003.

Anderberg, Y. (1978) Analytical Fire Engineering design of reinforced concrete structures based on real fire characteristics, *Proceedings of the Eighth Congress of the Fédération Internationale de la Précontrainte, London 1978, Concrete Society,* 112–123.

Bailey, C.J. (2002) Holistic Behaviour of Concrete Buildings in Fire, *Proceedings of the Institution of Civil Engineers, Buildings and Structures*, **152**, 199–212.

Bamforth, P.B. (2004) Enhancing Concrete Durability, *Technical Report 61*, Concrete Society.

Becker, J., Bizri, H. and Bresler, B. (1974) FIRES-T2 *A Computer Program for the Fire Response of Structures-Thermal,* Report No UCB-FRG 74-1, Department of Civil Engineering, University of Berkeley, California.

Beeby, A.W. (1978) Corrosion of reinforcing steel in concrete and its relation to cracking, *Structural Engineer*, **56A(3)**, 77–81.

Beeby, A.W. (1979) The prediction of crack widths in hardened concrete, *Structural Engineer*, **57A(1)**, 9–17.

Beeby, A.W., Scott, R.H. and Jones, A.E.K. (2005) Revised Code provisions for long-term deflection calculations, *Proceedings of the Institution of Civil Engineers, Structures and Buildings*, **158**, 71–75.

Bobrowski, J. and Bardhan-Roy, B.K. (1969) A method of calculating the ultimate strength of reinforced and prestressed concrete beams in combined flexure and shear, *Structural Engineer*, **47**, 3–15.

BRE (2004) Alkali-Silica Reaction in Concrete (Parts 1–4), Building Research Establishment, Garston.

BS EN 206-1:2000 Concrete – Part 1: Specification, performance, production and conformity, British Standards Institution/ Comité Européan de Normalisation.

Chana, P. and Price, W. (2003) The Cardington Fire Test, *Concrete*, **37(1)**, 28, 30–33.

Clark, L.A., Shammas-Toma, M.G.K., Seymour, E., Pallett, P.F. and Marsh, B.K. (1997) How can we get the cover we need? *Structural Engineer*, **75(17)**, 289–296.

Connolly, R.J (1995) The Spalling of Concrete in Fires, PhD Thesis, University of Aston.

Connolly, R.J. (1997) The spalling of concrete, *Fire Engineers Journal*, 38–40.

Dallard, P., Fitzpatrick, A.J., Flint, A., LeBourva, S., Ridstill Smith, R.M. and Willford, M. (2001) The London Millennium Footbridge, *Structural Engineer*, **79**(22), 17–33.

DOE (1991) Building Regulations: Approved Document B: Fire Safety (reissued 2000), Stationary Office.

Dotreppe, J.C. and Franssen, J.M. (undated) Fire Attack on concrete Columns and Design Rules under Fire Conditions, *Proceedings of a Symposium on Computational and Experimental Methods in Mechanical and Thermal Engineering* [100th Anniversary of the Laboratory of Machines and Machine Construction, Department of Mechanical and Thermal Engineering], Universeit Gent, unpaginated.

Dotreppe, J.C., Franssen, J.M., and Vanderzeypen, Y. (1999) Calculation Method for Design of Reinforced Concrete Columns under fire Conditions, American Concrete Institute, *Structural Journal*, **96**(1), 9–18.

Dunster, A.M. and Crammond, N.J. (2003) Deterioration of cement-based building materials: Lessons learnt, Information Paper IP 4/03, Building Research Establishment.

Ehm, H (1967) Ein Betrag zur reichnerischung Bemessung von brandbeanspruchten balkenartigen Stahlbetonbauteilen, Technische Hochschule Braunschweig.

Henderson, N.A., Baldwin, N.J.R., McKibbins, L.D., Winsor, D.S. and Shanghavi, H.B. (2002) Concrete technology for cast in-situ foundations, Report 569, Construction Industry Research and Information Association.

Hertz, K (1981) Simple Temperature Calculations of Fire Exposed Concrete Structures, Report No 159, Institute of Building Design, Technical University of Denmark.

Hertz, K (1985) Analyses of Prestressed Concrete Structures exposed to Fire, Report No 174, Institute of Building Design, Technical University of Denmark.

Hertz, K (1988) Stress Distribution Factors (2nd Ed), Report No. 190, Institute of Building Design, Technical University of Denmark.

Johnson, R.P. and Anderson, D. (2004) Designers' Guide to EN 1994-1-1. Eurocode 4; *Design of Composite steel and concrete structures*, Part 1.1: General Rules and Rules for buildings, Thomas Telford.

Kirby, B.R., Newman, G.M., Butterworth, N., Pagan, J. and English, C (2004) A new approach to specifying fire resistance periods. *Structural Engineer*, **82**(12), 34–37.

Lawson, R.M. (1985) Fire resistance of Ribbed Concrete floors, Report No 107, Construction Industry Research and Information Association, London.

Lennon, T. (2003) Precast Hollowcore slabs in Fire, Information Paper IP 5/03, Building Research Establishment.

Lennon, T. (2004a) Fire safety of concrete structures: Background to BS 8110 fire design, Report BR 468, Building Research Establishment.

Lennon, T. (2004b) Fire safety of concrete structures, *Structural Engineer*, **82**(6), 22–24

Malhotra, H.L. (1984) Spalling of Concrete in Fires, Technical Note No 118, Construction Industry Research and Information Association, London.

Meyer-Ottens, C (1975) Zur Frage der Abplatzungen an Bautelien aus Beton bei Brandbeanspruchungen, Deutsche Ausschuß für Stahlbeton, Heft 248.

Page, C.L. and Treadaway, K.W.J. (1982) Aspects of the electrochemistry of steel in concrete, *Nature*, **297**(5862), 109–115.

Persson, B. (2003) Self-compacting concrete at fire temperatures. Report No TVBM 3110, Lund Institute of Technology, Lund University, Denmark.

Purkiss, J.A (1996) Fire Safety Engineering Design of Structures, Butterworth-Heinemann.

Purkiss, J.A. and Du, Y. (2000) A parametric study of factors affecting the fire performance of reinforced concrete columns, In Concrete Communication Conference 2000 (10th Annual BCA Conference), Birmingham University 29–30th June 2000, pp. 267–275.

Purkiss, J.A., Claridge, S.L. and Durkin, P.S. (1989) Calibration of simple methods of calculating the fire resistance of flexural reinforced concrete members, *Fire Safety Journal*, **15**, 245–263.

Shorter, G.W. and Harmathy, T.Z. (1965) *Moisture clog spalling*. Proceedings of the Institution of Civil Engineers, **20**, 75–90.

Redfern, B. (2005) Lack of fire stops blamed for speed of Madrid tower inferno, New Civil Engineer, 17 February, 5–7.

Taumasite Experts Group (1999) The Structural implications of the taumasite form of sulfate attack, *Structural Engineer*, **77**(4), 10–11.

Chapter 6 / Reinforced Concrete Beams in Flexure

This chapter deals with the flexural design of both rectangular and flanged reinforced concrete beams.

Clause 6.1 of EN 1992-1-2 lays down the general principles which are:
Plane sections remain plane
The strain in the reinforcing steel is equal to that of the surrounding concrete
The tensile strength of the concrete is ignored
The stresses in the concrete are derived from
(a) Full stress–strain curve (Fig. 6.1 (a))
(b) Parabolic–rectangular stress block (Fig. 6.1 (b))
(c) Equivalent rectangular stress block (Fig. 6.1 (c)).
The stresses in the reinforcement are derived from the relationship shown in Fig. 6.2
The compressive strain of the concrete is limited to ε_{cu2} or ε_{cu3} as appropriate.

Note, the values of ε_{cu2} or ε_{cu3} are given in Table 3.1 of EN 1991-1-2. Moss and Webster (2004) indicate there is little difference between whichever of the three stress blocks are chosen for pure flexural design. As the rectangular stress block leads to a very simple approach to design, it is the assumption which will be used in this chapter.

At relatively low values of applied bending moment, beams are designed with tensile reinforcement only, and are said to be *singly reinforced*, and those requiring tensile and compression reinforcement to resist much higher applied loading are denoted as *doubly reinforced*. In addition, the shear induced by the applied loading also needs to be resisted. This resistance is commonly provided by the use of shear links, although other approaches such as bent-up bars or a combination of bent-up bars and shear links may be used. Even if a beam is singly reinforced, the shear links will need supporting in the compression zone by bars at each corner of any links. These bars provide anchorage for the link and enable the total reinforcement to be assembled as

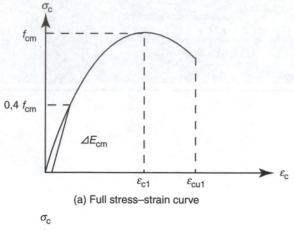

(a) Full stress–strain curve

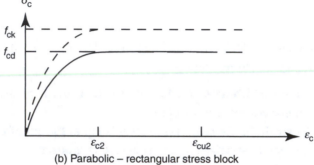

(b) Parabolic – rectangular stress block

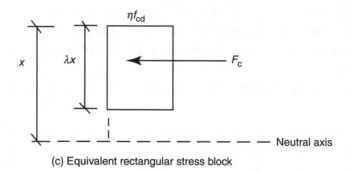

Neutral axis

(c) Equivalent rectangular stress block

FIGURE 6.1 Possible concrete stress blocks

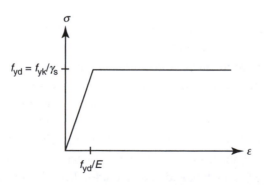

FIGURE 6.2 Reinforcement stress–strain behaviour

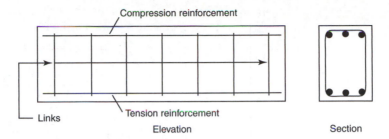

FIGURE 6.3 Arrangement of reinforcement in a beam

a reinforcing cage. In order to provide the cage with stability and, where the cage is prefabricated, to enable it to be lifted into position, the top steel should be not less than 16 mm high yield steel. A typical beam detail is given in Fig. 6.3. Where there is compression steel, the links have a secondary function which is to restrain the compression steel to prevent it from buckling.

6.1 BEHAVIOUR OF BEAMS IN FLEXURE

A set of theoretical moment-curvature diagrams are plotted in Fig. 6.4, albeit with a slightly different concrete stress–strain curve to that in EN 1992-1-1 and a reinforcement yield stress of 460 MPa. A similar set of diagrams would be obtained were the stress–strain curve in EN 1992-1-1 and a yield stress of 500 MPa to be used.

If Fig. 6.4 be examined, there appears two distinct patterns of behaviour. The first pattern is where there is a considerable increase in curvature after the attainment of the maximum moment and where the behaviour exhibits a characteristic close to elastic–perfectly plastic behaviour. The second type of behaviour is where failure occurs as the maximum moment is achieved with no subsequent ductility. The first type of behaviour clearly allows a large degree of rotation after yield, albeit accompanied by severe and increased cracking and deflection. This behaviour is exhibited by sections defined as *under-reinforced*, where the reinforcement yields before the concrete reaches its maximum strain capacity. The behaviour with no post-yield plateau is given by an *over-reinforced* section where the steel remains elastic and the concrete reaches its maximum strain before the steel yields. An over-reinforced section gives no prior warning of distress and therefore tends to be considered as unsafe. A section at which the steel yields at the same time as the concrete reaches its maximum strain is known as a *balanced* section.

At balance, the steel strain ε_{yd} is given by

$$\varepsilon_{yd} = \frac{f_{yk}}{E_s \gamma_s} = \frac{f_{yk}}{200 \times 10^3 \times 1{,}15} = \frac{f_{yk}}{0{,}23 \times 10^6} \quad (6.1)$$

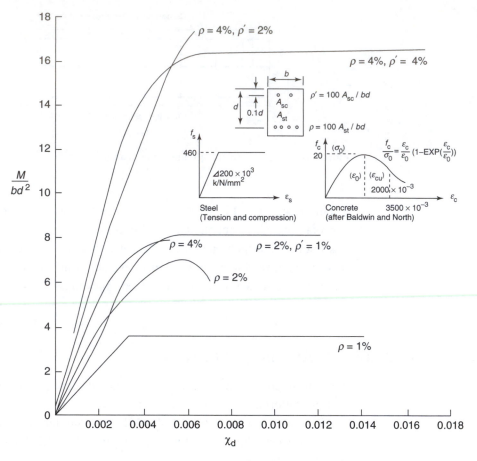

FIGURE 6.4 Moment-curvature diagrams showing effect of reinforcement percentage

the concrete strain ε_c is related to the steel strain by

$$\frac{\varepsilon_c}{(d-x)} = \frac{\varepsilon_c}{x} \tag{6.2}$$

where d the effective depth, x_{bal} is the neutral axis depth at balance, i.e. the point at which the steel strain equals its yield strain ε_{yd} and the concrete its ultimate strain ε_{cu3}. Thus x_{bal}/d is given by

$$\frac{x_{bal}}{d} = \frac{1}{(\varepsilon_{yd}/\varepsilon_{cu3}) + 1} \tag{6.3}$$

The concrete strain ε_{cu3} equals 3500 µstrain (numerically) for concrete of Grade C50 or less and $2{,}6 + 35[(90 - f_{ck})/100]^4 \times 10^3$ µstrain for concrete Grades C50 or above (Table 3.1, EN 1992-1-1), i.e. x_{bal}/d is given by

$$\frac{x_{bal}}{d} = \frac{1}{(f_{yk}/(0{,}23 \times 10^6))/\text{lesser of } [3500 \times 10^{-6} \text{ or } (2{,}6 + 35[(90 - f_{ck})/100]^4) \times 10^{-3}] + 1} \tag{6.4}$$

For Grade C50 concrete (or below) Eq. (6.4) reduces to

$$\frac{x_{bal}}{d} = \frac{1}{(f_{yk}/(0{,}23 \times 10^6))(1/(3500 \times 10^{-6})) + 1} = \frac{1}{(f_{yk}/805) + 1} \tag{6.5}$$

for $f_{yk} = 500$ MPa, $x_{bal}/d = 0{,}617$ and for $f_{yk} = 250$ MPa, $0{,}763$, respectively. In both these cases the balanced neutral axis depth is higher than that allowed by cl 5.5 (4).

6.1.1 Limits on Neutral Axis Depth

Where plastic analysis has been used to determine the internal forces in a member, such as slabs using yield line theory (Section 10.4), the limiting value of x_u/d is 0,25 for concretes of Grade 50 or below and 0,15 for concrete grades higher than C50 (cl 5.6.2 (2)). This is to ensure the section has sufficient ductility to allow the formation and rotation of plastic hinges.

There are also restrictions placed on the rotations at internal supports for both continuous beams and one-way spanning slabs (See Section 4.7).

Where an elastic analysis is used and the resultant moments redistributed to give simulated plastic analysis (Section 4.7) there are also limits on the x/d ratio to ensure ductility. These limits are

- for concrete of grade C50 or less

$$\delta \geq 0{,}44 + 1{,}25\left(0{,}6 + \frac{0{,}0014}{\varepsilon_{cu2}}\right)\frac{x}{d} \geq 0{,}7 \tag{6.6}$$

- for concrete above grade C50

$$\delta \geq 0{,}54 + 1{,}25\left(0{,}6 + \frac{0{,}0014}{\varepsilon_{cu2}}\right)\frac{x}{d} \geq 0{,}8 \tag{6.7}$$

where δ is defined as the ratio of the redistributed moment to the moment before redistribution. For concrete of grade C50 or less, $\varepsilon_{cu2} = 0{,}0035$, thus Eq. (6.7) reduces to

$$\delta \geq 0{,}44 + 1{,}25\frac{x}{d} \geq 0{,}7 \tag{6.8}$$

6.2 SINGLY REINFORCED SECTIONS

The forces acting on the section are shown in Fig. 6.5. The numerical value of the resultant compression force F_c from the rectangular concrete stress block is given by

$$F_c = \frac{\eta \alpha_{cc} f_{ck}}{\gamma_c}(\lambda x)b \tag{6.9}$$

and the tensile force F_s in the reinforcement by

$$F_s = A_s \frac{f_{yk}}{\gamma_s} \tag{6.10}$$

or

$$\frac{x}{d} = \frac{\gamma_c}{\gamma_s \eta \lambda \alpha_{cc}} \frac{A_s f_{yk}}{b\,d\,f_{ck}} \tag{6.11}$$

The ultimate moment of resistance of the section M_{Rd} is obtained by taking moments about the centroid of the compression zone, and is given by

$$M_{Rd} = \frac{A_s f_{yk}}{\gamma_s}\left(d - \frac{\lambda}{2}x\right) \tag{6.12}$$

or

$$\frac{M_{Rd}}{b\,d^2 f_{ck}} = \frac{1}{\gamma_s}\frac{A_s f_{yk}}{b\,d\,f_{ck}}\left(1 - \frac{\lambda}{2}\frac{x}{d}\right) \tag{6.13}$$

Eliminating x/d between Eqs (6.11) and (6.13) and rearranging gives

$$\left(\frac{A_s f_{yk}}{b\,d\,f_{ck}}\right)^2 - \frac{2\gamma_s \eta \alpha_{cc}}{\gamma_c}\frac{A_s f_{yk}}{b\,d\,f_{ck}} + \frac{2\gamma_s^2 \eta \alpha_{cc}}{\gamma_c}\frac{M_{Sd}}{b\,d^2 f_{ck}} = 0 \tag{6.14}$$

Solving gives

$$\frac{A_s f_{yk}}{b\,d\,f_{ck}} = \frac{\gamma_s \eta \alpha_{cc}}{\gamma_c} - \sqrt{\frac{\gamma_s^2 \eta^2 \alpha_{cc}^2}{\gamma_c^2} - \frac{2\eta \alpha_{cc}\gamma_s^2}{\gamma_c}\frac{M_{Sd}}{b\,d^2 f_{ck}}} \tag{6.15}$$

These equations hold irrespective of the values of the parameters γ_c, γ_s, η, λ and α_{cc}. For all concrete and steel grades $\gamma_s = 1,15$, $\gamma_c = 1,5$. The parameters η applied to the concrete strength and λ to the depth of the stress block are to give the correct equivalent values of the resultant compressive force and its point of action when the

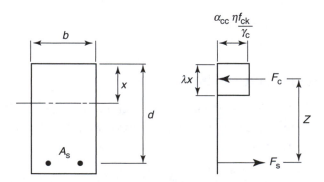

FIGURE 6.5   Forces on a singly reinforced section

rectangular stress block rather than the complete stress–strain curve is used. The values of λ are concrete strength dependant as the shape of the stress–strain curve changes with concrete strength as the absolute value of ultimate strain in the concrete decreases with strength and the strain at maximum stress increases with strength. If the concrete grade is restricted to C50/60 or below, and λ may be taken as constant with $\lambda = 0,8$ and $\eta = 1,0$, which allows Eq. (6.12) to (6.15) to be simplified. For concrete grades higher than C50, it is not possible to perform this operation as both η and λ are dependant upon the concrete strength,

$$\lambda = 0,8 - \frac{f_{ck} - 50}{400} \tag{6.16}$$

and

$$\eta = 1,0 - \frac{f_{ck} - 50}{200} \tag{6.17}$$

Equation (6.12) is independent of the value of α_{cc} and becomes for Grade C50 or below,

$$\frac{M_{Rd}}{b\,d^2 f_{ck}} = \frac{1}{1,15} \frac{A_s f_{yk}}{b\,d\,f_{ck}} \left(1 - 0,4 \frac{x}{d}\right) \tag{6.18}$$

The value of the coefficient α_{cc} which allows for long term effects on the compressive strength should lie between 0,8 and 1,0 (cl 3.1.6 (1)P), with a recommendation that it may be taken as 1,0. However, UK practice has conventionally taken a value of 0,85 for ULS design under ambient conditions. For any accidental limit state, such as fire, the need to consider creep effects can be ignored and α_{cc} set equal to 1.

It is therefore useful to derive flexural design equations for values of α_{cc} of both 1,0 and 0,85, although it should be noted that all the ambient design examples for reinforced concrete in this and subsequent chapters will be carried out with α_{cc} of 0,85.

- For $\alpha_{cc} = 0,85$,
  The neutral axis depth x/d is given by

$$\frac{x}{d} = \frac{1,5}{1,15 \times 1,0 \times 0,8 \times 0,85} \frac{A_s f_{yk}}{b\,d\,f_{ck}} = 1,918 \frac{A_s f_{yk}}{b\,d\,f_{ck}} \tag{6.19}$$

Equation (6.14) reduces to

$$\left(\frac{A_s f_{yk}}{b\,d\,f_{ck}}\right)^2 - \frac{1,15 \times 2 \times 0,85}{1,5} \frac{A_s f_{yk}}{b\,d\,f_{ck}} + \frac{1,15^2 \times 2 \times 0,85}{1,5} \frac{M_{Sd}}{bd^2 f_{ck}} = 0 \tag{6.20}$$

which becomes

$$\left(\frac{A_s f_{yk}}{b\,d\,f_{ck}}\right)^2 - 1{,}303\frac{A_s f_{yk}}{b\,d\,f_{ck}} + 1{,}534\frac{M_{Sd}}{bd^2 f_{ck}} = 0 \tag{6.21}$$

Solving gives (with a slight rounding of values)

$$\frac{A_s f_{yk}}{b\,d\,f_{ck}} = 0{,}652 - \sqrt{0{,}425 - 1{,}5\frac{M_{Sd}}{bd^2 f_{ck}}} \tag{6.22}$$

To determine M_{sd} in terms of A_s, Eq. (6.19) may be substituted into Eq. (6.18) to give

$$\frac{M_{Rd}}{bd^2 f_{ck}} = 0{,}87\frac{A_s f_{yk}}{b\,d\,f_{ck}}\left(1 - 0{,}767\frac{A_s f_{yk}}{b\,d\,f_{ck}}\right) \tag{6.23}$$

• For $\alpha_{cc} = 1{,}0$,

$$\frac{x}{d} = \frac{1{,}5}{1{,}15 \times 1{,}0 \times 0{,}8 \times 1{,}0}\frac{A_s f_{yk}}{b\,d\,f_{ck}} = 1{,}63\frac{A_s f_{yk}}{b\,d\,f_{ck}} \tag{6.24}$$

Equation (6.14) becomes

$$\left(\frac{A_s f_{yk}}{b\,d\,f_{ck}}\right)^2 - \frac{1{,}15 \times 2 \times 1{,}0}{1{,}5}\frac{A_s f_{yk}}{b\,d\,f_{ck}} + \frac{1{,}15^2 \times 2 \times 1{,}0}{1{,}5}\frac{M_{Sd}}{bd^2 f_{ck}} = 0 \tag{6.25}$$

or

$$\left(\frac{A_s f_{yk}}{b\,d\,f_{ck}}\right)^2 - 1{,}533\frac{A_s f_{yk}}{b\,d\,f_{ck}} + 1{,}763\frac{M_{Sd}}{bd^2 f_{ck}} = 0 \tag{6.26}$$

Solving gives

$$\frac{A_s f_{yk}}{b\,d\,f_{ck}} = 0{,}767 - \sqrt{0{,}588 - 1{,}76\frac{M_{Sd}}{b\,d^2 f_{ck}}} \tag{6.27}$$

To determine M_{sd} in terms of A_s, Eq. (6.24) may be substituted into Eq. (6.18) to give

$$\frac{M_{Rd}}{b\,d^2 f_{ck}} = 0{,}87\frac{A_s f_{yk}}{b\,d\,f_{ck}}\left(1 - 0{,}652\frac{A_s f_{yk}}{b\,d\,f_{ck}}\right) \tag{6.28}$$

It is instructive to compare the effects of the two values of α_{cc}. This is carried out in Table 6.1. It will be observed that there is virtually no difference in the values of $A_s f_{yk}/bdf_{ck}$ obtained using the two values of α_{cc}, any such reductions are likely to be

TABLE 6.1 Comparison between effects in values of α_{cc}.

M/bd^2f_{ck}	α_{cc}	A_sf_{yk}/bdf_{ck}		α_{cc}	x/d	
		0,85	*1,00*		*0,85*	*1,00*
0,05		0,060	0,060		0,115	0,098
0,10		0,128	0,125		0,247	0,204
0,15		0,205	0,198		0,393	0,323
0,20		0,298	0,281		0,393	0,323

TABLE 6.2 Values of limiting design parameters for concrete Grades less than or equal to C50.

α_{cc}	x_{max}/d	$A_{s.max}f_{yk}/bdf_{ck}$	$M_{Sd.max}/bd^2f_{ck}$
0,85	$(\delta - 0,44)/1,25$ for $\delta = 1$	$0,417\delta - 0,184$	$0,535\delta - 0,21 - 0,133\delta^2$
	0,45	0,233	0,168
	Plastic		
	0,25	0,13	0,117
1,00	$(\delta - 0,44)/1,25$ for $\delta = 1$	$0,491\delta - 0,216$	$0,63\delta - 0,246 - 0,157\delta^2$
	0,45	0,275	0,227
	Plastic		
	0,25	0,153	0,138

eliminated when the mechanical reinforcement ratio A_sf_{yk}/bdf_{ck} is converted to actual bar sizes. The main effect is that the higher value of α_{cc} will give lower values of x/d, hence a slightly higher moment capacity before compression reinforcement is required.

For $\alpha_{cc} = 0,85$, the limiting moment a singly reinforced section can take is obtained by substituting the maximum value of x/d from Eq. (6.8) into Eq. (6.19) to determine the limiting value of A_sf_{yk}/bdf_{ck}, and the coexistent value of M_{Sd}/bd^2f_{ck} from Eq. (6.21). A similar procedure may be adopted for $\alpha_{cc} = 1,0$ using Eqs (6.24) and (6.28). The results of these calculations are given in Table 6.2.

EXAMPLE 6.1 Design of a singly reinforced section. A singly reinforced beam has a width of 300 mm and an effective depth of 600 mm. The concrete is Grade C25/30 and the steel is Grade 500. Determine

(a) the maximum design moment of resistance of the section and the reinforcement necessary. If the maximum aggregate size is 20 mm and the cover to the main reinforcement is 40 mm, and

(b) the area of reinforcement required to resist a moment of 350 kNm.

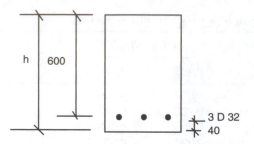

FIGURE 6.6   Detailing for Example 6.1

(a) From Table 6.2 (with M_{Sd} replaced by M_{Rd}) the maximum value of M_{Rd}/bd^2f_{ck} is 0,168, so

$$M_{Rd} = 0,168 \times 300 \times 600^2 \times 25 \times 10^{-6} = 453,6 \text{ kNm}.$$

$A_s f_{yk}/bdf_{ck} = 0,234$, so

$$A_s = 0,234 \times 300 \times 600 \times 25/500 = 2106 \text{ mm}^2$$

Fix 3D32 ($A_s = 2413 \text{ mm}^2$)
Minimum spacing between bars is greater of bar diameter or 20 mm.
Minimum width required $3 \times 32 + 2 \times 32 + 2 \times 40 = 240 \text{ mm} < 300 \text{ mm}$
therefore satisfactory (See Fig. 6.6).
Overall depth is $600 + 32/2 + 40 = 656 \text{ mm}$ (round to 660 mm).

(b) Steel area for moment of 350 kNm.

$$M_{Sd}/bd^2f_{ck} = 350 \times 10^6/(300 \times 600^2 \times 25) = 0,130.$$

From Eq. (6.22) determine $A_s f_{yk}/bdf_{ck}$,

$$A_s f_{yk}/bdf_{ck} = 0,652 - \sqrt{(0,425 - 1,5 \times 0,130)} = 0,172$$
$$A_s = 0,172 \times 600 \times 300 \times 25/500 = 1548 \text{ mm}^2.$$

Fix 2D32 in a single layer (1608 mm²).
Check x/d from (6.19):

$$x/d = 1,918 \times A_s f_{yk}/bdf_{ck} = 1,918 \times 0,172 = 0,330 < 0,45.$$

EXAMPLE 6.2   Ultimate resistance of a singly reinforced section. Determine the moment of resistance of a section whose width is 200 mm, effective depth 450 mm and is reinforced with 2D20 and 2D16 bars. The concrete is Grade C25/30.

A_s: Area of 2 No 20 mm and 2 No 16 mm bars is 1030 mm²

$$A_s f_{yk}/bdf_{ck} = 1030 \times 500/(200 \times 450 \times 25) = 0,229$$

Determine $M_{Rd}/bd^2 f_{ck}$ from Eq. (6.23)

$$M_{Rd}/bd^2 f_{ck} = 0{,}87 \times 0{,}229 \times (1 - 0{,}767 \times 229) = 0{,}164$$
$$M_{Rd} = 0{,}164 \times 200 \times 450^2 \times 25/10^6 = 166 \text{ kNm.}$$

From Table 6.2, $M_{Rd.max}/bd^2 f_{ck} = 0{,}168$ for $\delta = 1{,}0$.
Thus the moment is below the maximum for a singly reinforced beam.

6.3 DOUBLY REINFORCED SECTIONS

If the moment applied to the section is such that the limiting value of x/d is exceeded when it is attempted to design the section as singly reinforced, then compression reinforcement must be supplied in order to resist the excess moment. The configuration for a doubly reinforced beam is given in Fig. 6.7.

The applied moment on the section M_{Sd} is resisted in part by the limiting moment $M_{Rd,lim}$ determined from the maximum allowed value of x/d and the remainder by the couple induced by the forces in the compression reinforcement (and the additional equal force in the tension reinforcement).

Assume initially that the compression reinforcement has yielded. Equilibrium of the section is maintained with no shift in the centroidal axis position PROVIDED the force in the compression reinforcement F'_s is exactly balanced by an equivalent force in the tension reinforcement. The additional moment is simply the force in the compression steel F'_s times $d - d'$ the distance between the tension and compression reinforcement. The normalized moment capacity $M_{Sd}/bd^2 f_{ck}$ is then given by the limiting value $M_{Rd\,lim}/bd^2 f_{ck}$ in Table 6.2 plus an additional moment $M_{Rd.add}/bd^2 f_{ck}$ given by

$$\frac{M_{Rd.add}}{b\,d^2\,f_{ck}} = \frac{1}{1{,}15} \frac{A_{s'}\,f_{yk}}{bd\,f_{ck}} \left(1 - \frac{d'}{d}\right) \tag{6.29}$$

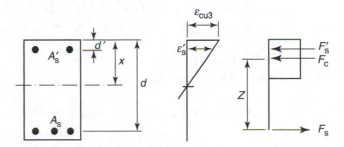

FIGURE 6.7   Doubly reinforced section

where d' is the depth from the compression face to the centroid of the compression reinforcement.

The tension reinforcement A_s is then given by $A_{s.\text{lim}} + A_s'$.

Check on steel yield:
From strain compatibility,

$$\varepsilon_{s'} = \frac{x - d'}{x} \varepsilon_c \tag{6.30}$$

The limiting steel strain is $f_{yk}/(1,15 \times 200 \times 10^3)$ (as for steel in tension), ε_c is 3500 μstrain, thus d'/d is given by

$$\frac{d'}{d} = \left(1 - \frac{f_{yk}}{805}\right) \frac{x_{\text{lim}}}{d} \tag{6.31}$$

If d'/d is greater than the value given by Eq. (6.31), then the stress f_s' in the compression reinforcement is given by

$$f_{s'} = \frac{805}{\gamma_s} \left(1 - \frac{d'}{d} \frac{d}{x_{\text{lim}}}\right) = 700 \left(1 - \frac{d'}{d} \frac{d}{x_{\text{lim}}}\right) \tag{6.32}$$

and the equivalent extra area of tension reinforcement is then given by $A_s' f_s / (f_{yk}/1,15)$.

EXAMPLE 6.3   Maximum capacity of a singly reinforced beam. A singly reinforced beam constructed from Grade C25/30 concrete has a width of 200 mm and an effective depth of 450 mm. The reinforcement has a characteristic strength of 500 MPa, determine the maximum capacity of the section and the reinforcement if the moments have been reduced by 30 per cent due to redistribution.

The ratio of the moment after redistribution to that before: $\delta = 0,7$.
Maximum allowable value of x_{max}/d from Table 6.2,

$$x_{\text{max}}/d = (\delta - 0,44)/1,25 = (0,7 - 0,44)/1,25 = 0,208$$

From Table 6.2,

$$A_{s.\text{lim}} f_{yk}/bdf_{ck} = 0,417\delta - 0,184 = 0,417 \times 0,7 - 0,184 = 0,108$$
$$A_{s.\text{lim}} = 0,108 \times 200 \times 450 \times 25/500 = 486 \text{ mm}^2$$

Fix 2D20 (628 mm^2)
From Table 6.2,

$$M_{Rd,\text{max}}/bd^2 f_{ck} = 0,535\delta - 0,21 - 0,133\delta^2$$
$$= 0,535 \times 0,7 - 0,21 - 0,133 \times 0,7^2 = 0,0993$$
$$M_{Rd,\text{max}} = 0,0993 \times 200 \times 450^2 \times 25 \times 10^{-6} = 101 \text{ kNm}$$

EXAMPLE 6.4   Design of a doubly reinforced section. A doubly reinforced beam constructed from Grade C25/30 concrete has a width of 250 mm and an effective depth of 450 mm. The reinforcement has a characteristic strength of 500 MPa, and the axis depth of the compression reinforcement, if required, is 50 mm. If the moment applied to the section is 200 kNm, determine the reinforcement requirements for (a) no redistribution and (b) a reduction of 30 per cent due to redistribution.

(a) Maximum $x/d = 0,45$ (Table 6.2)
Maximum moment without compression steel:

$$M_{Rd,max}/bd^2f_{ck} = 0,168 \times 250 \times 450^2 \times 25 \times 10^{-6} = 213 \, \text{kNm}$$

Beam is therefore singly reinforced.

$$M_{Sd}/bd^2f_{ck} = 200 \times 10^6/250 \times 450^2 \times 25 = 0,158$$

From Eq. (6.22)

$$A_s f_{yk}/bdf_{ck} = 0,652 - \sqrt{(0,425 - 1,5 \times 0,158)} = 0,218$$
$$A_s = 0,218 \times 250 \times 450 \times 25/500 = 1226 \, \text{mm}^2$$

Fix 4D20 (1257 mm^2)

(b) 30 per cent reduction due to redistribution ($\delta = 0,7$)
Tension steel requirement for limiting neutral axis depth (Table 6.2)

$$A_{s,lim} f_{yk}/bdf_{ck} = 0,417\delta - 0,184 = 0,417 \times 0,7 - 0,184 = 0,108$$
$$A_{s,lim} = 0,108 \times 250 \times 450 \times 25/500 = 608 \, \text{mm}^2$$
$$M_{Sd}/bd^2f_{ck} = 200 \times 10^6/250 \times 450^2 \times 25 = 0,158$$

Check limiting values from Table 6.2,

$$M_{Rd,lim}/bd^2f_{ck} = 0,535\delta - 0,21 - 0,133\delta^2 = 0,0993 < M_{Sd}/bd^2f_{ck}$$

Therefore compression steel is required, thus,

$$M_{Rd,add}/bd^2f_{ck} = M_{Sd}/bd^2f_{ck} - M_{Rd,lim}/bd^2f_{ck}$$
$$= 0,158 - 0,0993 = 0,0587$$

Check yield status of the compression reinforcement:

$$x/d = (\delta - 0,44)/1,25 = (0,7 - 0,44)/1,25 = 0,208$$

Critical value of d'/d is given by Eq. (6.31)

$$(d'/d)_{\text{crit}} = (x_{\text{lim}}/d)\,(1 - f_{\text{yk}}/805)$$
$$= 0{,}208 \times (1 - 500/805) = 0{,}079$$
$$\text{Actual } d'/d = 50/450 = 0{,}111.$$

The compression steel has not therefore yielded, and its stress level f_s is given by Eq. (6.32)

$$f_s = 700\,(1 - (d'/d)\,(d/x_{\text{lim}})) = 700(1 - 0{,}111/0{,}208) = 326\,\text{MPa}$$

From Eq. (6.29) with the steel yield stress replaced by its elastic stress,

$$A'_s f_s/bd f_{\text{ck}} = (M_{\text{Rd,add}}/bd^2 f_{\text{ck}})/(1 - d'/d)$$
$$= 0{,}0587/(1 - 0{,}111) = 0{,}066$$
$$A'_s = 0{,}066 \times 250 \times 450 \times 25/326 = 569\,\text{mm}^2$$

Fix 2D20 ($628\,\text{mm}^2$)

Two bars are preferred to three in order to give the maximum space possible between the top steel to allow a 75 mm diameter poker vibrator to be able to reach the complete cross section.

Additional tension steel, $A_{\text{s,add}}$:
Since the compression steel has not yielded then this must be given by an equivalent area of yielded steel, so

$$A_{\text{s,add}} = A'_s f_s/(f_{\text{yk}}/1{,}15) = 569 \times 326/(500/1{,}15) = 427\,\text{mm}^2$$
$$A_{\text{s,total}} = A_{\text{s, lim}} + A_{\text{s,add}} = 613 + 427 = 1040\,\text{mm}^2$$

Fix 4D20 ($1257\,\text{mm}^2$)

EXAMPLE 6.5   Determination of section capacity. The cross section of a doubly reinforced concrete beam is given in Fig. 6.8. The concrete is Grade C25 and the characteristic strength of the reinforcement is 500 MPa. There has been no redistribution of moments.

Since the tension steel is in two layers the effective depth must be determined from the centroid of area of the bars.

Bottom layer 3D40 ($3770\,\text{mm}^2$) at 60 mm from bottom face,
Top layer 2D32 ($1608\,\text{mm}^2$) at 136 mm from bottom face, so effective centroid is at

$$(60 \times 3770 + 136 \times 1608)/(3770 + 1608) = 82{,}7\,\text{mm (round to 83)}$$

Effective depth $= 1000 - 83 = 917\,\text{mm}$

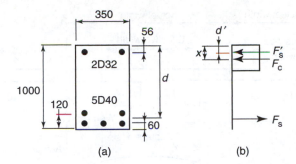

FIGURE 6.8 Data for Example 6.5

Consider equilibrium of forces to determine depth of neutral axis, assuming the compression steel has yielded (Fig. 6.8(b)):

$$F_s = A_s f_{yk}/1,15 = (3770 + 1608) \times 500 \times 10^{-3}/1,15 = 2338 \, \text{kN}.$$
$$F'_s = A'_s f_{yk}/1,15 = 1608 \times 500 \times 10^{-3}/1,15 = 699 \, \text{kN}.$$
$$F_c = b(0,8x)\,(0,85 f_{ck}/1,5)$$
$$= 350 \times (0,8x)(0,85 \times 25/1,5) \times 10^{-3} = 3,977x$$

Equating forces gives

$$3,977x + 699 = 2338, \text{ or } x = 412 \, \text{mm}.$$

Limiting value of $x_{max}/d = 0,45$, so $x_{max} = 0,45 \times 917 = 413$ mm.
This is satisfactory.

$$x/d = 412/917 = 0,449$$

Check limiting value of d'/d Eq. (6.31):

$$(d'/d)_{lim} = (x/d)\,(1 - f_{yk}/805) = 0,449(1 - 500/805) = 0,170$$

Actual value $= 56/917 = 0,061$. This is less than the limiting value, thus the assumption of yield was valid.
To determine M_{Rd} take moments about the centroid of the tension steel:

$$M_{Rd} = F'_s\,(d - d') + F_c(d - 0,4x)$$
$$= 699(917 - 56) + 3,977 \times 412(917 - 0,4 \times 412) \times 10^{-3} = 1835 \, \text{kNm}$$

6.4 FLANGED BEAMS

Flanged beams occur where it is necessary to deepen a section in order to carry the reactions from one or two spanning slabs and any additional loads from cladding or permanent partitions. Such beams are cast as a part of the slab system, and the

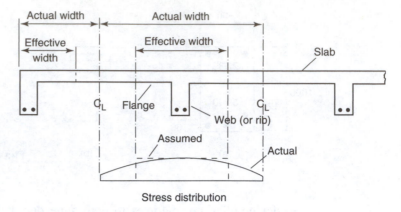

FIGURE 6.9   Effective width of a flanged beam

flexural resistance is calculated with an allowance of part of the slab acting integrally with the beam. For one way spanning slabs half the total loading from the slab is taken by each beam and slab system. For two spanning slabs the load is determined either on the basis of a 45° dispersion or where the Hillerborg strip method has been used, the reaction from each strip (Section 10.6). The full width of the flange of a 'T' beam is not effective in resisting the applied loads as the stresses are not constant across the width owing to a phenomenon known as shear lag (also See Section 4.5.3.2). The full width is therefore reduced in such a manner that the total force in the flange is given by a constant stress times the effective width. Strictly the effective width should vary over the span (Beeby and Taylor, 1978), although in practice it is sufficient to consider the reduced effective width as a constant over the whole span (Fig. 6.9).

6.4.1 Determination of Effective Width (cl 5.3.2.1)

For a 'T' beam, the effective width b_{eff} is given by

$$b_{\text{eff}} = \Sigma b_{\text{eff,j}} + b_{\text{w}} \leq b \tag{6.33}$$

where,

$$b_{\text{eff,j}} = 0{,}2\,b_{\text{i}} + 0{,}1\,l_0 \leq 0{,}2\,l_0 \tag{6.34}$$

subject to the condition

$$b_{\text{eff,j}} \leq b_i \tag{6.35}$$

where b_{w} is the width of the web, b the actual width of the flange, b_1 and b_2 are half widths of the flange outstands either side of the web (see Fig. 6.10), and l_0 is the

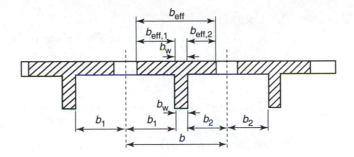

FIGURE 6.10   Determination of effective widths

distance between points of contraflexure in the beam. For an 'L' beam either b_1 or b_2 equals zero.

In the absence of a more accurate determination, the distance between the points of contraflexure may be taken as 0,85 times the span for a beam continuous over one support, 0,7 times the span for a beam continuous over both supports and for a cantilever, the clear span plus 0,15 times the adjacent span. These approximate data may only be used where adjacent spans are in the ratio of between 2/3 and 1,5 (cl 5.3.2.2).

EXAMPLE 6.6   Determination of effective widths. A section through a beam and slab floor assembly is given in Fig. 6.11. Determine the effective widths for the flanges if the beams are (a) simply supported over a span of 8,0 m, and (b) continuous over the same span.

Beam A:

(a) Simply supported:

$$l_0 = 8000 \text{ mm}$$
$$b_1 = 0, b_{\text{eff},1} = 0$$
$$b_2 = (3500 - 250)/2 = 1625 \text{ mm}$$
$$b_{\text{eff},2} = 0,2b_i + 0,1l_0 = 0,2 \times 1625 + 0,1 \times 8000 = 1125 \text{ mm}$$

This is less than the limiting value of b_2 ($= 1625$ mm) and $0,2l_0$ ($= 0,2 \times 8000 = 1600$ mm)

so from Eq. (6.33)

$$b_{\text{eff}} = \sum b_{\text{eff,j}} + b_w = 1255 + 250 = 1505 \text{ mm}$$

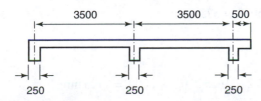

FIGURE 6.11   Data for Example 6.6

(b) Continuous:

$l_0 = 0.7 \times 8000 = 5600$ mm. The values of b_1, b_2 and $b_{\text{eff},1}$ are same as above.

$$b_{\text{eff},2} = 0.2 b_i + 0.1 l_0 = 0.2 \times 1625 + 0.1 \times 5600 = 885 \text{ mm}$$

This is less than the limiting value of b_2 ($= 1625$ mm) and $0.2 l_0$ ($= 0.2 \times 5600 = 1120$ mm)

so from Eq. (6.33)

$$b_{\text{eff}} = \sum b_{\text{eff},j} + b_w = 885 + 250 = 1135 \text{ mm}$$

Beam B:

(a) Simply supported:

As the beam is symmetric about its centre line, $b_1 = b_2$, and $b_{\text{eff},1} = b_{\text{eff},2}$

$b = 3500$ mm

$b_1 = b_2 = (3500 - 250)/2 = 1625$ mm

$b_{\text{eff},1} = 0.2 b_1 + 0.1 l_0 = 0.2 \times 1625 + 0.1 \times 8000 = 1125$ mm

This is less than the limits of b_1 or $0.2 l_0$.

$$b_{\text{eff}} = b_{\text{eff},1} + b_{\text{eff},2} + b_w = 2 \times 1125 + 250 = 2500 \text{ mm}.$$

This is less than the actual width b.

(b) Continuous:

As the beam is symmetric about its centre line, $b_1 = b_2$, and $b_{\text{eff},1} = b_{\text{eff},2}$

$b = 3500$ mm

$b_1 = b_2 = (3500 - 250)/2 = 1625$ mm

$b_{\text{eff},1} = 0.2 b_1 + 0.1 l_0 = 0.2 \times 1625 + 0.1 \times 5600 = 885$ mm

This is less than the limits of b_1 or $0.2 l_0$.

$$b_{\text{eff}} = b_{\text{eff},1} + b_{\text{eff},2} + b_w = 2 \times 885 + 250 = 2020 \text{ mm}.$$

This is less than the actual width b.

Beam C:

$b_1 = 1625$ mm (as in the previous calculations)

$b_2 = 500 - 250/2 = 375$ mm

(a) Simply supported:

$$b_{\text{eff},1} = 0.2 \times 1625 + 0.1 \times 8000 = 1125 \text{ mm}$$

This is less than the limiting values

$$b_{\text{eff},2} = 0{,}2 \times 375 + 0{,}1 \times 8000 = 875\,\text{mm}$$

Limiting values are b_2 (=375 mm) or $0{,}2 \times l_0$ (=1600 mm)
Thus $b_{\text{eff},2} = 375\,\text{mm}$

From Eq. (6.33)

$$b_{\text{eff}} = b_{\text{eff},1} + b_{\text{eff},2} + b_w = 1125 + 375 + 250 = 1750\,\text{mm}$$

(b) Continuous:

$$b_{\text{eff},1} = 0{,}2 \times 1625 + 0{,}1 \times 5600 = 885\,\text{mm}$$

This is less than the limiting values.

$$b_{\text{eff},2} = 0{,}2 \times 375 + 0{,}1 \times 5600 = 635\,\text{mm}$$

Limiting values are b_2 (=375 mm) or $0{,}2 \times l_0$ (=1120 mm)
Thus $b_{\text{eff},2} = 375\,\text{mm}$
From Eq. (6.33)

$$b_{\text{eff}} = b_{\text{eff},1} + b_{\text{eff},2} + b_w = 885 + 375 + 250 = 1510\,\text{mm}$$

6.4.2 Design of a Flanged Beam

Under hogging moments the flange is in tension, and is thus cracked. The beam is then designed as a rectangular beam (with the tension reinforcement in the top of the beam, spaced to allow a vibrating poker to compact the concrete adequately, i.e. a clear space of at least 75 mm should be left between the bars).

Under sagging moments two cases must be considered (Fig. 6.12):

(a) where the neutral axis is in the web, and
(b) where the neutral axis is in the flange.

6.4.2.1 Sagging Moment: Neutral Axis in the Flange

For this case the design follows the calculations for rectangular beams with the width of the beam taken as the effective width of the flange.

6.4.2.2 Sagging Moment: Neutral Axis in the Web

It is sufficiently accurate to design the tension reinforcement on the basis that the entire flange is in compression with the centre of compression at the mid-height of the flange (with any concrete in compression in the web being ignored). Thus A_s may

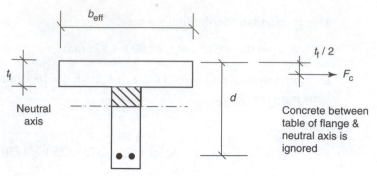

(a) Neutral axis in web

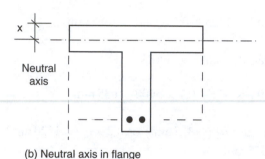

(b) Neutral axis in flange

FIGURE 6.12 'T' Beam design

be determined from

$$M_{Rd} = \frac{A_s f_{yk}}{1,15}\left(d - \frac{t_f}{2}\right) \tag{6.36}$$

where t_f is the thickness of the flange. However the moment applied cannot exceed the value $M_{Sd,max}$ based on the capacity of the concrete flange,

$$M_{Sd,max} = 0,85\frac{f_{ck}}{\gamma_c}b_{eff}\,t_f\left(d - \frac{t_f}{2}\right) \tag{6.37}$$

EXAMPLE 6.7 Design of a 'T' beam. A 'T' beam has an effective flange width of 1370mm, a flange thickness of 150 mm, a web width of 250 mm and an overall depth of 500 mm. The concrete Grade is C25 and the steel 500 MPa. The axis distance from the reinforcement to any face is 50 mm. Design suitable reinforcement for the following moments (there is no redistribution):

(i) 400 kNm (sagging), and
(ii) 300 kNm (hogging).

(a) effective depth $d = 500 - 50 = 450$ mm
Check the moment capacity of the beam assuming the neutral axis is in the web with the complete flange available to resist compression using Eq. (6.37),

$$M_{Sd} = (0,85 \times 25 \times 1370 \times 150/1,5)\,(450 - 150/2)10^{-6} = 1091\,\text{kNm.}$$

Neutral axis is therefore in the flange as this value exceeds the applied moment. So use Eq. (6.22) (with b replaced by b_{eff}):

$$M_{sd}/b_{eff}d^2f_{ck} = 400 \times 10^6/1370 \times 450^2 \times 25 = 0{,}0577$$
$$A_s f_{yk}/b_{eff}df_{ck} = 0{,}652 - \sqrt{(0{,}425 - 1{,}5 \times 0{,}0577)} = 0{,}070$$
$$A_s = 0{,}070 \times 1370 \times 450 \times 25/500 = 2158\,\mathrm{mm^2}$$

Fix 3D32 (2413 mm²)
Check x/d:

$$x/d = 1{,}918 A_s f_{yk}/b_{eff}df_{ck} = 1{,}918 \times 0{,}070 = 0{,}134 < 0{,}45$$

(b) $d = 450\,\mathrm{mm}$ (as above): $b = b_w = 250\,\mathrm{mm}$
$M_{Sd.max}/bd^2f_{ck}$ for a singly reinforced beam is 0,168 (Table 6.2).

$$M_{Rd.\,max} = 0{,}168 \times 250 \times 450^2 \times 25 \times 10^{-6} = 213\,\mathrm{kNm} < M_{Sd}$$

Beam requires compression reinforcement.

$$A_{s,\,lim} = 0{,}234 \times 450 \times 250 \times 25/500 = 1294\,\mathrm{mm^2}$$
$$M_{Sd}/bd^2f_{ck} = 300 \times 10^6/250 \times 450^2 \times 25 = 0{,}237$$
$$d'/d = 50/450 = 0{,}111.$$
$$x_{lim}/d = 0{,}45.$$

Check d'/d Eq. (6.31):

$$d'/d = (x_{lim}/d)(1 - f_{yk}/805) = 0{,}45(1 - 500/805) = 0{,}170.$$

Actual d'/d is less than the limiting value, so compression steel has yielded.

$$M_{Rd,add}/bd^2f_{ck} = 0{,}237 - 0{,}168 = 0{,}069$$

Calculate A'_s from Eq. (6.29)

$$A'_s f_{yk}/bdf_{ck} = 0{,}069/(1 - 0{,}111) = 0{,}0776$$
$$A_sl = 0{,}0776 \times 250 \times 450 \times 25/500 = 437\,\mathrm{mm^2}$$

Fix 2D20 (628 mm²)

$$A_s = A_{s,\,lim} + A'_s = 1249 + 437 = 1686\,\mathrm{mm^2}$$

Fix 2D40 (2513 mm²) (this arrangement gives a clear enough space between the bars to allow full compaction of the concrete. An alternative would be to either bundle bars or to place steel in two layers with a subsequent recalculation of moment capacity).

EXAMPLE 6.8 Design of a T beam. A beam with a flange width of 1500 mm, flange thickness 150 mm, effective depth 800 mm and web width 350 mm carries a moment of 2000 kNm. Concrete Grade C25 and steel strength 500 MPa. Design suitable reinforcement.

Initially calculate the moment capacity assuming the neutral axis is at the top of the web (or the underside of the flange).
$x/d = 150/800 = 0,1875$

Using Eq. (6.19)

$$\frac{A_s\, f_{yk}}{b\, d\, f_{ck}} = 1,918\frac{x}{d} = 1,918 \times 0,1875 = 0,36 \tag{6.38}$$

Using Eq. (6.18)

$$\frac{M_{Rd}}{b\, d^2 f_{ck}} = \frac{A_s f_{yk}}{b\, d\, f_{ck}}\left(1 - 0,4\frac{x}{d}\right)$$

$$= 0,36(1 - 0,4 \times 0,1875) = 0,333 \tag{6.39}$$

or, $M_{Rd} = 0,333 \times 1500 \times 800^2 \times 25 \times 10^{-6} = 7992$ kNm.

This is greater than the applied moment of 2000 kNm, thus the neutral axis is in the flange.

As a check, the moment capacity M assuming the neutral axis is the web with the concrete fully stressed is given by

$$M = b_f t_f \left(\alpha_{cc}\frac{f_{ck}}{\gamma_c}\right)\left(d - \frac{t_f}{2}\right)$$

$$= 1500 \times 150 \times \left(0,85 \times \frac{25}{1,5}\right)\left(800 - \frac{150}{2}\right) \times 10^{-6}$$

$$= 2311 \text{ kNm} \tag{6.40}$$

M_{Sd} is less than either of these limiting values, thus beam is singly reinforced with the neutral axis in the flange. $M_{Sd}/b_f d^2 f_{ck} = 2000 \times 10^6/1500 \times 800^2 \times 25 = 0,0833$
Use Eq. (6.22) to determine $A_s f_{yk}/b_w d f_{ck}$:

$$\frac{A_s\, f_{ck}}{b\, d\, f_{ck}} = 0,652 - \sqrt{0,425 - 1,5\frac{M_{Sd}}{b\, d^2 f_{ck}}}$$

$$= 0,652 - \sqrt{0,425 - 1,5 \times 0,0833} = 0,104 \tag{6.41}$$

$A_s = 0,104 \times 1500 \times 800 \times 25/500 = 6240$ mm^2

Fix 6D40 [7540 mm^2] in TWO layers. This will reduce the effective depth, but the reinforcement is overprovided.
Check x/d from Eq. (6.19):

$$\frac{x}{d} = \frac{1}{1,918}\frac{A_s f_{yk}}{b\, d\, f_{ck}} = \frac{0,104}{1,918} = 0,054 \tag{6.42}$$

$x = 0,054d = 0,054 \times 800 \times = 43,2$ mm (this is within the flange).
Fix 8D40 (10,500 mm^2) in TWO layers.

6.5 HIGH STRENGTH CONCRETE

Although, high strength concrete is not commonly used in cases where flexure is dominant as it is the steel strength which tends to control behaviour, it will allow slightly less shear reinforcement and give improved deflection behaviour as the empirical rules are dependant on the concrete strength.

The design procedure is essentially similar to that for more normal strength concretes except the parameters needed with the flexural design Eq. (6.12) to (6.15) need evaluating for each concrete strength.

Two examples (based on Examples 6.1 and 6.4) will be carried out to demonstrate the differences.

EXAMPLE 6.9   Design of a singly reinforced section. A singly reinforced beam has a width of 300 mm and an effective depth of 600 mm. The concrete is Grade C70 and the steel is Grade 500. Determine

(a) the maximum design moment of resistance of the section and determine the reinforcement necessary. If the maximum aggregate size is 20 mm and the cover to the main reinforcement is 40 mm, and,
(b) the area of reinforcement required to resist a moment of 350 kNm.

Limiting neutral axis depth is determined from Eq. (6.7). However it is first necessary to determine ε_{cu2}:

$$
\begin{aligned}
\varepsilon_{cu2} &= \left(2,6 + 35\left[\frac{90 - f_{ck}}{100}\right]^4\right) \times 10^{-3} \\
&= \left(2,6 + 35\left[\frac{90 - 70}{100}\right]^4\right) \times 10^{-3} \\
&= 2620 \times 10^{-6}
\end{aligned}
\tag{6.43}
$$

Equation (6.7) may be rewritten as

$$
\begin{aligned}
\frac{x}{d} &= \frac{\delta - 0,54}{1,25\left(0,6 + \dfrac{0,0014}{\varepsilon_{cu2}}\right)} \\
&= \frac{1 - 0,54}{1,25\left(0,6 + \dfrac{0,0014}{2620 \times 10^{-6}}\right)} = 0,324
\end{aligned}
\tag{6.44}
$$

(as there is no redistribution, $\delta = 1,0$) Determine η from (6.17)

$$\eta = 1,0 - \frac{f_{ck} - 50}{200} = 1 - \frac{70 - 50}{200} = 0,9 \tag{6.45}$$

and λ from (6.16)

$$\lambda = 0,8 - \frac{f_{ck} - 50}{400} = 0,8 - \frac{70 - 50}{400} = 0,75 \tag{6.46}$$

Use Eq. (6.11) to determine the relationship between $A_s f_{yk}/bdf_{ck}$ and x/d,

$$\frac{A_s f_{yk}}{b d f_{ck}} = \frac{\gamma_s \eta \lambda \, \alpha_{cc}}{\gamma_c} \frac{x}{d}$$

$$= \frac{1,15 \times 0,9 \times 0,75 \times 0,85}{1,5} \frac{x}{d} = 0,44 \frac{x}{d} \tag{6.47}$$

With the limiting value of x/d,

$$A_s f_{yk}/bdf_{ck} = 0,44 \times 0,324 = 0,143$$

Use Eq. (6.13) to determine M/bd^2f_{ck}:

$$\frac{M_{Sd}}{b d^2 f_{ck}} = \frac{A_s f_{yk}}{b d f_{ck}} \left[1 - \frac{\lambda}{2} \frac{x}{d} \right]$$

$$= 0,143 \left[1 - \frac{0,75}{2} \times 0,324 \right] = 0,126 \tag{6.48}$$

$M_{Sd} = 0,126 \times 300 \times 600^2 \times 70 \times 10^{-6} = 952\,\text{kNm}$.
$A_s f_{yk}/b d f_{ck} = 0,143$, so

$$A_s = 0,143 \times 300 \times 600 \times 70/500 = 3604\,\text{mm}^2$$

Fix 2D32 and 2D40 ($A_s = 4122\,\text{mm}^2$)
Minimum spacing between bars is greater of bar diameter or 20 mm.

$$\text{Minimum width required} = 2 \times 32 + 3 \times 40 + 2 \times 40 + 2 \times 40$$
$$= 284\,\text{mm} < 300\,\text{mm}$$

Therefore satisfactory overall depth is $600 + 40/2 + 40 = 660\,\text{mm}$.

(b) Steel area for moment of 350 kNm.

$$M_{Sd}/bd^2f_{ck} = 350 \times 10^6/(300 \times 600^2 \times 70) = 0,0463.$$

Use Eq. (6.15) to determine $A_s f_{yk}/bdf_{ck}$, however the coefficients in this equation need to be calculated first

$$\frac{\gamma_s \eta \, \alpha_{cc}}{\gamma_c} = \frac{1,15 \times 0,9 \times 0,85}{1,5} = 0,587 \tag{6.49}$$

$$\frac{2\gamma_s^2 \eta \alpha_{cc}}{\gamma_c} = \frac{2 \times 1,15^2 \times 0,9 \times 0,85}{1,5} = 1,35 \qquad (6.50)$$

So Eq. (6.15) becomes

$$\frac{A_s f_{yk}}{b\,d\,f_{ck}} = 0,587 - \sqrt{0,587^2 - 1,35\frac{M}{bd^2 f_{ck}}}$$

$$= 0,587 - \sqrt{0,587^2 - 1,35 \times 0,0463} = 0,0555 \qquad (6.51)$$

$A_s = 0,0555 \times 600 \times 300 \times 70/500 = 1399\,\text{mm}^2$.
Fix 3D25 in a single layer ($1473\,\text{mm}^2$).
Check x/d from Eq. (6.11):

$x/d = (1/0,44)A_s f_{yk}/bdf_{ck} = 2,273 \times 0,0555 = 0,126 < 0,324$ (determined in (a)).

EXAMPLE 6.10 Design of a doubly reinforced section. A doubly reinforced beam constructed from Grade C70 concrete has a width of 250 mm and an effective depth of 450 mm. The reinforcement has a characteristic strength of 500 MPa, and the axis depth of the compression reinforcement, if required, is 50 mm. If the moment applied to the section is 400 kNm, determine the reinforcement requirements for (a) no redistribution and (b) a reduction of 20 per cent due to redistribution.

The necessary design, Eqs (6.44), (6.47) and (6.51) have already been derived in Example 6.9 and will not be repeated.

(a) Maximum $x/d = 0,324$ (from Example 6.9)
Maximum moment without compression steel:

$M_{Rd,max}/bd^2 f_{ck} = 0,126$ as before

$$M_{Rd,max} = 0,126 \times 250 \times 450^2 \times 70 \times 10^{-6} = 447\,\text{kNm}$$

As M_{Sd} is less than the maximum value, the beam is singly reinforced.

$$M_{Sd}/bd^2 f_{ck} = 400 \times 10^6/250 \times 450^2 \times 70 = 0,113$$

Using Eq. (6.51)

$$A_s f_{yk}/bdf_{ck} = 0,587 - \sqrt{(0,587^2 - 1,35 \times 0,113)} = 0,149$$
$$A_s = 0,149 \times 250 \times 450 \times 70/500 = 2347\,\text{mm}^2$$

Fix 3D32 ($2412\,\text{mm}^2$)

(b) 20 per cent reduction due to redistribution ($\delta=0{,}8$)
Use Eq. (6.44) to determine limiting x/d value

$$\frac{x}{d} = \frac{\delta - 0{,}54}{1{,}25\left(0{,}6 + \dfrac{0{,}0014}{\varepsilon_{cu2}}\right)}$$

$$= \frac{0{,}8 - 0{,}54}{1{,}25\left(0{,}6 + \dfrac{0{,}0014}{2620 \times 10^{-6}}\right)} = 0{,}184 \tag{6.52}$$

Limiting value of $A_s f_{yk}/bdf_{ck}$ from Eq. (6.47)

$$\left(\frac{A_s f_{yk}}{b d f_{ck}}\right)_{lim} = 0{,}44 \times 0{,}184 = 0{,}0810 \tag{6.53}$$

$A_{s,lim} = 0{,}0810 \times 500 \times 250 \times 70/500 = 1418\,\text{mm}^2$
Limiting value of $M/bd^2 f_{ck}$ is given by Eq. (6.13)

$$\frac{M_{Sd}}{b d^2 f_{ck}} = \frac{A_s f_{yk}}{b d f_{ck}}\left[1 - \frac{\lambda x}{2 d}\right]$$

$$= 0{,}0810\left[1 - \frac{0{,}75}{2} \times 0{,}184\right] = 0{,}0754 \tag{6.54}$$

$M_{Sd}/bd^2 f_{ck} = 400 \times 10^6/250 \times 450^2 \times 70 = 0{,}113$
This is greater than the limiting value, therefore compression steel is required, so,

$$M_{Rd,add}/bd^2 f_{ck} = M_{Sd}/bd^2 f_{ck} - M_{Rd,lim}/bd^2 f_{ck}$$
$$= 0{,}113 - 0{,}0754 = 0{,}0376$$

Check yield status of the compression reinforcement with $x/d=0{,}185$
Critical value of d'/d is given by a similar equation to Eq. (6.31) but with the constant replaced by

$$1{,}15 \times 200 \times 10^3 \varepsilon_{cu2} = 1{,}15 \times 200 \times 10^3 \times 2660 \times 10^{-6} = 612 \, (d'/d)_{crit}$$
$$= (x_{lim}/d)\,(1 - f_{yk}/612)$$
$$= 0{,}184 \times (1 - 500/612) = 0{,}034$$

Actual $d'/d = 50/450 = 0{,}111$.
The compression steel has not therefore yielded, and its stress level f_s is given by Eq. (6.32) with the constant replaced by $612/1{,}15 = 532$,

$$f_s = 532(1 - (d'/d)\,(d/x_{lim})) = 532(1 - 0{,}111/0{,}184) = 211\,\text{MPa}$$

From Eq. (6.29) with the steel yield stress replaced by its elastic stress,

$$A'_s f_s / b d f_{ck} = (M_{Rd,add}/bd^2 f_{ck})/(1 - d'/d)$$
$$= 0{,}0376/(1 - 0{,}111) = 0{,}0334$$
$$A'_s = 0{,}0334 \times 250 \times 450 \times 70/211 = 1247\,\text{mm}^2.$$

Fix 2D32 ($1608\,\text{mm}^2$)

Two bars are preferred to three in order to give the maximum space possible between the top steel to allow a 75 mm diameter poker vibrator to be able to reach the complete cross section. Additional tension steel, $A_{s,add}$.

Since the compression steel has not yielded then this must be given by an equivalent area of yielded steel, so

$$A_{s,add} = A'_s f_s /(f_{yk}/1{,}15) = 1247 \times 211/(500/1{,}15) = 606\,\text{mm}^2$$
$$A_{s,total} = A_{s,lim} + A_{s,add} = 1418 + 606 = 2024\,\text{mm}^2$$

Fix 4D32 ($3217\,\text{mm}^2$).

6.6 DESIGN NOTES

6.6.1 Sizing of Sections

The size of sections are generally constrained by the situation in which they are being used, e.g. the width of a beam must be sufficient to provide the requisite bearing lengths for precast floor units, or sufficient to support both leaves of a cavity wall. The construction depth should be the least possible, except that it is cheaper and easier to construct formwork for a beam layout when all the soffits are at the same level and the beams are no narrower than the columns into which they frame.

Where there are no limitations the most effective section is one in which the width is around half the depth. The depth for a preliminary design can be taken between 1/10 to 1/15 of the span, with the higher figure being used for longer spans.

6.6.2. Procedure for Design of Beams

(1) Check the allowable x/d ratio (with consideration being taken of redistribution, if appropriate).
(2) Determine the maximum moment $M_{Rd,lim}$ capacity at the limiting x/d ratio.
(3) Determine whether the applied moment M_{Sd} is greater or less than $M_{Rd,lim}$.
(4) If less, the beam is singly reinforced and A_s determined from Eq. (6.22) for $\alpha_{cc} = 0{,}85$ and Eq. (6.27) for $\alpha_{cc} = 1{,}0$.
(5) If greater, beam needs compression reinforcement with A'_s determined from Eq. (6.29). Note, compression steel needs checking for yield.
(6) In all cases minimum and maximum percentages of steel should be checked. (The minimum steel requirement will often govern for slab design).

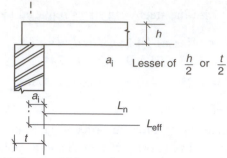

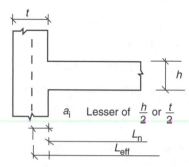

(a) Wall support (also applies to continuous members)

(b) Restrained support (or cantilever)

FIGURE 6.13 Determination of effective spans

(7) Design shear reinforcement.

(8) Check bar spacing and span/effective depth ratio for deflection and cracking. Note reinforcement in the top face of a beam should have clear spacing of at least 75 mm to allow the insertion of a poker vibrator.

6.7 EFFECTIVE SPANS (cl 5.3.2.2)

The effective span l_{eff} used to determine the internal forces in a member should be taken as

$$l_{eff} = l_n + a_1 + a_2 \tag{6.55}$$

where l_n is the clear distance between the supports and a_1 and a_2 are the end distances at either end of the element and are determined from Fig. 6.13. For beams supported on bearings the effective span is the distance between the centre lines of the bearings. For columns the effective heights are those between beam centre lines.

REFERENCES AND FURTHER READING

Baldwin, R. and North, M.A. (1973) A stress–strain relationship for concrete at high temperatures. *Magazine of Concrete Research*, **24**, 99–101.

Beeby, A.W. and Taylor, H.P.J. (1978) The use of simplified methods in CP110- is rigour necessary? *Structural Engineer*, **56A(8)**, 209–215.

Moss, R. and Webster, R. (2004) EC2 and BS8110 compared. *The Structural Engineer*, **82(6)**, 33–38.

Chapter 7 / Shear and Torsion

7.1 SHEAR RESISTANCE OF REINFORCED CONCRETE

7.1.1 Introduction (cl 6.2, EN)

The dimensions of the cross section of a reinforced concrete beam and the area of longitudinal steel are usually determined from calculations which consider resistance to the bending at the ultimate limit state. In situations of minor importance, e.g. a lintel, these dimensions and reinforcement will be adequate to resist small transverse shear forces. In other cases at least minimum shear reinforcement is required, and where shear forces in beams are large shear reinforcement must be added to resist all of the shear force.

7.1.2 Types of Shear Failure

The mechanism of shear failure in reinforced concrete is complicated and despite extensive experimental and theoretical research work semi-empirical solutions are currently still in use. The appearance of a beam failing in shear is shown in Fig. 7.1, and may be described as cracked concrete interacting with, and held together by the reinforcement. Prior to failure, vertical cracks are produced by the bending moment but later these are linked to diagonal cracks produced by the shear forces. Broadly there are five types of failure for reinforced concrete beams as shown in Fig. 7.1. The diagrams show the left hand support for a simply supported beam carrying a central point load. The first four types of failure are related to the shear span ratio (a/d). As the shear span ratio increases shear resistance decreases, but for values of $a/d > 3$ the decrease is relatively small. The longitudinal steel does not generally reach yield.

Type I ($a/d > 6$) The bending moment is large in comparison to the shear force and the mode of failure is similar to that in pure bending, i.e. rotation about the compression zone. The initial vertical bending cracks become inclined due to the action of the shear stresses and failure occurs in the compression zone (Fig. 7.1(a)). The stresses in the tensile steel are close to yield and only minimum shear reinforcement is required in design situations.

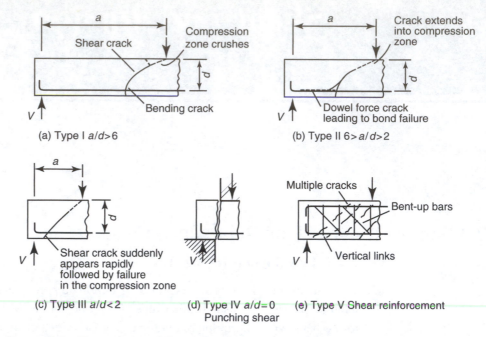

FIGURE 7.1 Types of shear failure in reinforced concrete beams

Type II ($6 > a/d > 2$) The initial bending cracks become inclined early in the loading sequence and failure is associated more with vertical shear deformation and less with vertical shear deformation and less with rotation about the compression zone. At collapse, horizontal cracks form running along the line of the tensile reinforcement (Fig. 7.1(b)) which reduce the shear resistance of the section by destroying the dowel force, which reduces the bond stresses between the steel and concrete. Finally the compression zone fails when subject to shear and compression.

Type III ($a/d < 2$) Bending cracks are small and do not develop but shear cracks, connecting a load point and the support, suddenly appear and run through to the compression zone producing collapse (Fig. 7.1(c)).

Type IV ($a_v/d = 0$) Punching shear failure occurs when the plane of failure is forced to run parallel to the shear forces as shown in Fig. 7.1(d). This can occur when the opposing shear forces are close together or, if vertical shear links have been added, when a failure plane forms which does not intercept the shear links. When this type of failure occurs the shear resistance of the section is at a maximum. The addition of shear reinforcement in the form of horizontal or inclined links increases the shear resistance.

Type V The addition of shear reinforcement increases the shear resistance provided that the shear reinforcement intercepts the shear cracks. Numerous diagonal cracks develop as shown in Fig. 7.1(e) and at failure the shear reinforcement reaches yield, provided that the shear reinforcement is anchored and not excessive.

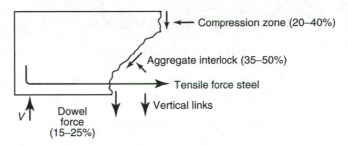

FIGURE 7.2   Components of shear resistance for a singly reinforced beam

7.1.3 Components of Shear Resistance

The shear resistance of a section may be represented by component strength parts as shown in Fig. 7.2. The approximate percentage of the total shear load resisted by each component, obtained experimentally (Taylor, 1974) for a singly reinforced concrete beam without shear reinforcement, is indicated.

The shear resistance of a beam can be expressed as the sum of the resistance of the parts, but the exact interaction of these shear components is not fully understood, and theories which are reasonably accurate are often too complicated for use in practical situations. Experiments (ASCE-ACI, 1973) show that the shear resistance increases with an increase in, area of the section, area of the tensile steel, direct compressive stress and strength of the concrete. These variables are included in Eq. (7.1). Except for small percentage of steel, the steel does not yield and therefore the yield strength of the longitudinal steel is not a variable.

7.2 MEMBERS NOT REQUIRING SHEAR REINFORCEMENT (cl 6.2.2, EN)

Shear reinforcement is not required in slabs of less than 200 mm nor for members of minor importance, e.g. a lintel. The reasons are that loading on a lintel is low, while for a slab the depth is often insufficient to accommodate shear reinforcement. The following equation is largely based on empirical evidence.

Design shear resistance (Eq. 6.2a, EN)

$$V_{Rd,c} = [C_{Rd,c}k(100\rho_1 f_{ck})^{1/3} + k_1\sigma_{cp}]b_w d \tag{7.1}$$

where

$$V_{Rd,c} > (v_{min} + k_1\sigma_{cp})b_w d \text{ and } v_{min} = 0{,}035k_1^{1,5}f_{ck}^{0,5}$$
$$C_{Rd,c} = 0{,}18/\gamma_c$$

b_w is the minimum width of the section in the tensile area

$k = 1 + (200/d)^{0,5}$ where d is in mm

$\rho_1 = A_{sl}/(b_w d) \leq 0,02$

A_{sl} = the area of tension reinforcement extending not less than $d + l_{bd}$ beyond the section considered

$\sigma_{cp} = N_{Ed}/A_c < 0,2 f_{cd}$ MPa

N_{Ed} = axial force in the section due to loading or prestressing (compression positive)

$k_1 = 0,15$

Where shear cracks are restrained by the loading point and support, and where $0,5d \leq a_v \leq 2d$, the shear force may be reduced by $\beta = a_v/2d$. For $a_v \leq 0,5d$, $a_v = 0,5d$ (cl 6.2.2(6), EN). The longitudinal reinforcement must be anchored securely at the support.

The factor k indicates that shear resistance decreases with increase in depth, which is justified by experiments (Chana, 1981 and Shioya *et al.*, 1989). The reason is thought to be due to larger crack widths in deep beams (ASCE-ACI, 1998). The resistance of the longitudinal steel is related to the steel ratio and does not include the yield strength.

The maximum shear force, calculated without the reduction factor β, that a section can resist (Eq. 6.5, EN) is related to the crushing strength of the concrete

$$V_{Rd} \leq 0,5 b_w d \nu f_{cd} \tag{7.2}$$

where the strength reduction factor for concrete cracked in shear (cl 6.6, EN)

$\nu = 0.6 (1 - f_{ck}/250)$

EXAMPLE 7.1 Shear resistance of a singly reinforced concrete lintel beam. A simple supported lintel beam spans 2 m and carries a total uniformly distributed design load of 15,8 kN. The beam section is 230 mm square ($d = 200$) with 2T10 bars in tension which are sufficient to resist bending at midspan. Determine the shear resistance as a singly reinforced beam assuming C20/25 Grade concrete.

Reduced design shear force at a distance d from the support (cl 6.2.1(8), EN)

$$V_{Ed} = (L/2 - d)/(L/2) \times W/2 = (2000/2 - 200)/(2000/2) \times 15,8/2 = 6,3 \text{ kN}$$

Shear force resistance of a singly reinforced section (Eq. 6.2a, EN)

$$\begin{aligned} V_{Rd,c} &= [C_{Rd,c} k (100 \rho_1 f_{ck})^{1/3} + 0,15 \sigma_{cp}] b_w d \\ &= [0,18/1,5 \times 2(100 \times 0,00343 \times 20)^{1/3} + 0] \times 230 \times 200/1E3 \\ &= 21,0 > V_{Ed}(6,3) \text{ kN, therefore satisfactory.} \end{aligned}$$

Minimum shear resistance

$$v_{min} = 0,035\,k^{1,5}f_{ck}^{0,5} = 0,035 \times 2^{1,5} \times 20^{0,5} = 0,443$$
$$v_{min} = (v_{min} + k_1\sigma_{cp})b_w d$$
$$= (0,443 + 0) \times 230 \times 200/1E3 = 20,4 < V_{Rd,c}(21,0)\text{ kN, satisfactory.}$$

Longitudinal tensile reinforcement ratio (cl 6.2.2(1), EN)

$$\rho_l = A_s/(b_w d) = 2 \times 79/(230 \times 200) = 0,00343 < 0,02\text{, acceptable.}$$

Minimum area of the longitudinal steel (cl 9.2.1.1, EN) which includes $f_{ctm} = 2,2$ MPa from Table 3.1, EN (Annex A2)

$$\rho_{l,min} = 0,26f_{ctm}/f_{yk} = 0,26 \times 2,2/500 = 0,00114 < \rho_l(0,00343)\text{, satisfactory.}$$

Depth factor

$$k = 1 + (200/d)^{0,5} = 1 + (200/200)^{0,5} = 2$$

This is a member of minor importance and shear reinforcement is not required (cl 6.2.1(4), EN).

Maximum shear resistance (Eq. 6.5, EN), generally not critical for a lintel beam

$$V_{Rd,max} = 0,5b_w dvf_{cd} = 0,5 \times 230 \times 200 \times 0,552 \times (20/1,5)/1E3$$
$$= 169,3 > V_{Ed,max}(7,9)\text{ kN, therefore satisfactory.}$$

where the strength reduction factor (Eq. 6.6 EN)

$$v = 0,6(1 - f_{ck}/250) = 0,6(1 - 20/250) = 0,552$$

7.3 MEMBERS REQUIRING SHEAR REINFORCEMENT (cl 6.2.3, EN)

At the ultimate limit state, where the design shear force is greater than the resistance of a singly reinforced beam, then shear reinforcement is required to resist all of the applied shear force. The most common type of shear reinforcement is in the form of vertical stirrups, or links, spaced at intervals along the length of a member. The links should be well anchored (cl 8.5, EN). The shape of the link, shown in Fig. 7.3(b), is preferred because it is closed, but the other shapes shown in Fig. 7.3(c) and (d) are also used where there are practical difficulties in fixing the reinforcement. See also Fig. 9.5, EN. Vertical links strengthen the beam web and resist dowel failure (Fig. 7.1(b)).

The European Code method for the shear resistance of a beam with vertical links is based on the traditional truss analogy (Ritter, 1899). The truss consists of a tie BC at

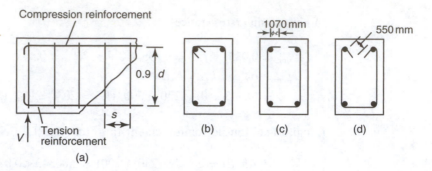

FIGURE 7.3   Types of vertical links

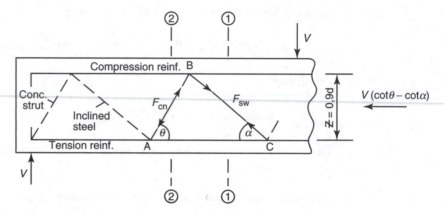

FIGURE 7.4   Truss analogy

an angle α which represents the links acting over a length AC, a compression strut AB representing a concrete compression at an angle θ, together with longitudinal tension and compression reinforcement (Fig. 7.4). The truss analogy applies to vertical and inclined links or bars.

Resolving forces vertically at section 1-1 (Fig. 7.4) to check for strength of a link (Eq. 6.13, EN).

$$V_{Rd,s} = F_{sw} \sin \alpha = A_{sw} f_{ywd}(AC/s) \sin \alpha = A_{sw}/s f_{ywd} z (\cot \theta + \cot \alpha) \sin \alpha \qquad (7.3)$$

To check for the crushing of the equivalent concrete strut resolve forces vertically at section 2-2 (Eq. 6.14, EN).

$$V_{Rd, max} = F_{cn} \sin \theta = f_{ck} b_w (AC \sin \theta) \sin \theta = \alpha_c f_{cd} b_w z (\cot \theta + \cot \alpha) \sin^2 \theta$$
$$= \alpha_c f_{cd} b_w z (\cot \theta + \cot \alpha)/(1 + \cot^2 \theta) \qquad (7.4)$$

The factor α_c is introduced for design purposes to allow for the effect of shear, cracking and prestressing on concrete. The strength of the steel tie BC should be less than the strength of the concrete strut AB.

The additional tensile force in the tension steel due to shear is obtained from the equilibrium of horizontal forces (Eq. 6.18, EN)

$$2\Delta F_{td} = F_{cn}\cos\theta - F_{sw}\cos\alpha = V_{Ed}/\tan\theta - V_{Ed}/\tan\alpha$$
$$\Delta F_{td} = 0{,}5V_{Ed}(\cot\theta - \cot\alpha) \tag{7.5}$$

Total force in the tension steel

$$M_{Ed}/z + \Delta F_{td} \leq M_{Ed,max}/z \tag{7.6}$$

These last two equations are used to determine the longitudinal compression steel and to check that there is sufficient tension steel at sections where the bending moment and shear force are large.

7.3.1 Minimum, or Nominal, Vertical Links (cl 9.2.2 EN)

In most building construction work the smallest diameter link is 6 mm and the largest diameter is generally 12 mm. 12 mm diameter links are not often used because of the difficulties involved in bending and because of the displacement of the longitudinal bars at the corners of links when large corner radii are used.

If too little shear reinforcement is provided it is not effective (Yoon *et al.*, 1996). It is also required to control cracking and to provide ductility. This is estimated from the shear ratio (Eq. 9.4, EN)

$$\rho_w = A_{sw}/(sb_w\sin\alpha) \tag{7.7}$$

Minimum values of the shear ratio (Eq. 9.5, EN)

$$\rho_{w,min} = 0{,}08f_{ck}^{0,5}/f_{yk} \tag{7.8}$$

7.3.2 Spacing of Links (cl 9.2.2, EN)

In practice longitudinal spacings of links are generally in intervals to 20 mm. If the longitudinal spacing of links is greater than approximately 0,75d then the links will not intercept cracks, nor the distance AC in Fig. 7.4 and consequently will not be effective in resisting the applied shear force.

Recommendations for minimum longitudinal spacing for links are related to the requirement for the concrete to flow round the reinforcement during casting. Hence a space of at least the diameter of the maximum size of aggregate plus 5 mm, or bar diameter if greater, is required.

For vertical links the maximum longitudinal spacing (Eq. 9.6, EN)

$$s_{l,max} = 0{,}75d(1 + \cot\alpha) \tag{7.9}$$

For bent-up bars to intercept AC (Fig. 7.4) the maximum longitudinal spacing (cl 9.2.2(7), EN)

$$s_{b,max} = (0{,}9d/1{,}5)(\cot\theta + \cot\alpha) = 0{,}6d(1 + \cot\alpha) \tag{7.10}$$

Maximum transverse spacing of legs in a series of shear links (cl 9.2.2(8), EN)

$$s_{t,max} = 0{,}75d \le 600\,\text{mm} \tag{7.11}$$

7.3.3 Anchorage of Links (cls 8.5 and 9.2.2, EN)

Generally, it is not possible to anchor a link by bond alone because of the limited bond length available, and therefore it is necessary to provide longitudinal bars in the bottom and top of a beam which are enclosed by the link and used as an anchorage. This arrangement produces a cage of reinforcement which is easier for the steel fixer to construct, as a separate unit and place in the mould prior to casting a beam. The longitudinal bars should be at least equal in diameter to the link and generally are not less than 12 mm, although some designers may use larger sizes. The shape of link shown in Fig. 7.3(b) is the most effective and common form used in practice.

7.3.4 Crack Control for Links (cl 7.3, EN)

Cracks are normal in reinforced concrete structures that are stressed, and if they do not impair the functioning of the structure they are accepted. If calculations are made acceptable values are $0{,}2 < w < 0{,}4$ mm. Cracking is controlled without calculation by limiting the tensile stress in the steel and by limiting bar spacing (cl 7.3.3, EN). See Chapter 5 for further information.

EXAMPLE 7.2   Beam with vertical links. A simply supported beam, span $20d$, carries a uniformly distributed design load of 475 kN which includes self weight. The beam cross section in Fig. 7.5 shows the tension reinforcement required to resist the bending moment at midspan at the ultimate limit state. Design the cross section for shear using C30/35 concrete assuming no bar curtailment.

Reduced design shear force at distance d supports (cl 6.2.1(8), EN)

$$\begin{aligned}
V_{Ed} &= (10d - d)/(10d) \times W/2 \\
&= (10 \times 0{,}55 - 0{,}55)/(10 \times 0{,}55) \times 475/2 = 213{,}75\,\text{kN}
\end{aligned}$$

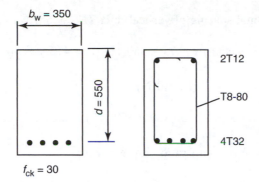

FIGURE 7.5 Example: beam with vertical links

Stage (a) Check shear force resistance as singly reinforced (Eq. 6.2a, EN)

$V_{Rd,c} = [C_{Rd,c}k(100\rho_l f_{ck})^{1/3} + 0,15\sigma_{cp}]b_w d = [0,18/1,5 \times 1,6(100 \times 0,0167 \times 30)^{1/3} + 0] \times$
$350 \times 550/1E3 = 136,1 < V_{Ed}(213,75)$ kN, add links to resist all of the shear force.

Minimum shear resistance

$$v_{min} = 0,035k^{1,5}f_{ck}^{0,5} = 0,035 \times 1,6^{1,5} \times 30^{0,5} = 0,388$$
$$V_{min} = (v_{min} + k_1\sigma_{cp})b_w d$$
$$= (0,388 + 0) \times 350 \times 550/1E3$$
$$= 74,7 < V_{Rd,c}(136,1) \text{ kN, satisfactory.}$$

Depth factor

$$k = 1 + (200/d)^{0,5} = 1 + (200/550)^{0,5} = 1,6$$

Longitudinal reinforcement ratio (cl 6.2.2(1), EN)

$$\rho_l = A_s/(b_w d) = 3217/(350 \times 550) = 0,0167 < 0,02$$

Minimum longitudinal reinforcement ratio (cl 9.2.1.1, EN)

$$\rho_{l,min} = 0,26f_{ctm}/f_{yk} = 0,26 \times 2,9/500 = 0,0051 < \rho_l(0,0167), \text{ acceptable.}$$

Stage (b) Solution with single vertical links to resist all of the shear force ($\theta = 45°$, $\alpha = 90°$) (Eq. 6.8, EN)

$$A_{sw}/s_l = V_{Ed}/(zf_{ywd}\cot\theta) = (213,75E3/(0,9 \times 550 \times 0,8 \times 500 \times 1) = 1,08 \text{ mm}$$

Assuming T8 mm diameter single vertical links, longitudinal spacing

$$s_l = A_{sw}/1,08 = 2 \times 50/1,08 = 92,6 \text{ mm, use } s_l = 80 \text{ mm.}$$

Maximum shear resistance of the concrete (Eq. 6.9, EN) with $v = 0,6$ for ≤ 60 MPa

$$V_{Rd,max} = \alpha_{cw}b_w zvf_{cd}/(\cot\theta + \tan\theta)$$
$$= 1,0 \times 350 \times 0,9 \times 550 \times 0,6 \times 30/1,5/(1+1) \times 1E\text{-}3$$
$$= 1040 > V_{Ed,max}(237,5) \text{ kN, therefore satisfactory.}$$

Maximum longitudinal spacing of vertical links (Eq. 9.6, EN)

$$s_{1,\max} = 0.75\,d(1 + \cot\alpha) = 0.75 \times 550(1 + 0) = 412.5 > s_1(80)\,\text{mm}$$

Shear reinforcement ratio (Eq. 9.4, EN)

$$\rho_w = A_{sw}/(sb_w\sin\alpha) = 2 \times 50/(80 \times 350 \times 1) = 0.00357$$

Minimum shear reinforcement ratio (Eq. 9.5, EN)

$$\rho_{w,\min} = 0.08f_{ck}^{0.5}/f_{yk} = 0.08 \times 30^{0.5}/500 = 0.00159 < \rho_w(0.00357),\ \text{satisfactory.}$$

Compression reinforcement related to the shear force (Eq. 6.18, EN)

$$A_{sl} = 0.5V_{Ed}(\cot\theta - \cot\alpha)/f_{yd} = 0.5 \times 213.75\text{E}3 \times (1 - 0)/(500/1.15)$$
$$= 246\,\text{mm}^2.\ \text{Use}\ 2\text{T}12 = 226\,\text{mm}^2.$$

The balancing horizontal force is included in the longitudinal steel in tension which close to the support is not fully stressed.

Use single vertical links T8-80 mm centres, with 2T12 as compression reinforcement (Fig. 7.5).

If, in a different situation, the maximum shear force occurs at the maximum bending moment, e.g. a central point load on a simply supported beam, then check that the design stress in the tensile steel is not exceeded (Eq. 6.18, EN), i.e.

$$(M_{Ed,\max}/z + \Delta F_{td}) = [M_{Ed,\max}/z + 0.5V_{Ed}(\cot\theta - \cot\alpha)] < A_{sl}f_{yd}.$$

Stage (c) Alternative solution using single vertical T8 links ($\alpha = 90°$) and $\cot\theta = 2.5$ (Eq. 6.8, EN)

$$A_{sw}/s_1 = V_{Ed}/(zf_{ywd}\cot\theta) = 213.75\text{E}3/(0.9 \times 550 \times 0.8 \times 500 \times 2.5) = 0.432\,\text{mm}$$

Assuming T8 vertical links, longitudinal spacing

$$s_1 = A_{sw}/0.432 = 2 \times 50/0.432 = 232\,\text{mm, use}\ s_1 = 200\,\text{mm.}$$

Maximum shear resistance of the concrete (Eq. 6.9, EN)

$$V_{Rd,\max} = \alpha_{cw}b_wzvf_{cd}/(\cot\theta + \tan\theta)$$
$$= 1.0 \times 350 \times 0.9 \times 550 \times 0.6 \times 30/1.5/(2.5 + 1/2.5)/1\text{E}3$$
$$= 717 > V_{Ed,\max}(237.5)\,\text{kN, therefore satisfactory.}$$

Maximum longitudinal spacing of vertical links (Eq. 9.6, EN)

$$s_{1,\max} = 0.75d(1 + \cot\alpha) = 0.75 \times 550(1 + 0) = 412.5 > s_1(220)\,\text{mm}$$

Shear reinforcement ratio (Eq. 9.4, EN)

$$\rho_w = A_{sw}/(s_1 b_w \sin \alpha) = 2 \times 50/(220 \times 350 \times 1) = 0{,}0013$$

Minimum shear reinforcement ratio (Eq. 9.5, EN)

$$\rho_{w,min} = 0{,}08 f_{ck}^{0,5}/f_{yk} = 0{,}08 \times 30^{0,5}/500 = 0{,}00159 > \rho_w (0{,}0013), \text{ not acceptable}$$

Reduce longitudinal link spacing to satisfy shear ratio

$$s_1 = 0{,}0013/0{,}00159 \times 220 = 180 \text{ mm, use } s_1 = 180 \text{ mm}$$

Compression reinforcement (Eq. 6.18, EN)

$$A_{sl} = 0{,}5 V_{Ed}(\cot \theta - \cot \alpha)/f_{yd} = 0{,}5 \times 213{,}75E3 \times 2{,}5/(500/1{,}15)$$
$$= 615 \text{ mm}^2, \text{ use } 2T20 = 628 \text{ mm}^2.$$

Use single vertical links T8-180 mm centres, with 2T20 as compression reinforcement.

The solution with $\cot\theta = 2{,}5$ has increased the spacing of the T8 links from 80 to 180 mm and thus economises on shear steel but increases the longitudinal compression steel from 2T10 to 2T20. It can be shown theoretically that the minimum steel is used when $\cot\theta = 2{,}5$ but this may be complicated by practical bar sizes, link spacing, percentage reinforcement limits and costs.

EXAMPLE 7.3 Floor beam loaded through the bottom flange. Reinforce for shear the section of a simply supported floor beam shown in Fig. 7.6. The beam which spans 3 m is loaded through the bottom flange with a uniformly distributed design load of 8 kN/m (which includes self weight) at the ultimate limit state. Assume Grade C30/35 concrete.

Shear force at distance d from the supports (cl 6.2.1(8), EN)

$$V_{Ed} = (L/2 - d)/(L/2)wl/2 = (1{,}5 - 0{,}184)/1{,}5 \times 8 \times 3/2 = 11 \text{ kN}$$

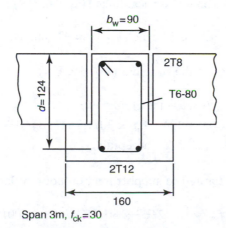

Span 3m, $f_{ck}=30$

FIGURE 7.6 Example: floor beam loaded through the bottom flange

Check shear force resistance as a singly reinforced beam (Eq. 6.2a, EN)

$$V_{Rd,c} = [C_{Rd,c}k(100\rho_l f_{ck})^{1/3} + 0,15\sigma_{cp}]b_w d$$
$$= [0,18/1,5 \times 2(100 \times 0,02 \times 30)^{1/3} + 0] \times 90 \times 124/1E3$$
$$= 10,47 < V_{Ed}(11)\,\text{kN, unsatisfactory and links to resist all of the shear force.}$$

Minimum shear resistance

$$v_{min} = 0,035k^{1,5}f_{ck}^{0,5} = 0,035 \times 2^{1,5} \times 30^{0,5} = 0,542$$
$$V_{min} = (v_{min} + k_1\sigma_{cp})b_w d$$
$$= (0,542 + 0) \times 90 \times 124/1E3 = 6,05 < V_{Rd,c}(10,47) \text{ kN, satisfactory.}$$

Depth factor

$$k = 1 + (200/d)^{0,5} = 1 + (200/124)^{0,5} = 2,27 > 2, \text{ use } k = 2$$

Longitudinal steel ratio (cl 6.2.2(1), EN)

$$\rho_l = A_s/(b_w d) = 226/(90 \times 124) = 0,02, \text{ acceptable}$$

Minimum longitudinal steel ratio (cl 9.2.1.1, EN)

$$\rho_{l,min} = 0,26f_{ctm}/f_{yk} = 0,26 \times 2,9/500 = 0,0015 < \rho_l(0,02), \text{ acceptable.}$$

Shear reinforcement ($\theta = 45°$ and $\alpha = 90°$) (Eq. 6.8, EN)

$$A_{sw}/s_1 = V_{Ed}/(z f_{ywd}) = 11E3/(0,9 \times 124 \times 0,8 \times 500) = 0,246\,\text{mm}$$

Assuming T6 vertical links, longitudinal spacing

$$s_1 = A_{sw}/0,246 = 2 \times 28/0,246 = 227,2\,\text{mm}$$

Maximum longitudinal spacing of vertical links (Eq. 9.6, EN)

$$s_{l,max} = 0,75d(1 + \cot\alpha) = 0,75 \times 124(1 + 0) = 93\,\text{mm, use } s_1 = 80\,\text{mm.}$$

Maximum shear resistance of the concrete (Eq. 6.9, EN)

$$V_{Rd,max} = \alpha_{cw}b_w z \, v f_{cd}/(\cot\theta + \tan\theta)$$
$$= 1,0 \times 90 \times 0,9 \times 124 \times 0,6 \times 30/1,5/(1 + 1)/1E3$$
$$= 60,3 > V_{Ed,max}(12)\,\text{kN acceptable.}$$

Additional area of link required to support load on bottom flange ($s_1 = 80\,\text{mm}$)

$$A_{sw(extra)} = (Ws/L)/(f_{yw}/\gamma_s) = (24E3 \times 80/3E3)/(0,8 \times 500) = 1,6\,\text{mm}^2$$

Total area of link required (cl 6.2.1(9), EN)

$$= A_{sw(shear)} + A_{sw(extra)} = 0,246 \times 80 + 1,6 = 21,3 \, \text{mm}^2.$$

Single T6 links, $A_{sw} = 2 \times 28 = 56 > 21,3 \, \text{mm}^2$, satisfactory.

Shear reinforcement ratio (Eq. 9.4, EN)

$$\rho_w = A_{sw}/(sb_w \sin\alpha) = 2 \times 28/(80 \times 90 \times 1) = 0,00778$$

Minimum shear reinforcement ratio (Eq. 9.5, EN)

$$\rho_{w,min} = 0,08 f_{ck}^{0,5}/f_{yk} = 0,08 \times 30^{0,5}/500 = 0,0016 < \rho_w(0,00778), \text{ acceptable.}$$

Compression streel required based on shear (Eq. 6.18, EN)

$$A_{sl} = 0,5 V_{Ed}(\cot\theta - \cot\alpha)/f_{yd} = 0,5 \times 11E3 \times 1/(500/1,15) = 12,65 \, \text{mm}^2, \text{ use}$$
$$2T8 = 101 \, \text{mm}^2.$$

Use single vertical links, T6-80, with 2T8 as compression reinforcement (Fig. 7.6).

7.3.5 Shear Resistance of Bent-up Bars

In situations where all the longitudinal tensile reinforcement is not required to resist the bending moment at the ultimate limit state, some of the bars may be bent-up to form shear reinforcement (Fig. 7.7). The tensile bars are shown at different levels for ease in identification, but in practice they are all at the same level as shown in the cross section. If the tensile reinforcement is used in this way it must be adequately anchored at the top of the beam to prevent bond failure (See Chapter 8).

Generally, there are insufficient bent-up bars available to resist the total shear force and vertical links are therefore added. Only a maximum of 50 per cent of the design shear force (V_{Ed}) is allowed (cl 9.2.2(4), EN) to be resisted by the bent-up bars, the other 50 per cent is to be provided by vertical links. When bent-up bars and vertical

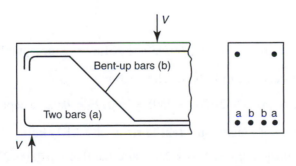

FIGURE 7.7   Bent-up bars

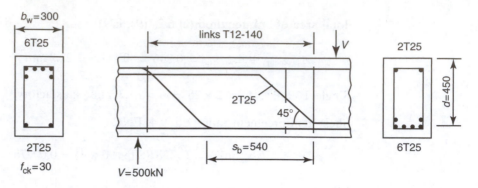

FIGURE 7.8 Example: bent-up bars

links are used together their effect is added (Evans, 1967 and Kong, 1978). It should be appreciated that if tensile bars are bent-up then there is a reduction in tensile steel, and as a consequence a reduction in bending and shear resistance of the beam.

EXAMPLE 7.4 Beam containing bent-up bars and vertical links. Determine the shear resistance of the bent-up bars in the beam shown in Fig. 7.8 and add vertical links to resist a design shear force of $V_{Ed} = 500$ kN. Assume C30/35 Grade concrete.

Stage (1) Shear force resistance as a singly reinforced beam at a section with 2T25 bars (Eq. 6.2a, EN)

$$V_{Rd,c} = [C_{Rd,c}k(100\rho_l f_{ck})^{1/3} + 0,15\sigma_{cp}]b_w d$$
$$= [0,18/1,5 \times 1,67(100 \times 0,00727 \times 30)^{1/3} + 0] \times 300 \times 450/1E3$$
$$= 75,51 < V_{Ed}(500)\,\text{kN, add shear reinforcement to resist all of the}$$
shear force.

Minimum shear force resistance

$$V_{min} = (0,035k^{1,5}f_{ck}^{0,5} + k_1\sigma_{cp})b_w d$$
$$= (0,035 \times 1,67^{1,5} \times 30^{0,5} + 0) \times 300 \times 450/1E3$$
$$= 55,85 < V_{Rd,c}(75,51)\,\text{kN, acceptable.}$$

Depth factor

$$k = 1 + (200/d)^{0,5} = 1 + (200/450)^{0,5} = 1,67 < 2, \text{acceptable}$$

Tensile reinforcement ratio (cl 6.2.2(1), EN)

$$\rho_1 = A_s/(b_w d) = 982/(300 \times 450) = 0,00727 < 0,02, \text{acceptable}$$

Minimum ratio of longitudinal tensile steel (cl 9.2.1.1, EN)

$$\rho_{l,min} = 0,26f_{ctm}/f_{yk} = 0,26 \times 2,9/500 = 0,00151 < \rho_l(0,00727), \text{acceptable.}$$

Stage (2) For bent-up bars with $\alpha = \theta = 45°$.

Maximum longitudinal spacing of bent-up bars (Eq. 9.7, EN)

$$s_{b,max} = 0,6d(1 + \cot \alpha) = 0,6 \times 450(1 + 1) = 540 \, mm, \text{ use } s_b = 540 \, mm.$$

Design shear resistance of 2T25 bent-up bars at 45° (Eq. 6.13, EN)

$$V_{Rd,s} = (A_{sw}/s)z f_{ywd}(\cot \theta + \cot \alpha) \sin \alpha$$
$$= (982/540) \times 0,9 \times 450 \times 0,8 \times 500(1 + 1) \times 0,707/1E3 = 417 > (V_{Ed}/2) \, kN$$

Add vertical links to resist half of the shear force (Eq. 6.8, EN)

$$A_{sw}/s = (V_{Ed}/2)/(z f_{ywd}) = (500E3/2)/(0,9 \times 450 \times 0,8 \times 500) = 1,54 \, mm$$

For single T12 vertical links the longitudinal spacing

$$s_l = A_{sw}/1,54 = 2 \times 113/1,54 = 147 \, mm, \text{ use } s_l = 140 \, mm.$$

Maximum longitudinal spacing of vertical links (cl 9.2.2(6), EN)

$$s_{l,max} = 0,75d(1 + \cot \alpha) = 0,75 \times 450(1 + 0) = 270 > s_l = 140 \, mm.$$

Check for maximum shear resistance (Eq. 6.9, EN)

$$V_{Rd,max} = \alpha_{cw} b_w z v f_{cd}/(\cot \theta + \cot \alpha)$$
$$= 1 \times 300 \times 0,9 \times 450 \times 0,6 \times (30/1,5)/(1 + 1)/1E3$$
$$= 729 > V_{Ed}(500) \, kN, \text{ therefore satisfactory.}$$

Shear reinforcement ratio for vertical links (Eq. 9.4, EN)

$$\rho_w = A_{sw}/(s b_w \sin \alpha) = 2 \times 113/(140 \times 300 \times 1) = 0,00538$$

Shear reinforcement ratio for bent-up bars (Eq. 9.4, EN)

$$\rho_w = A_{sw}/(s b_w \sin \alpha) = 2 \times 491/(540 \times 300 \times 0,707) = 0,00857$$

Minimum shear reinforcement ratio (Eq. 9.5, EN)

$$\rho_{w,min} = 0,08 f_{ck}^{0,5}/f_{yk} = 0,08 \times 30^{0,5}/500 = 0,00159 < \rho_w(0,00538), \text{ acceptable.}$$

Maximum effective area of vertical links (Eq. 6.12, EN)

$$A_{sw,max} f_{ywd}/(b_w s_l) = 2 \times 113 \times 0,8 \times 500/(300 \times 140) = 2,15 \, MPa$$
$$0,5 \alpha_c v f_{cd} = 0,5 \times 1 \times 0,6 \times 30/1,5 = 6 > 2,15 \, MPa, \text{ therefore satisfactory.}$$

Maximum effective area of inclined bars (Eq. 6.13, EN)

$$A_{sw,max} f_{ywd} z/(b d s_l)(\cot \theta + \cot \alpha) \sin \alpha = 2 \times 491 \times 0,8 \times 500 \times 0,9/(300 \times 540)$$
$$(1 + 1) \times 0,707 = 3,09 \, MPa$$
$$0,5 \alpha_c v f_{cd}(z/d)(\cot \theta + \cot \alpha)/(1 + \cot^2 \theta) = 0,5 \times 1 \times 0,6 \times 30/1,5 \times 0,9(1 + 1)/(1 + 1)$$
$$= 5,4 > 3,09 \, MPa, \text{ acceptable.}$$

Compression reinforcement (Eq. 6.18, EN)

$$A_{sl} = 0,5V_{Ed}(\cot\theta - \cot\alpha)/f_{yd}$$
$$= 0,5(500E3/2) \times (1-0)/(500/1,15)$$
$$= 575\,mm^2, \text{ use } 2T25 = 982\,mm^2.$$

Balancing half of this longitudinal reinforcement is included in the tension steel.

For a complete design, check that the extra longitudinal steel does not exceed that required for pure bending (Eq. 6.18, EN), check anchorage for the bent-up bars as shown in Chapter 8, and check bar diameters and link spacing for control of crack width.

Use 2T25-540 as bent-up bars, T12-140 mm centres as links with 2T25 as compression reinforcement (Fig. 7.8).

7.3.6 Horizontal Shear Between Web and Flange for a 'T' Beam (cl 6.2.4, EN)

If a 'T' beam is subjected to a change in bending moment along its length then this produces shear stresses between the flange and the web at sections 1-1 and 1-2. (Fig. 7.9).

Horizontal design shear stress (Eq. 6.20, EN)

$$v_{Ed} = \Delta F_d/(h_f \Delta x) \tag{7.12}$$

where ΔF_d is the range in normal force over a distance Δx

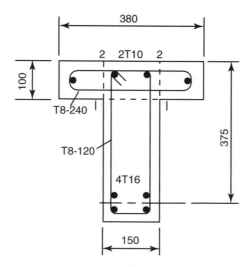

FIGURE 7.9 Example: horizontal and vertical shear stresses in a 'T' beam

A minimum amount of reinforcement must be provided in the form of vertical links. The shear resistance per unit length of concrete is assumed to be resisted only by the links and the cracked concrete resistance is zero (Eq. 6.21, EN)

$$A_{st}f_{yd}/s_f > v_{Ed}h_f/\cot\theta_f \qquad\qquad (7.13)$$

Maximum horizontal shear resistance per unit length of concrete reinforced with links to prevent crushing of the compression struts in the flange (Eq. 6.22, EN)

$$v_{Ed} < vf_{cd}\sin\theta_f\cos\theta_f \qquad\qquad (7.14)$$

Experiments (Saemann and Washa (1964), Loov and Patnaik (1994)) justify these expressions.

EXAMPLE 7.5 Horizontal and vertical shear stresses in a 'T' beam. Determine the size and spacing of links for the 'T' beam shown in Fig. 7.9. The beam is simply supported over a span of 6 m and carries a uniformly distributed design load of 136 kN. Assume C30/35 Grade concrete.

Vertical design shear force at a distance d from the support (cl 6.2.1 (8), EN)

$$V_{Ed} = (L/2 - d)/(L/2)V_{max} = (1 - 2d/L)V_{max} = (1 - 2 \times 375/3E3)136/2 = 51\,kN$$

Stage (1) Vertical shear force resistance as a singly reinforced T beam at a section with 4T16 bars in tension (Eq. 6.2a, EN)

$$\begin{aligned}
V_{Rd,c} &= [C_{Rd,c}k(100\rho_l f_{ck})^{1/3} + 0{,}15\sigma_{cp}]b_w d \\
&= [0{,}18/1{,}5 \times 1{,}73(100 \times 0{,}0143 \times 30)^{1/3} + 0] \times 150 \times 375/1E3 \\
&= 40{,}89 < V_{Ed}\,(51)\,kN, \text{ add stirrups to resist all of the shear force.}
\end{aligned}$$

Minimum shear force resistance

$$\begin{aligned}
V_{min} &= (0{,}035k^{1,5}f_{ck}^{0,5} + k_1\sigma_{cp})b_w d \\
&= (0{,}035 \times 1{,}73^{1,5} \times 30^{0,5} + 0) \times 150 \times 375/1E3 \\
&= 24{,}5 < V_{Rd,c}(40{,}89)\,kN, \text{ acceptable.}
\end{aligned}$$

Depth factor

$$k = 1 + (200/d)^{0,5} = 1 + (200/375)^{0,5} = 1{,}73 < 2, \text{ acceptable}$$

Tensile reinforcement ratio (cl 6.2.2(1), EN)

$$\rho_l = A_s/(b_w d) = 804/(150 \times 375) = 0{,}0143 < 0{,}02, \text{ acceptable.}$$

Minimum ratio of longitudinal tensile steel (cl 9.2.1.1, EN)

$$\rho_{l,min} = 0{,}26f_{ctm}/f_{yk} = 0{,}26 \times 2{,}9/500 = 0{,}0015 < \rho_l(0{,}0143), \text{ acceptable.}$$

Stage (2) Combine reinforcement to resist vertical shear stress in the web and horizontal shear stresses on plane 1-1 (Fig. 7.9)

Change in normal force in the flange over a distance $L/4$ at the end of the beam (Eq. 6.20, EN)

$$\Delta F_d = (wL^2/8 - wL^2/32)/d = 3/4\,W$$

Shear stress on horizontal failure plane 1-1 (Fig. 7.9)

$$v_{Ed} = (3/4W)/(b_w z) = (3/4 \times 136E3)/(150 \times 0,9 \times 375) = 2,01\,\text{MPa}$$

Shear stress on the vertical plane

$$v_{Ed} = V_{Ed}/(b_w d) = 51E3/(150 \times 375) = 0,907\,\text{MPa}$$

Spacing of T8 single vertical links ($\theta = 45°$ and $\alpha = 90°$) to resist horizontal shear stresses

$$s_1 = A_{sf} f_{yd}/(v_{Ed} b_w)$$
$$= 2 \times 50 \times 0,8 \times 500/(2,01 \times 150) = 133\,\text{mm, use } s_1 = 120\,\text{mm.}$$

Check crushing of the compression strut in the flange, $\cot\theta_f = 1$ (Eq. 6.22, EN)

$$v_{Ed} < v f_{cd} \sin\theta_f \cos\theta_f = 0,6 \times 30/1,5 \times 0,707 \times 0,707 = 6 > 2,01\,\text{MPa, acceptable.}$$

Maximum longitudinal spacing of vertical links (Eq. 9.6, EN)

$$s_{l,max} = 0,75d(1 + \cot\alpha) = 0,75 \times 375 \times (1 + 0) = 281 < 300\,\text{mm, use } s_l = 120\,\text{mm.}$$

Shear reinforcement ratio (Eq. 9.4, EN)

$$\rho_w = A_{sw}/(s\,b_w \sin\alpha) = 2 \times 50/(120 \times 150 \times 1) = 0,0056$$

Minimum shear reinforcement ratio (Eq. 9.5, EN)

$$\rho_{w,min} = 0,08 f_{ck}^{0,5}/f_{yk} = 0,08 \times 30^{0,5}/500 = 0,00159 < \rho_w(0,0056), \text{ satisfactory.}$$

Compression reinforcement (Eq. 6.18, EN)

$$A_{sl} = 0,5V_{Ed}(\cot\theta - \cot\alpha)/f_{yd} = 0,5 \times 51E3 \times (1 - 0)/(500/1,15)$$
$$= 58,65\,\text{mm}^2, \text{ use 2T10} = 157\,\text{mm}^2.$$

Stage (3) Consider the two vertical shear failure planes 1-2 (Fig. 7.9). The change in normal force in the flange over a length $L/4$ at the end of the beam (cl 6.2.4(3), EN)

$$\Delta F_d = (wL^2/8 - wL^2/32)/d$$

Longitudinal shear stress (Eq. 6.20, EN) on one side of the flange

$$v_{\text{Ed}} = \Delta F_{\text{d}}/(h_{\text{f}}\Delta x) = 0,5[(wL^2/8 - wL^2/32)/d]/(h_{\text{f}}L/4) = (3/8W)/(h_{\text{f}}z)$$
$$= (3/8 \times 136\text{E}3)/(100 \times 0,9 \times 375) = 1,51\,\text{MPa}$$

Rearranging (Eq. 6.21, EN), assuming $\cot\theta_{\text{f}}=1$, to determine the transverse reinforcement per unit length. Assume T8 as a single horizontal link

$$s_{\text{f}} = A_{\text{sf}}f_{\text{yd}}\cot\theta_{\text{f}}/(v_{\text{Ed}}h_{\text{f}}) = 2 \times 50 \times 0,8 \times 500 \times 1/(1,51 \times 100) = 265\,\text{mm, use}$$
$$s_{\text{f}} = 240\,\text{mm}.$$

Use T8-120 mm centres as single vertical links, T8-240 mm centres as single horizontal links and 2T10 as compression reinforcement (Fig. 7.9).

7.4 Shear Resistance of Solid Slabs (cls 6.4 and 9.4.3, EN)

The shear resistance of slabs and foundations subject to concentrated loads is related to a 'punching shear' form of failure. This takes the form of a truncated cone with cracks at approximately 30° (Regan, 2004) and occurs when concentrated loads, such as columns and piles, produce shear stresses on defined perimeters. The perimeters are immediate to the loaded area and at a distance of $2d$ (cl 6.4.2(1), EN) outside the loaded area.

If the applied design shear stress (v_{Ed}) is less than the design shear resistance ($v_{\text{Rd,c}}$) then no shear reinforcement is required, not even the minimum required in beams. This decision is based on the practical consideration that it is difficult to bend and fix shear reinforcement in slabs with a depth of ≤ 200 mm. If additional shear capacity is required for a slab without shear reinforcement it can be obtained by increasing one or all of the following, percentage of tensile reinforcement, concrete strength or depth of the slab.

If shear reinforcement is required in slabs with a thickness greater than 200 mm, then it is introduced in the form of castellated links (Fig. 7.10), or bent-up bars.

7.4.1 Slabs and Foundations without Shear Reinforcement (cl 6.4.4, EN)

Punching shear stress resistance (cl 6.4.4(1), EN)

$$v_{\text{Rd,c}} = C_{\text{Rd,c}}k(100\rho_{\text{l}}f_{\text{ck}})^{1/3} + 0,15\sigma_{\text{cp}} \geq (v_{\text{min}} + k_1\sigma_{\text{cp}}) \qquad (7.15)$$

Figure 7.10 Castellated link (5 legs)

where

$$C_{Rd,c} = 0{,}18/\gamma_c$$

Minimum shear resistance

$$(v_{min} + k_1\sigma_{cp}) = 0{,}035k^{1,5}f_{ck}^{0,5} + k_1\sigma_{cp}$$

Depth factor

$$k = 1 + (200/d)^{0,5} \leq 2$$

Longitudinal tensile reinforcement ratio

$$\rho_l = (\rho_{ly}\,\rho_{lz})^{0,5} \leq 0{,}02$$

These expressions are similar to those for singly reinforced beams except for the tensile reinforcement ratio, which is for steel in both directions.

7.4.2 Slabs or Foundations with Shear Reinforcement (cl 6.4.5, EN)

For slabs of thickness greater than 200 mm which are subject to concentrated loads, where shear reinforcement is required, links are used in castellated form (Fig. 7.10), or bent-up bars. The shear resistance is part concrete and part steel shear reinforcement for a perimeter of length u_1.

Shear stress resistance of a slab with shear reinforcement in the form of links (Eq. 6.52, EN)

$$v_{Rd,cs} = 0{,}75v_{Rd,c} + 1{,}5(d/s_r)A_{sw}f_{ywd,ef}(1/(u_1d))\sin\alpha \tag{7.16}$$

where

s_r is the radial spacing of perimeters of shear reinforcement (mm) and $f_{ywd,ef} = 250 + 0{,}25d \leq f_{ywd}$ MPa.

It should be noted that part of the applied design shear stress is resisted by the concrete and part by the steel. This is in contrast to singly reinforced beams with shear reinforcement where all of the applied design shear stress is resisted by the shear reinforcement.

The design process for a concentrated column load consists of making checks for shear resistance at a basic control perimeters (Fig. 7.11). The first perimeter is around the loaded area, the second $2d$ from the loaded area, and then usually at successive perimeters at increments of $2d$ until shear reinforcement is no

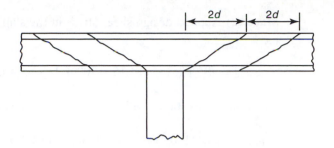

FIGURE 7.11 Shear failure of a slab subjected to a concentrated column load

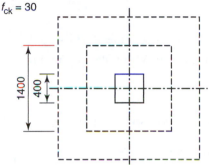

FIGURE 7.12 Example: slab with a concentrated load

longer required. Critical second perimeters for various shapes of loaded areas are shown in cl 6.4.2(1), EN.

EXAMPLE 7.6 Flat slab subjected to a concentrated load. Check the flat slab shown in Fig. 7.12 for shear failure. The internal column design load $N = 1000\,\text{kN}$ and the concrete slab is 300 mm deep, with $d = 250\,\text{mm}$, and reinforced with T16-200 mm centres in both directions at the top and bottom of the slab. Assume Grade C30/35 concrete.

Stage (1) At the column face for an interior column, $\beta = 1{,}15$ (cl 6.4.3(6), EN), the applied punching design shear stress in the slab (Eq. 6.53, EN)

$$v_{\text{Ed}} = \beta V_{\text{Ed}}/(u_{\text{o}}d) = 1{,}5 \times 1000\text{E}3/(4 \times 400 \times 250) = 2{,}88\,\text{MPa}.$$

Maximum shear stress resistance allowed at the column face (Eq. 6.53, EN)

$$v_{\text{Rd,max}} = 0{,}5vf_{\text{cd}} = 0{,}5 \times 0{,}528 \times 30/1{,}5 = 5{,}28 > v_{\text{Ed}}(2{,}88)\,\text{MPa acceptable},$$

where strength reduction factor (Eq. 6.6, EN)

$$v = 0{,}6(1 - f_{\text{ck}}/250) = 0{,}6(1 - 30/250) = 0{,}528.$$

Stage (2) Applied design shear stress in the slab for perimeter $2d$ from the column face

$$v_{Ed} = \beta V_{Ed}/(u_1 d) = 1{,}15 \times 1000E3/(4 \times (400 + 4 \times 250) \times 250) = 0{,}821 \, \text{MPa}.$$

Shear stress resistance without shear reinforcement (Eq. 6.47, EN)

$$v_{Rd,c} = C_{Rd,c}k(100\rho_l f_{ck})^{1/3} + 0{,}15\sigma_{cp} = 0{,}18/1{,}5 \times 1{,}89 \times (100 \times 0{,}00446 \times 30)^{1/3} + 0$$
$$= 0{,}538 < v_{Ed}(0{,}821) \, \text{MPa},$$

therefore shear reinforcement required.

Minimum shear resistance

$$(v_{min} + k_1\sigma_{cp}) = 0{,}035k^{1,5}f_{ck}^{0,5} + k_1\sigma_{cp} = 0{,}035 \times 1{,}89^{1,5} \times 30^{0,5} + 0$$
$$= 0{,}36 < v_{Rd,c}(0{,}538) \, \text{MPa, acceptable.}$$

Depth factor

$$k = 1 + (200/d)^{0,5} = 1 + (200/250)^{0,5} = 1{,}89 < 2.$$

Tensile reinforcement ratio (cl 6.4.4, EN)

$$\rho_l = (\rho_{ly} \, \rho_{lz})^{0,5} = 1115/(1E3 \times 250) = 0{,}00446 < 0{,}02, \text{ acceptable.}$$

Shear stress resistance with vertical links (T10-180) (Eq. 6.52, EN)

$$v_{Rd,cs} = 0{,}75 v_{Rdc} + 1{,}5(d/s_r) \, A_{sw} f_{ywd,ef}(1/(u_1 d)) \sin \alpha$$
$$= 0{,}75 \times 0{,}538 + 1{,}5 \times (250/180) \times 12 \times 79 \times 312{,}5/$$
$$(4 \times (400 + 4 \times 250) \times 250) \times 1$$
$$= 0{,}846 > v_{Ed}(0{,}821) \, \text{MPa, acceptable}$$

where

$$f_{ywd,ef} = 250 + 0{,}25d = 250 + 0{,}25 \times 250 = 312{,}5 < f_{ywd}(0{,}8 \times 500) \, \text{MPa}.$$

Minimum area of a link leg with $s_r = s_t = 180 \, \text{mm}$ (cl 9.4.3(2), EN)

$$A_{sw,\,min}(1{,}5 \sin \alpha + \cos \alpha)/(s_r s_t) = 79 \times (1{,}5 \times 1 + 0)/(180 \times 180) = 0{,}00366$$
$$0{,}08 f_{ck}^{0,5}/f_{yk} = 0{,}08 \times 30^{0,5}/500 = 0{,}000825 < 0{,}00366, \text{ acceptable.}$$

Repeat calculations at increasing radii until the control perimeter at which shear reinforcement is not required (Eq. 6.54, EN). Details of punching shear reinforcement are given in cl 9.4.3, EN.

$$u_{out,ef} = \beta V_{Ed}/(v_{Rd,c}d) = 1{,}15 \times 1000E3/(0{,}538 \times 250) = 8550 \, \text{mm}.$$

7.4.3 Shear Resistance of a Column Base without Shear Reinforcement (cl 6.4.4, EN)

The design of a pad foundation for shear is similar to that for concentrated loads on slabs. Checks for punching shear resistance are made at control perimeters within $2d$ from the column face (cl 6.4.4(2), EN). The lowest value controls the design. Similar criteria are used to decide whether shear reinforcement is necessary. Often, in pad foundations, the depth of the foundation is increased rather than introduce shear reinforcement.

EXAMPLE 7.7   Shear resistance of pad foundation. Design for shear the pad foundation shown in Fig. 7.13. The steel required to resist the bending moment is shown. Assume C20/25 Grade of concrete.

Stage (1) Perimeter at the face of the column
Bending moment at the column face increases the shear stress factor (Eqs. 6.39 and 6.51, EN)

$$\beta = 1 + kM_{Ed}u_1/(V_{Ed,red}W_1)$$
$$= 1 + 0.6 \times 3333E6 \times 4E3/(6500E3 \times 81.67E6) = 1.015$$
$$c_1/c_2 = 1 \text{ and hence } k = 0.6 \quad \text{(Table 6.1, EN)}$$

From Eq. 6.41, EN

$$W_1 = c_1^2/2 + c_1c_2 + 4c_2d + 16d^2 + 2\pi dc_1$$
$$= 0.5 + 1 + 4 \times 1 \times 1.94 + 16 \times 1.94^2 + 2\pi \times 1.94 \times 1 = 81,67\,\text{m}^2.$$

Applied design punching shear stress at the face of the column (Eq. 6.38, EN)

$$v_{Ed} = \beta V_{Ed}/(u_i d) = 1.015 \times 6500E3/(4E3 \times 1.94E3) = 0.85\,\text{MPa}.$$

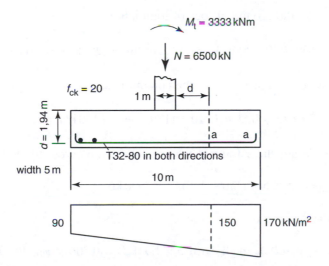

FIGURE 7.13   Example: a pad foundation

Maximum shear stress allowed (Eq. 6.53, EN)

$$v_{Rd, max} = 0.5 v f_{cd} = 0.5 \times 0.552 \times 20/1.5 = 3.68 > v_{Ed}(0.85)\,\text{MPa, acceptable,}$$

where the strength reduction factor (Eq. 6.6, EN)

$$v = 0.6(1 - f_{ck}/250) = 0.6(1 - 20/250) = 0.552$$

Stage (2) The $2d$ perimeter lies outside the base. Check a section across the base at a distance d from the face of the column.

Shear stress resistance of a singly reinforced base at a section distance d from the face of the column (Eq. 6.50, EN).

$$v_{Rd,c} = C_{Rd,c}k(100\rho_1 f_{ck})^{1/3} \times 2d/a$$
$$= 0.18/1.5 \times 1.32(100 \times 0.00518 \times 20)^{1/3} \times 2 = 0.345 \times 2 = 0.69\,\text{MPa.}$$

Minimum shear stress resistance

$$v_{min}(2\,d/a) = 0.035k^{1.5}f_{ck}^{0.5}(2d/a)$$
$$= 0.035 \times 1.32^{1.5} \times 20^{0.5} \times 2 = 0.376 < v_{Rd,c}(0.69)\,\text{MPa, acceptable.}$$

Depth factor

$$k = 1 + (200/d)^{0.5} = 1 + (200/1940)^{0.5} = 1.32 < 2.$$

Tensile reinforcement ratio (cl 6.4.4, EN)

$$\rho_1 = (\rho_{ly}\,\rho_{lz})^{0.5} = 10053/(1E3 \times 1940) = 0.00518 < 0.02, \text{ acceptable.}$$

To determine the applied shear stress at a distance d from the column face use the soil pressure diagram (Fig. 7.13).

Pressure on the ground at toes of foundation

$$p = 6500/(10 \times 5) \pm 3333/(5 \times 10^2/6) = 130 \pm 40 = +170 \text{ and } +90 \text{ kN/m}^2.$$

Pressure at a distance d from the column face

$$p = 90 + (5 + B/2 + d)/10 \times (170 - 90) = 150\,\text{kN/m}^2.$$

Applied design shear force at the critical section across the full width of the base

$$V_{Ed} = 2.5 \times 5 \times (170 + 150)/2 = 2000\,\text{kN.}$$

Design shear stress

$$v_{Ed} = V_{Ed}/(ud) = 2000E3/(5E3 \times 1.94E3) = 0.206 < v_{Rd,c}(0.69)\,\text{MPa acceptable.}$$

Alternatively calculate the applied design shear stress from (Eq. 6.51, EN)

$$v_{Ed} = V_{Ed}/(ud)(1 + kM_{Ed}u/V_{Ed}W)$$
$$= 6500E3/(2 \times 5E3 \times 1,94E3)(1 + 0,6 \times 3333E6 \times 2 \times 5E3/(6500E3 \times 81,67E6))$$
$$= 0,335 \times (1 + 0,0377) = 0,348 < v_{Rd,c}(0,69)\,\text{MPa also satisfactory.}$$

This applied design shear stress (0,348 MPa) is considerably higher than previously (0,206 MPa) and is therefore more conservative.

Maximum punching shear stress resistance (Eq. 6.53, EN)

$$v_{Rd,max} = 0,5vf_{cd} = 0,5 \times 0,6(1 - 20/250) \times 20/1,5$$
$$= 3,68 > v_{Ed}(0,206)\,\text{MPa, satisfactory.}$$

Stage (3) Check bond length (a-a) (Fig. 7.13) (cl 9.8.2.2, EN).

Force in the tensile reinforcement (Eq. 9.13, EN)

$$F_s = Rz_e/z_i = 1600 \times 3,847/(0,9 \times 1,94) = 3195\,\text{kN}$$

where

$$e = M/N = 3333/6500 = 0,513\,\text{m}$$
$$R = 160 \times 2 \times 5 = 1600\,\text{kN (approx.)}$$
$$z_e = 5 - e - 1 = 5 - 0,513 - 1 = 3,487\,\text{m}$$
$$\sigma_{sd} = F_s/A_s = 3195E3/(5E3/80 \times 804) = 63,6\,\text{MPa.}$$

Bond length required (Eq. 8.3, EN)

$$l_{b,req} = (\phi/4)(\sigma_{sd}/f_{bd}) = (32/4) \times (63,6/2,25) = 226\,\text{mm.}$$

Design value of the bond stress (Eq. 8.2, EN)

$$f_{bd} = 2,25f_{ctk,0,5}/\gamma_c = 2,25 \times 1,5/1,5 = 2,25\,\text{MPa.}$$

Design anchorage length (Eq. 8.4, EN)

$$l_{bd} = \alpha_1\alpha_2\alpha_3\alpha_4 l_{b,req} = 0,7 \times 226 = 158\,\text{mm.}$$

Bond length available

$$= 2000 - \text{cover} = 2000 - 50 = 1950 > l_{bd}(158)\,\text{mm, acceptable.}$$

The procedure should be repeated at other perimeters to check whether there is one more critical. The magnitude of the applied shear stress ($v_{Ed} = 0,206$ MPa) could be reduced by deducting the self weight of the base (Eq. 6.48, EN) but in this case it is not critical.

7.4.4 Shear Resistance of Pile Caps (cl 9.8, EN)

Pile caps are required to transfer a column load to a pile group (Fig. 7.14). Toe bearing piles are used in situations where a load bearing stratum is available. Alternatively friction piles are used. Piles are generally in groups of three or more, but two can be used if stabilized by tying to another foundation. Shear reinforcement is generally added to resist high shear stresses, and formed into a steel cage with the longitudinal reinforcement which facilitates placing during construction. Further information is given in Chapter 11.

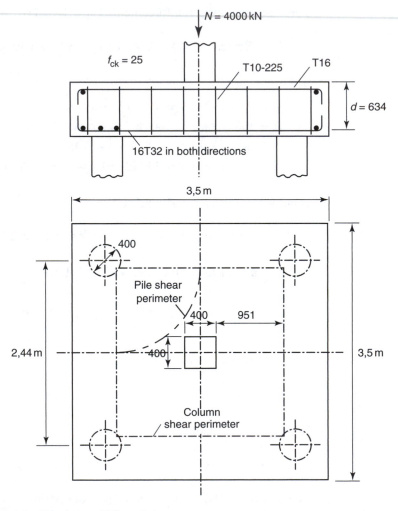

FIGURE 7.14   Example: pile foundation

EXAMPLE 7.8 *Shear resistance of a pile cap.* Check the pile cap shown in Fig. 7.14 for shear resistance and reinforce if necessary. Assume an internal column and Grade C25/30 concrete.

Stage (1) Check for punching shear failure round the column.
Applied shear stress at column perimeter (Eq. 6.53, EN)

$$v_{Ed} = \beta V_{Ed}/(u_i d) = 1{,}15 \times 4000E3/(4 \times 400 \times 634) = 4{,}53 \text{ MPa}.$$

Maximum shear stress resistance (Eq. 6.5, EN)

$$v_{Rd,max} = 0{,}5vf_{cd} = 0{,}5 \times 0{,}6(1 - 25/250) \times 25/1{,}5 = 4{,}5 < 4{,}53 \text{ MPa, acceptable}.$$

Applied design shear stress round shear perimeter (Fig. 7.14)

$$v_{Ed} = 4000E3/(4 \times (2 \times 951 + 400) \times 634) = 0{,}685 \text{ MPa}.$$

Shear stress resistance singly reinforced (Eq. 6.50, EN).

$$
\begin{aligned}
v_{Rd,c} &= C_{Rd,c} k (100 \rho_1 f_{ck})^{1/3} \times 2d/a \\
&= 0{,}18/1{,}5 \times 1{,}56(100 \times 0{,}0058 \times 25)^{1/3} \times 2 \times 634/951 \\
&= 0{,}608 < v_{Ed}(0{,}685) \text{ MPa},
\end{aligned}
$$

therefore shear reinforcement required.

$$
\begin{aligned}
v_{min} &= 0{,}035k^{1,5}f_{ck}^{0,5} \times 2d/a = 0{,}035 \times 1{,}56^{1,5} \times 20^{0,5} \times 1{,}33 \\
&= 0{,}455 < v_{Rd,c}(0{,}608) \text{ MPa, satisfactory}.
\end{aligned}
$$

Depth factor

$$k = 1 + (200/d)^{0,5} = 1 + (200/634)^{0,5} = 1{,}56 < 2.$$

Reinforcement ratio (cl 6.4.4(1), EN)

$$\rho_1 = (\rho_{ly} \rho_{lz})^{0,5} = 16 \times 804/(3500 \times 634) = 0{,}00518 < 0{,}02, \text{ satisfactory}.$$

Shear stress resistance of T12 vertical links assuming a radial spacing $s_r = 225$ mm (Eq. 6.52, EN)

$$
\begin{aligned}
v_{Rd,cs} &= 0{,}75V_{Rd,c} + 1{,}5(d/s_r) A_{sw} f_{ywd,ef}(1/(u_1 d)) \sin \alpha \\
&= 0{,}75 \times 0{,}608 + 1{,}5 \times (634/225) \times 4 \times 3 \times 113 \times 400/ \\
&\quad (4(400 + 2 \times 951) \times 634) \times 1 \\
&= 0{,}849 > v_{Ed}(0{,}685) \text{ MPa satisfactory}.
\end{aligned}
$$

$$f_{ywd,ef} = 250 + 0{,}25d = 250 + 0{,}25 \times 634 = 408{,}5 > 0{,}8 \times 500 = 400 \text{ MPa, use}$$
$$f_{ywd,ef} = 400 \text{ MPa}.$$

Maximum longitudinal spacing of the vertical links (Eq. 9.6, EN)

$$s_{l,max} = 0{,}75d(1 + \cot\alpha) = 0{,}75 \times 634 = 475{,}5 > 225 \text{ mm, satisfactory}.$$

Shear reinforcement ratio (Eq. 9.4, EN)

$$\rho_w = A_{sw}/(s\,b_w \sin\alpha) = 4 \times 3 \times 113/(225 \times 4 \times 400 \times 1) = 0{,}00377.$$

Minimum shear reinforcement ratio (Eq. 9.5, EN)

$$\rho_{w,\min} = 0{,}08 f_{ck}^{0,5}/f_{yk} = 0{,}08 \times 250{,}5/408{,}5 = 0{,}00098 < \rho_w(0{,}00377), \text{ acceptable}$$

Stage (2) Check for shear resistance round a pile.
Applied punching shear stress round a pile at edge of slab (Eq. 6.53, EN)

$$v_{Ed} = \beta V_{Ed}/(u\,d) = 1{,}5 \times 1000E3/(\pi \times 400 \times 634) = 1{,}88\,\text{MPa}.$$

Maximum shear stress resistance (Eq. 6.5, EN)

$$v_{Rd,\max} = 0{,}5 v f_{cd} = 0{,}5 \times 0{,}6(1 - 25/30) \times 25/1{,}5 = 4{,}5 > v_{Ed}(1{,}88)\,\text{MPa acceptable.}$$

Applied shear stress at the pile shear perimeter (Fig. 7.14)

$$v_{Ed} = \beta V_{Ed}/(u\,d) = 1{,}5 \times 1000E3/[(2 \times 530 + \pi/4 \times 1220) \times 634] = 1{,}17\,\text{MPa}.$$

Shear stress resistance (see previous calculations) $v_{Rd,c} = 0{,}608 < v_{Ed}(1{,}17)$ MPa therefore shear reinforcement required.

Shear stress resistance with link T12-225 (Eq. 6.52, EN)

$$v_{Rd,cs} = 0{,}75 V_{Rdc} + 1{,}5(d/s_r) A_{sw} f_{ywd,ef}(1/(u_1 d)) \sin\alpha$$
$$= 0{,}75 \times 0{,}608 + 1{,}5 \times (634/225) \times 3 \times 4 \times 113 \times 400/(1220(2 + \pi/2) \times 634) \times 1$$
$$= 1{,}29 > v_{Ed}(1{,}17)\,\text{MPa, acceptable.}$$

7.4.5 Corbels (cl J3, Figs 6.4 (b) and 6.6, EN)

Corbels are brackets on columns supporting loads as shown in Fig. 7.15. Generally the ratio $a_c/d < 2$ and the shear resistance is enhanced. Design consists in checking for combined shear force and bending moment, and bond lengths. See also Section 8.4.7.

EXAMPLE 7.9 Design of a corbel. Reinforce the corbel (Fig. 7.15(a)) for the design loads shown. Assume C30/40 Grade concrete.

Ratio

$a_c/h_c = 160/300 = 0{,}53 > 0{,}5$, therefore use vertical links (see Fig. J6(b), EN).

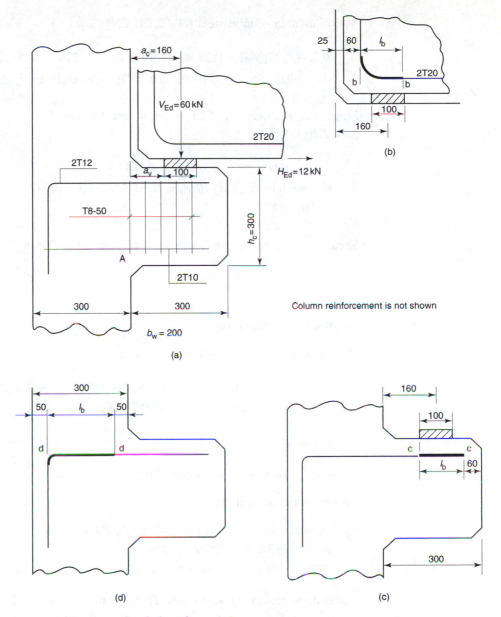

FIGURE 7.15   Example: design of a corbel

Stage (1) Force in the tension steel from moments of forces about A

$$F_{ld} = \text{bending} + \text{horizontal force} + \text{shear} = V_{Ed}a_c/z + H_{Ed} + 0{,}5V_{Ed}$$
$$= 60 \times 160/(0{,}9 \times 270) + 0{,}2 \times 60 + 0{,}5 \times 60 = 81{,}5 \text{ kN.}$$

Area of steel required to resist the force F_{ld}

$$A_s = F_{ld}/f_{yd} = 81{,}5\text{E3}/(500/1{,}15) = 187{,}5 \text{ mm}^2, \text{ use 2T12} = 226 \text{ mm}^2.$$

Minimum area of mild steel (cl 9.2.1.1, EN)

$$0,26(f_{ctm}/f_{yk})b_t d = 0,26 \times (2,9/500) \times 200 \times 270 = 81,5 < 226\,\text{mm}^2$$
$$0,0013 b_t d = 0,0013 \times 200 \times 270 = 70 < 226\,\text{mm}^2,\ \text{acceptable.}$$

Stage (2) Reduced applied shear stress for small values of shear/span ratio (cl 6.2.2(6), EN)

$$a_v/(2d) = (160 - 50)/(2 \times 270) = 0,204$$
$$v_{Ed} = 60E3(1 - a_v/(2d))/(b\,d) = 60E3(1 - (160 - 50)/(2 \times 270))/(200 \times 270)$$
$$= 0,885\,\text{MPa.}$$

Shear stress resistance without shear reinforcement (Eq. 6.2a, EN)

$$v_{Rd,c} = C_{Rd,c}k(100\rho_l f_{ck})^{1/3} + 0,15\sigma_{cp}$$
$$= 0,18/1,5 \times 1,86(100 \times 0,00418 \times 30)^{1/3} - 0,15 \times 0,2 = 0,488 < v_{Ed}(0,885)\,\text{MPa,}$$
therefore links required.

Minimum shear stress resistance

$$(v_{min} + k_1\sigma_{cp}) = 0,035k^{1,5}f_{ck}^{0,5} + k_1\sigma_{cp} = 0,035 \times 1,56^{1,5} \times 25^{0,5} - 0,15 \times 0,2$$
$$= 0,311 < 0,488\,\text{MPa, acceptable.}$$

Depth factor

$$k = 1 + (200/d)^{0,5} = 1 + (200/270)^{0,5} = 1,86 < 2.$$

Tensile reinforcement ratio

$$\rho_1 = A_s/(b_w d) = 226/(200 \times 270) = 0,00418$$
$$\sigma_{cp} = 12E3/(200 \times 300) = 0,2\,\text{MPa}$$
$$0,2f_{cd} = 0,2 \times 30/1,5 = 4 > 0,2\,\text{MPa, satisfactory.}$$

Shear stress resistance with single T8-50 vertical links (Eq. 6.19, EN)

$$v_{Rs} = A_{sw}(0,75a_v/s)f_{ywd}\sin\alpha/(b_w d)$$
$$= 2 \times 50 \times 0,75 \times ((160 - 50)/50) \times 0,8 \times 500/(200 \times 270)$$
$$= 1,22 > v_{Ed}(0,885)\,\text{MPa, acceptable.}$$

Maximum shear stress resistance (Eq. 6.5, EN)

$$v_{Rd,\,max} = 0,5vf_{cd} = 0,5 \times 0,528 \times 30/1,5 = 5,28 > v_{Ed,\,max}\,(1,11)\,\text{MPa, acceptable.}$$

Strength reduction factor (Eq. 6.6, EN)

$$v = 0,6(1 - f_{ck}/250) = 0,6(1 - 30/250) = 0,528 > 0,5.$$

Shear reinforcement ratio (Eq. 9.4, EN)

$$\rho_w = A_{sw}/(s\,b_w \sin\alpha) = 2 \times 50/(50 \times 200 \times 1) = 0{,}01.$$

Minimum shear reinforcement (Eq. 9.5, EN)

$$\rho_{w,min} = 0{,}08f_{ck}^{0,5}/f_{yk} = 0{,}08 \times 30^{0,5}/500 = 0{,}00087 < \rho_w(0{,}01), \text{ acceptable.}$$

Maximum spacing of links (Eq. 9.6, EN)

$$s_{max} = 0{,}75d(1+0) = 0{,}75 \times 270 = 203 > s(50)\,\text{mm, therefore satisfactory.}$$

Compression reinforcement (Eq. 6.18, EN)

$$A_{sl} = 0{,}5V_{Ed}(\cot\theta - \cot\alpha)/f_{yd} = 0{,}5 \times 60E3 \times 1/(500/1{,}15) = 16{,}6\,\text{mm}^2, \text{ use}$$
$$2T10 = 157\,\text{mm}^2.$$

Stage (3) Bond length (b-b) required for tension bars (2T20) end of beam (Fig. 7.15(b)). Axial stress in reinforcement

$$\sigma_{sd} = (H_{Ed} + 0{,}5V_{Ed})/A_s = (12E3 + 0{,}5 \times 60E3)/628 = 67\,\text{MPa}.$$

Basic anchorage length required (Eq. 8.3, EN)

$$l_{b,req} = (\phi/4)(\sigma_{sd}/f_{bd}) = (20/4) \times (67/3) = 112\,\text{mm}.$$

Design value of the bond stress (Eq. 8.2, EN)

$$f_{bd} = 2{,}25f_{ctk,0,5}/\gamma_c = 2{,}25 \times 2/1{,}5 = 3\,\text{MPa}.$$

Design anchorage length (Eq. 8.4, EN)

$$l_{bd} = \alpha_1\alpha_2\alpha_3\alpha_4 l_{b,req} = 0{,}7 \times 112 = 78\,\text{mm}.$$

Bond length available

$$a_c + 50 - \text{clearance} - \text{cover} = 160 + 50 - 25 - 60 = 125 > l_{bd}(78)\,\text{mm, acceptable.}$$

Bond length (c-c) required for tension bars (2T12) at end of the bracket (Fig. 7.15(c)). Axial stress in reinforcement

$$\sigma_{sd} = (A_{s,req}/A_s)f_{yd} = (187{,}5/226) \times (500/1{,}5) = 277\,\text{MPa}.$$

Basic anchorage length required (Eq. 8.3, EN)

$$l_{b,req} = (\phi/4)(\sigma_{sd}/f_{bd}) = (12/4) \times (277/3) = 277\,\text{mm}.$$

Design value of the bond stress (Eq. 8.2, EN)

$$f_{bd} = 2{,}25 f_{ctk,0,5}/\gamma_c = 2{,}25 \times 2/1{,}5 = 3\,\text{MPa}.$$

Design anchorage length (Eq. 8.4, EN)

$$l_{bd} = \alpha_1 \alpha_2 \alpha_3 \alpha_4 l_{b,req} = 0{,}7 \times 277 = 194\,\text{mm}.$$

Bond length available $= 300 - 160 + 50 = 190 = l_{bd}$ (194) mm approx. acceptable. Bond length (d-d) required for tension bars (2T12) embedded in the column (Fig. 7.15(d)). Axial stress in reinforcement

$$\sigma_{sd} = (A_{s,req}/A_s)f_{yd} = (187{,}5/226) \times (500/1{,}5) = 277\,\text{MPa}.$$

Bond length required (Eq. 8.3, EN)

$$l_{b,req} = (\phi/4)(\sigma_{sd}/f_{bd}) = (12/4) \times (277/3) = 277\,\text{mm}.$$

Design value of the bond stress (Eq. 8.2, EN)

$$f_{bd} = 2{,}25 f_{ctk,0,5}/\gamma_c = 2{,}25 \times 2/1{,}5 = 3\,\text{MPa}.$$

Design anchorage length (Eq. 8.4, EN)

$$l_{bd} = \alpha_1 \alpha_2 \alpha_3 \alpha_4 l_{b,req} = 0{,}7 \times 277 = 194\,\text{mm}.$$

Bond length available

$$= h - 2 \times \text{cover} = 300 - 2 \times 50 = 200 > l_{bd}(194)\,\text{mm, acceptable.}$$

An alternative method of design for corbels using a strut and tie model is given in cl J3, EN.

7.5 SHEAR RESISTANCE OF PRESTRESSED CONCRETE BEAMS (cls 6.2.2 AND 6.2.3 EN)

7.5.1 Introduction

The modes of shear failure and design requirements for prestressed beams are similar to those for reinforced concrete. Prestressing and increased concrete strength delays the formation of cracks and increases shear resistance. However the lower percentage of tensile reinforcement reduces the shear resistance. There are two situations to check. The first is as for a singly reinforced section at a section where

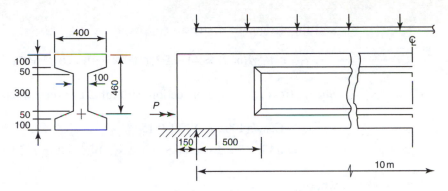

FIGURE 7.16   Example: shear resistance of a prestressed beam

there are bending cracks. The second is where there are no bending cracks but the principal tensile stress in the web is high.

EXAMPLE 7.10   Shear resistance of a pretensioned beam. Design for shear at the ultimate limit state the pretensioned prestressed concrete beam shown in Fig. 7.16. The beam is simply supported over a span of 10 m and carries a uniformly distributed design load of 160 kN at the ultimate limit state. The beam has been designed to meet the requirements for bending. Assume the prestressing force is transferred when the concrete strength is Grade C30/40 and is mature at concrete Grade C50/60. The area of prestressing steel is 645 mm^2 with an initial prestressing force of 1000 kN with 20 per cent losses, and 7 mm wires stressed to 1550 MPa.

Stage 1 Check as a cracked singly reinforced concrete beam.
Determine the section closest to the support at a distance x where cracking occurs at the bottom of the beam.

Bending stress = effective prestress + concrete tensile strength

$$0,5wLx(1-x/L)/Z = P_e/A + P_ee/Z + f_{ctm}/\gamma_c$$
$$0,5 \times 16 \times 10E3x(1-x/10E3)/20,4E6$$
$$= 800E3/135E3 + 800E3 \times 160/20,4E6 + 4,1/1,5$$
$$x^2 - 16E3x + 30464E3 = 0$$
$$x = 2209 \text{ mm from the support.}$$

where

$$A = BD - bd = 400 \times 600 - 300 \times 350 = 135E3 \text{ mm}^2$$
$$I = BD^3/12 - bd^3/12 = (400 \times 600^3 - 300 \times 350^3)/12 = 6128E6 \text{ mm}^4$$
$$Z = I/(D/2) = 6128E6/300 = 20,4E6 \text{ mm}^3$$

Design shear force at 2,209 m from the support

$$V_{Ed} = (5 - 2,209)/5 \times 80 = 44,66 \text{ kN}$$

and the design shear stress

$$v_{Ed} = V_{Ed}/(b_w d) = 44{,}66E3/(100 \times 460) = 0{,}97\,\text{MPa}.$$

Shear stress resistance of section without shear reinforcement (Eq. 6.2a, EN)

$$v_{Rd,c} = C_{Rd,c}k(100\rho_l f_{ck})^{1/3} + 0{,}15\sigma_{cp}$$
$$= 0{,}18/1{,}5 \times 1{,}66(100 \times 0{,}014 \times 50)^{1/3} + 0{,}15 \times 5{,}93 = 1{,}71 > v_{Ed}(0{,}97)\,\text{MPa}$$

where

$$\sigma_{cp} = N_{Ed}/A_c = 800E3/135E3 = 5{,}93\,\text{MPa}$$
$$0{,}2f_{cd} = 0{,}2 \times 50/1{,}5 = 6{,}67 > \sigma_{cp}(5{,}93)\,\text{MPa satisfactory.}$$

Depth factor

$$k = 1 + (200/d)^{0{,}5} = 1 + (200/460)^{0{,}5} = 1{,}66 < 2.$$

Longitudinal tensile reinforcement ratio (cl 6.2.2(1), EN)

$$\rho_l = 645/(100 \times 460) = 0{,}014 < 0{,}02, \text{ satisfactory.}$$

Minimum shear stress resistance

$$(v_{min} + k_1\sigma_{cp}) = 0{,}035k^{1{,}5}f_{ck}^{0{,}5} + 0{,}15 \times 5{,}93$$
$$= 0{,}035 \times 1{,}66^{1{,}5} \times 50^{0{,}5} + 0{,}15 \times 5{,}93$$
$$= 1{,}42 < v_{Rd,c}(1{,}71)\,\text{MPa, acceptable.}$$

Add minimum shear reinforcement (cl 6.2.1(4), EN).

Minimum shear ratio (Eq. 9.5, EN)

$$\rho_{w,min} = 0{,}08f_{ck}^{0{,}5}/f_{yk} = 0{,}08 \times 50^{0{,}5}/500 = 0{,}00113.$$

Longitudinal spacing for single T8 vertical links (Eq. 9.4, EN)

$$s_l = A_{sw}/(\rho_{w,min}b_w) = 2 \times 50/(0{,}00113 \times 100) = 885\,\text{mm}.$$

Maximum longitudinal spacing of vertical links (Eq. 9.6, EN)

$$s_{l,max} = 0{,}75d(1 + \cot\alpha) = 0{,}75 \times 460(1+0) = 345, \text{ use } s_l = 340\,\text{mm}$$

Maximum shear force resistance of T8-340 vertical links (Eq. 6.8, EN)

$$V_{Rd,s} = (A_{sw}/s)zf_{ywd} = (2 \times 50/340) \times 0{,}9 \times 460 \times 0{,}8 \times 500/1E3$$
$$= 48{,}7 > V_{Ed}(44{,}66)\,\text{kN, acceptable.}$$

Maximum shear force resistance of the concrete (Eq. 6.9, EN)

$$0 < \sigma_{cp}(5{,}93) < 0{,}25 f_{cd}(8{,}33)\,\text{MPa}$$
$$\alpha_{cw} = 1 + \sigma_{cp}/f_{cd} = 1 + 5{,}93/(50/1{,}5) = 1{,}18$$
$$V_{Rd,max} = \alpha_{cw} b_w z\, \nu_1 f_{cd}/(\cot\theta + \cot\alpha)$$
$$= (1{,}18 \times 100 \times 0{,}9 \times 460 \times 0{,}6 \times 50/1{,}5)/(1+0)/1\text{E}3$$
$$= 977 > V_{Ed,max}(80)\,\text{kN, acceptable.}$$

Stage 2 Check shear resistance of the uncracked web 0,5 m from the support. Design shear force 0,5 m from the support

$$V_{Ed} = 80(5 - 0{,}5)/0{,}5 = 72\,\text{kN}$$

Shear resistance of the web based on the principal tensile stress criterion (Eq. 6.4, EN)

$$V_{Rd,c} = I b_w/S(f_{ctd}^2 + \alpha_1 \sigma_{cp} f_{ctd})^{0{,}5}$$
$$= 6128\text{E}6 \times 100/13{,}14\text{E}6(2{,}73^2 + 0{,}729 \times 5{,}93 \times 2{,}73)^{0{,}5}/1\text{E}3$$
$$= 200{,}5 > V_{Ed}(72)\,\text{kN, acceptable}$$

where

$$s = B(D/2)(D/4) + (B - b_w)(D - t_f - 25)^2/2$$
$$= 400 \times 300 \times 150 - 300(300 - 100 - 25)^2/2 = 13{,}41\text{E}6\,\text{(approx.)}$$

To find $\alpha_1 = l_x/l_{pt2}$, $l_x = 650\,\text{mm}$ (Fig. 7.16) and from Eq. 8.16, EN

$$l_{pt} = \alpha_1 \alpha_2 \phi \sigma_{pmo}/f_{bpt} = 1{,}0 \times 0{,}25 \times 7 \times 1550/3{,}65 = 743\,\text{mm}$$

where (Eq. 8.15, EN)

$$f_{bpt} = \eta_{p1} \eta_1 f_{ctd(t)} = 2{,}7 \times 1{,}0 \times 1{,}35 = 3{,}65\,\text{MPa}$$

and from Eq. 8.18, EN

$$f_{ctd(t)} = \alpha_{ct} 0{,}7 f_{ctm(t)}/\gamma_c = 1{,}0 \times 0{,}7 \times 2{,}9/1{,}5 = 1{,}35\,\text{MPa}$$
$$l_{pt2} = 1{,}2 l_{pt} = 1{,}2 \times 743 = 892\,\text{mm}$$
$$\alpha_1 = l_x/l_{pt2} = 650/892 = 0{,}729$$

Compression reinforcement to resist shear at the cracked section (Eq. 6.18, EN)

$$A_{sl} = 0{,}5\, V_{Ed}(\cot\theta - \cot\alpha)/f_{yd} = 0{,}5 \times 44{,}66\text{E}3 \times (1-0)/(500/1{,}15)$$
$$= 51{,}4\,\text{mm}^2, \text{ use 2T12 (226 mm}^2\text{)}.$$

Use vertical links T8-340 with 2T12 as compression reinforcement.

7.6 Torsional Resistance of Reinforced and Prestressed Concrete (cl 6.3, EN)

7.6.1 Introduction

Small torsional moments occur in many structures, e.g. corners of simply supported slabs, but these are of minor importance and can be strengthened by adding additional longitudinal steel. In a few situations however torsion is of major importance, e.g. edge beams supporting built-in floor slabs and helical staircases. Calculations are required to determine the additional reinforcement required to resist the torsional moments.

In general, torsional moments occur in conjunction with bending moments and shear forces. The section size is determined from consideration of resistance to the bending moment and separate, apparently independent calculations, are made in order to add reinforcement to resist the shear force and torsional moment. In reality the moments and forces interact but it is convenient in design to consider them separately.

Despite recent research the interaction of torsion, bending and shear is not fully understood and design recommendations are therefore semi-empirical. The torsional strength of prestressed members is greater than that of ordinary reinforced concrete members, and the European Code recommends that the methods of design use for ordinary reinforced concrete should also be applied to prestressed concrete.

7.6.2 Modes of Failure in Torsion

The mode of failure in torsion depends on the arrangement of reinforcement and the relative magnitudes of torsional moment, bending moment and shear force.

The action of a pure torsional moment on a plain concrete member is to produce diagonal cracks at approximately 45° which spiral along the member and result in failure on a skewed failure plane. Failure occurs as soon as major cracks form and plain concrete is of little use for structural members.

The addition of longitudinal reinforcement strengthens the member when the ratio of bending moment to torsional moment is high ($M/T > 4$), but has little strengthening effect when the ratio is low.

The addition of closed stirrups to the longitudinal reinforcement however strengthens the member for the full range of M/T ratios and it is this fact which forms the basis for design recommendations in the European Code. There are three main modes of torsional bending failure for members reinforced with longitudinal steel and links (Lessig, 1959 and Walsh *et al.*, 1966) as follows.

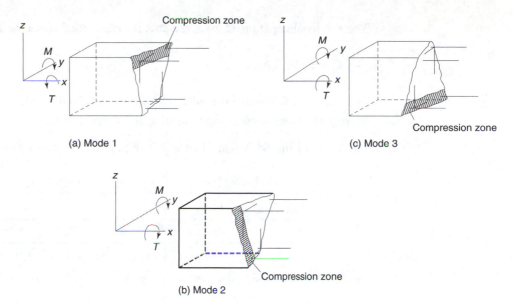

FIGURE 7.17 Modes of failure in torsion

Mode 1 type failure – This mode occurs when the M/T ratio is greater than 4 approximately. The mode of failure is comparable to that for pure bending, but the failure plane is skewed to the main longitudinal axis (See Fig. 7.17(a)).

Mode 2 type failure – This mode occurs when the M/T ratio is less than 4 approximately. This also is a type of bending failure but with the compression zone at the side of the member as shown in Fig. 7.17(b).

Mode 3 type failure – This mode occurs when the M/T ratio is less than 2 approximately. This also is a type of bending failure but with the compression zone at the bottom of the member as shown in Fig. 7.17(c). A requirement for this form of failure to occur is that the strength of the top longitudinal reinforcement is less than that for the bottom. These forms of failure are not identified by the European Code but they are useful as a guide for arranging reinforcement.

7.6.3 Torsional Stiffness of a Member (cl 6.3.1 EN)

The elastic analysis of a hyperstatic, or statically indeterminate, structure involves the stiffness of the members. For a concrete member the bending stiffness is E/lL and the torsional stiffness is GC/L where

G = elastic shear modulus and is equal to $0,42E$
L = length of a member between joint centres
C = torsional constant equal to half the St. Venant value for plain concrete.

The torsional constant C for a rectangular reinforced concrete section

$$C = 0{,}5\beta h_{min}^3 h_{max}$$

The factor of 0,5 incorporated in the value of C makes allowance for warping by halving the polar second moment of area of the section.

The value of the St. Venant factor β is obtained from the following table:

TABLE 7.1   St. Venant torsional factors for plain concrete.

h_{max}/h_{min}	1	1,5	2	3	5	>5
β	0,14	0,20	0,23	0,26	0,29	0,33

The St. Venant torsional stiffness of a non-rectangular section may be obtained by dividing the section into a series of rectangles and summing the torsional stiffness of these rectangles (cl 6.3.1(5), EN). The division of the rectangles is arranged to maximize the calculated stiffness. This will be generally achieved if the widest rectangle is made as long as possible.

EXAMPLE 7.11   Torsional stiffness of a member. Determine the torsional stiffness of a member with a rectangular cross section 150×300 mm, length between joint centres 4 m. Assume Grade C25/30 concrete.

Static modulus of elasticity of concrete (Table 3.1, EN)

$$E_{cm} = 22[(f_{cm} + 8)/10]^{0{,}3} = 22[(25 + 8)/10]^{0{,}3} = 31{,}5 \text{ GPa}.$$

Shear modulus

$$G = 0{,}42 E_{cm} = 0{,}42 \times 31{,}5 = 13{,}2 \text{ GPa}.$$

Torstional constant, $\beta = 0{,}23$ from Table 7.1

$$C = 0{,}5\beta h_{min}^3 h_{max} = 0{,}5 \times 0{,}23 \times 150^3 \times 300 = 116{,}4\text{E6 mm}^4$$

Torstional stiffness of the member

$$GC/L = 13{,}2 \times 116{,}4\text{E6}/(4\text{E3} \times 1\text{E3}) = 384 \text{ kNm}.$$

7.6.4 Equivalent Thin Walled Section (cl 6.3.1(3), EN)

In practice, there are solid and hollow sections which are reinforced to resist bending and torsion. Experiments (Hsu, 1968 and Lampert, 1971) show that the core area does not contribute significantly to the torsional resistance. For design purposes

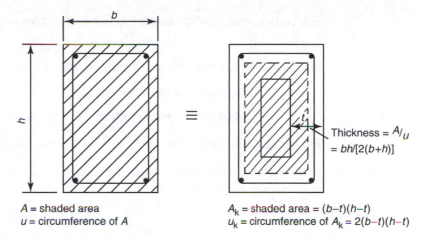

A = shaded area
u = circumference of A

A_k = shaded area = $(b-t)(h-t)$
u_k = circumference of A_k = $2(b-t)(h-t)$

FIGURE 7.18 Equivalent thin walled section

solid sections are replaced by thin walled sections (Fig. 7.18) where, for a rectangular section, the effective wall thickness

$$t_{ef,i} = \text{total area/outer circumference}$$
$$= A/u = bh/[2(b+h)] < \text{actual wall thickness.} \tag{7.17}$$

The wall thickness should be at least twice the cover to the longitudinal reinforcement plus the bar size.

Area enclosed by the centre lines of the connecting walls including the inner hollow areas

$$A_k = (b-t)(h-t) \tag{7.18}$$

7.6.5 Torsion Reinforcement (cl 9.2.3, EN)

Plain concrete is able to resist a small torsional moment, but because in practice the concrete may crack the section is always reinforced. The torsional reinforcement is generally in the form of closed links (Fig. 9.6, EN) similar in shape and additional to those required for shear reinforcement.

The truss analogy, used to develop theory for shear resistance, can also be used for torsional resistance. If each wall of a rectangular hollow section is considered as shown in Fig. 7.4, then Eq. (7.3) can be modified to form an equation to express the torsional resistance of the shear reinforcement about the longitudinal axis

$$T_{Rd2} = 2F_{sw} \sin \alpha A_k = 2A_{sw}f_{ywd}(AC/s) \sin \alpha A_k$$
$$= 2A_{sw}f_{ywd}(\cot \theta + \cot \alpha) \sin \alpha A_k \tag{7.19}$$

where A_k is the area enclosed by the centre lines of the connecting walls.

For vertical links $\alpha = 90°$ and assuming $\theta = 45°$ then this equation is equivalent to Eq. 6.28, EN.

Similarly the torsional resistance of the concrete about the longitudinal axis

$$T_{Rd1} = 2F_{cn}\sin\theta A_k = 2t(AC\sin\theta)f_{crushing}\sin\theta A_k$$
$$= 2tA_k\nu f_{cd}(\cot\theta + \cot\alpha)\sin^2\theta \tag{7.20}$$

For vertical stirrups $\alpha = 90°$ then this is comparable with Eq. 6.30, EN.

Equating total axial forces at section 2-2 (Fig. 7.4), where A_{sl} is the total area of axial steel to resist torsion

$$4F_c\cos\theta = A_{sl}f_{yld} \tag{i}$$

Resolving forces vertically at B

$$F_c\sin\theta = A_{sw}f_{ywd}(AC/s)\sin\alpha \tag{ii}$$

Equating values of F_c from Eqs (i) and (ii)

$$F_c = A_{sl}f_{yld}/\cos\theta = u_k(A_{sw}/s)f_{ywd}AC\sin\alpha/\sin\theta$$
$$A_{sl}f_{yld} = u_k\cos\theta(A_{sw}/s)f_{ywd}(\cot\theta + \cot\alpha)\sin\alpha/\sin\theta$$
$$= u_k\cos\theta[T_{Rd2}/(2A_k\cot\theta)](\cot\theta + \cot\alpha)\sin\alpha/\sin\theta \tag{iii}$$
$$A_{sl}f_{yld} = u_k\cot\theta[T_{Rd2}/(2A_k)] \tag{7.21}$$

From Eq. (iii) a relationship for $\tan\theta$ may be obtained

$$A_{sl}f_{yld} = u_k(A_{sw}/s)f_{ywd}\cot^2\theta$$
$$\tan^2\theta = u_k(A_{sv}f_{ywd}/s)/(A_{sl}f_{yld}) \tag{7.22}$$

Combining Eqs (7.21) and (7.22)

$$T_{Rd2} = 2A_k[(A_{sv}f_{ywd}/s)(A_{sl}f_{yld}/u_k)]^{0,5} \tag{7.23}$$

Recommended values of $\cot\theta$ vary between 1 and 2,5 (cl 6.3.2(2), EN) the same range as for shear resistance. When considering combined actions, the angle θ is the same for torsion and shear design. Notice that than θ is related to the ratio of links to the longitudinal steel (Eq. 7.22).

7.6.6 Arrangement of Torsion Reinforcement (cl 9.2.3, EN)

Careful detailing of the torsion reinforcement is required if it is to be effective. To intercept the diagonal cracks in the concrete the spacing of the links (s) is similar to that for shear resistance.

The longitudinal torsional reinforcement must be distributed evenly round the inside perimeter of the links to prevent a mode 3 type failure as shown in Fig. 7.17(c).

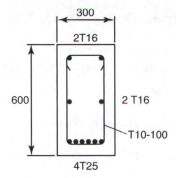

300

2T16

600

2 T16

T10-100

4T25

FIGURE 7.19   Example: torsion reinforcement

The distance between bars should not exceed 350 mm (cl 9.2.3(4), EN) and at least four bars, one in each corner should be used.

The torsion reinforcement should extend a distance equal to the largest dimension of the section beyond where it theoretically ceases to be required. This requirement is because of additional axial stresses introduced due to torsion, cracks spiralling for considerable lengths along a member, and to avoid abrupt changes of reinforcement.

EXAMPLE 7.12   Torsion reinforcement. A rectangular section of a beam is subject to a design torsional moment $T_{Ed} = 50$ kNm (Fig. 7.19). The cross section is 600 mm deep and 300 mm wide and made from C25/30 concrete. Determine the torsion reinforcement required. Assume that the section has been designed for bending and that the area of tensile reinforcement required from bending calculations is $A_{sl} = 1200$ mm^2.

Equivalent wall thickness of the section (Fig. 7.18) from Eq. (7.17)

$$t = bh/[2(b+h)] = 300 \times 600/[2(300+600)] = 100 \text{ mm}.$$

Area enclosed by the centre lines of enclosing walls

$$A_k = (b-t)(h-t) = (300-100)(600-100) = 100E3 \text{ mm}^2$$
$$u_k = 2[(b-t)+(h-t)] = 2[(300-100)+(600-100)] = 1400 \text{ mm}.$$

Shear stress in the longer wall of the section from torsion (Eq. 6.26, EN)

$$\tau_{t,i}t_{ef,i} = T_{Ed}/(2A_k) = 50E6/(2 \times 100E3) = 250 \text{ MPa}.$$

Shear force in the longer wall of the section (Eq. 6.27, EN)

$$V_{Ed,i} = \tau_{t,i}t_{ef,i}z_i = 250 \times (600 - 2 \times 30)/1E3 = 135 \text{ kN}$$
$$A_{sw}/s = V_{Ed,i}/(zf_{ywd}) = 135E3/(0{,}9 \times 550 \times 0{,}8 \times 500) = 0{,}682 \text{ mm}.$$

Longitudinal spacing of T10 vertical links

$$s_l = 79/0{,}682 = 115{,}8 \text{ mm, use } s_l = 100 \text{ mm}.$$

Maximum torsional moment of resistance ($\theta = 45°$) (Eq. 6.30, EN)

$$T_{\text{Rd, max}} = 2\nu\,\alpha_c f_{cd} A_k t_{ef,i} \sin\theta\cos\theta$$
$$= (2 \times 0{,}54 \times 25/1{,}5 \times 100\text{E}3 \times 100 \times 0{,}707 \times 0{,}707)/1\text{E}6$$
$$= 90 > T_{\text{Ed}}(50)\,\text{kNm, satisfactory.}$$

Strength reduction factor for concrete (Eq. 6.6, EN)

$$\nu = 0{,}6(1 - f_{ck}/250) = 0{,}6(1 - 25/250) = 0{,}54.$$

Area of the longitudinal reinforcement (Eq. 6.28, EN)

$$\Sigma A_{sl} = T_{\text{Ed}} \cot\theta\, u_k/(2 A_K f_{yd})$$
$$= 50\text{E}6 \times 1 \times 1400/(2 \times 100\text{E}3 \times 500/1{,}15) = 805\,\text{mm}^2$$

Use 6 longitudinal bars. Two of these bars can be combined with the tension reinforcement (cl 6.3.2(3), EN)

$A_{sl} = 1200 + 805/3 = 1468\,\text{mm}^2$. Tension reinforcement 4T25 ($1964\,\text{mm}^2$).

Area of other four bars $= 2/3 \times 805 = 537\,\text{mm}^2$. Use 4T16 $= 804\,\text{mm}^2$.

Maximum longitudinal spacing of links (Eq. 9.6, EN)

$$s_{l,\text{max}} = 0{,}75d(1 + \cot\alpha) = 0{,}75 \times 540 \times (1 + 0) = 405 > s_l(100)\,\text{mm.}$$

Also (cl 9.2.3, EN)

$$s_{l,\text{max}} = u/8 = 2(b+d)/8 = 2 \times (300 + 600)/8 = 225 < b(300)\,\text{mm, use}$$
$$s_l = 100\,\text{mm.}$$

Shear reinforcement ratio (Eq. 9.4, EN)

$$\rho_w = A_{sv}/(s\,b_w \sin\alpha) = 2 \times 79/(100 \times 300 \times 1) = 0{,}00527.$$

Minimum shear reinforcement ratio (Eq. 9.5, EN)

$$\rho_{\text{min}} = 0{,}08 f_{ck}^{0,5}/f_{yk} = 0{,}08 \times 25^{0,5}/500 = 0{,}0008 < \rho_w(0{,}00527),\text{ acceptable.}$$

REFERENCES AND FURTHER READING

ASCE-ACI Committee 445 (1998) Recent approaches to shear design of structural concrete, *Jnl Struct Eng* **124**, No 12, Dec, 1373–1417.

ASCE-ACI Committee 426 (1973) The shear strength of reinforced concrete structures, *Proc. ASCE*, **99**, NO ST6, 1091–1187.

Baumann, T. and Rusch, H. (1970) Tests studying the dowel action to the flexural tensile reinforcement of reinforced concrete beams, *Berlin, William Ernst und Sohn. Deutscher Ausschuss fur Stahlbeton*, Heft **210**, 42–82.

Chana, P.S. (July 1981) Some aspects of modelling the behaviour of reinforced concrete under shear loading, *Cement and Concrete Association*, Technical Report 543, 22.

Evans, R.H. and Kong, F.K. (1967) Shear design and British Code CP 114. *Structural Engineer*, **45(4)**, 153–187.

Hsu, T.T.C. (1968) Torsion of structural concrete - behaviour of reinforced rectangular members. *Special Publication, No 18-10, ACI*, 261–306.

Kong, F.K. (1978) *Bending, Shear and Torsion, Chapter 1, Developments in Prestressed Concrete*. Applied Science, London, Vol. 1, 1–68.

Lampert, P. (1971) Torsion and bending in reinforced and prestressed concrete beams. *Proc. ICE*, Dec, Vol. **50**, 487–505.

Lessig, N.N. (1959) The determination of the load bearing capacity of rectangular reinforced concrete sections subject to combined torsion and bending. *Study No 5, Concrete and Reinforced Concrete Inst., Moscow*, 5–28.

Loov, R.E. and Patnaik, A.K. (1994) *Horizontal shear strength of composite concrete beams with a rough interface*. PCIJ, **39(1)**, 48–109.

Placas, A. and Regan, P.E. (1971) Shear failure of reinforced concrete beams. *Proc. ACIJ*. October, 763–773.

Reynolds, G.C., Clark, J.L. and Taylor, H.P.J. (1974) Shear provisions for prestressed concrete in the unified code CP 110(1972). *Technical Report 42.500, Cement and Concrete Association*, London, 16.

Regan, P.E. (2004) Punching of slabs under highly concentrated loads. *Proc. ICE, Struct. and Bu.*, 157, April, 165–171.

Ritter, W. (1899) Die bauweiser hennebique. Schweizerische Bauzeitung, **33(7)**, 59–61.

Saemann, J.C. and Washa, G.W. (1964) Horizontal shear connections between precast beams and cast-in-place slabs. *Proc. ACI*, V61, 11, Nov, 1383–1408.

Shioya, T., Iguro, M., Nojiri, Y., Akiayma, H. and Okada, T. (1989) Shear strength of large concrete beams, fracture mechanics: application to concrete. *SP-118, ACI, Detroit*, 259–279.

Taylor, H.P.J. (1974) The fundamental behaviour of reinforced concrete beams in bending and shear. *Proc. ACI - ASCE Shear Symposium, Ottawa, ACI* Publication SP **42**, ACIJ, Detroit, 43–77.

Walraven, J.C., Vos, E. and Reinhardt, H.W. (1978) Experiments on shear and transfer in cracks in concrete, Pt 1, description of results. *Report 5-79-3, Stevin Laboratory, Delft University of Technology, The Netherlands*, 84.

Walsh, P.E., Collins, M.P., Archer, F.E. and Hall, A.S. (1966) The ultimate strength design of rectangular reinforced concrete beams subjected to combined torsion, bending and shear. *Trans. ICE, Australia*, 143–157.

Yoon Y.S., Cook W.D. and Mitchell D. (1996). Minimum shear reinforcement in normal, medium and high strength concrete beams. *ACIJ*, **93(5)**, 576–584.

Chapter 8 / Anchorage, Curtailment and Member Connections

8.1 ANCHORAGE (cl 8.4, EN)

8.1.1 Introduction

Reinforcing bars are produced in stock lengths, e.g. approximately 6 m for 8 mm diameter, and 12 m for 32 mm diameter. These stock lengths may be cut or terminated for the following reasons:

(a) to fit the member,
(b) to economise in steel, e.g. curtailment of bars in order to use fewer bars where the bending moment reduces along the length of a member and
(c) to make construction easier.

A bar may be terminated as straight, hook, bend or loop (Fig. 8.1, EN). Equivalent anchorage lengths for design purposes are shown (Fig. 8.1).

A reinforcing bar must be adequately anchored to prevent withdrawal from the concrete before it has reached its full tensile strength. The greater the angle change at the end of the bar the more effective the anchorage, but this is offset by the greater cost for shaping. Some situations in practice that require calculations for bond length (shown as a heavy black line) are shown in Fig. 8.2.

8.1.2 Anchorage of Bars (cl 8.4, EN)

The withdrawal of straight bar, with adequate cover and embedded in concrete, is prevented by the bond shear stresses parallel to the surface of the bar and the concrete. Bond is due to the combined effects of adhesion, friction and, in the case of deformed bars, the bearing of the projecting ribs. Plain bars withdraw from the concrete, but ribbed bars introduce radial tensile stresses in the concrete which generally produce splitting and cracking of the concrete.

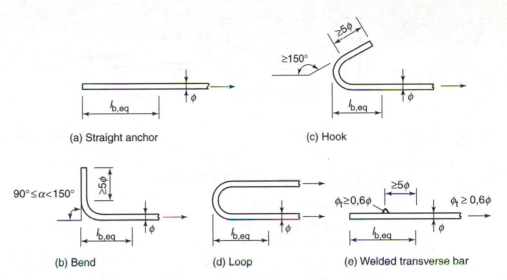

FIGURE 8.1   Equivalent anchorage lengths

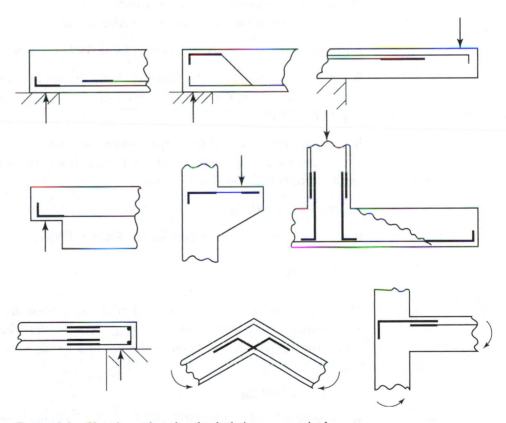

FIGURE 8.2   Situations where bond calculations are required

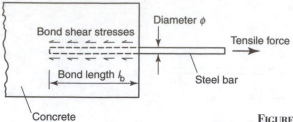

FIGURE 8.3 Anchorage of a straight bar

The factors that affect the bond are:

(a) concrete strength,
(b) axial stress in the bar,
(c) surface condition of the bar,
(d) straight or hooked bar,
(e) welding as anchorage,
(f) transverse reinforcement,
(g) pressure transverse to the bar,
(h) tension or compression in the bar and
(i) dowel forces at right angles to the reinforcement.

All these factors, except (i), are allowed for in Table 8.2, EN.

When a steel bar is pulled out of concrete bond stresses arise due to the differences in the axial stresses in the steel bar. This results in varying bond stresses along the length of the bar.

If for simplicity, the ultimate bond shear stresses are assumed to be uniformly distributed as shown in Fig. 8.3, then from the equilibrium of forces, tensile force in bar = bond shear force

$$\pi\phi^2/4\sigma_{sd} = \pi\phi l_b f_{bd}$$

Rearranging, the basic anchorage length (Eq. 8.3, EN)

$$l_{b,req} = (\phi/4)(\sigma_{sd}/f_{bd}).$$

The design axial stress in the bar (σ_{sd}) is at the position from where anchorage is measured. The design bond stress (f_{bd}) between the steel and the concrete varies with the surface condition of the bar and tensile strength of the concrete (Eq. 8.2, EN).

$$f_{bd} = 2{,}25n_1n_2f_{ctd}$$

n_1 is related to bond position of the bar, n_2 is related to bar diameter and the tensile strength of the concrete (cl 8.4.2(2), EN)

$$f_{ctd} = \alpha_{ct}f_{ctk0,05}/\gamma_c$$

FIGURE 8.4   Deformed bar

The 5 per cent fractile tensile strengths of concrete ($f_{ctk0,05}$) are obtained from Table 3.1, EN (Annex A2).

Ribbed bars are more difficult to withdraw from concrete because their surfaces are rougher (Fig. 8.4). At bond failure, a ribbed bar splits and cracks the surrounding concrete producing a volume change in the concrete (Reynolds, 1982). Despite the different failure mechanisms the ultimate bond strength is also related to the tensile strength of the concrete, although as expected, the values are greater than for plain bars.

Welded mesh reinforcement is difficult to withdraw from concrete because of the strength of the weld(F_{wd}) to the transverse reinforcement (Eq. 8.9, EN).

$$F_{btd} = F_{wd} \leq 16A_s f_{cd} \phi_t / \phi_l$$

The basic anchorage length ($l_{b,req}$) is factored (Table 8.2, EN), to allow for practical conditions e.g. hooks, concrete cover and confinement. The design anchorage length (Eq. 8.4, EN)

$$l_{bd} = \alpha_1 \alpha_2 \alpha_3 \alpha_4 l_{b,req} \geq l_{b, min}$$

Where there is insufficient length available for anchorage of a straight bar then bends, hooks or loops are used as shown in Fig. 8.1. To prevent overstraining and cracking of the bar the radii of the bends should not be too small. This is controlled by the mandrel diameter (Table 8.1, EN). For $\phi \leq 16\,\text{mm}$ the mandrel diameter is 4ϕ. For $\phi > 16\,\text{mm}$ the mandrel diameter is 7ϕ.

In practice, the use of bends and hooks is kept to a minimum because of the cost involved in their formation. Although improved bond characteristics of straight bars has reduced the use of bends and hooks they are required where bond length is restricted, e.g. supports for beams, starter bars in bases, links, bent-up shear reinforcement and member connections.

8.2 Bar Splices

8.2.1 Introduction (cl 8.7, EN)

In reinforced concrete construction splices between the ends of bars are required to extend their length or to facilitate construction. Splices transferring stress are generally lapped, but may also be welded or joined with mechanical devices (Paterson and Ravenshill, 1981). Splices should be placed, wherever possible, away from points of high stress and preferably should be staggered. Laps in fabric may be

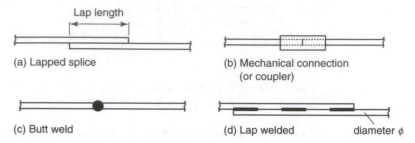

FIGURE 8.5   Bar splices

layered or nested to maintain the lapped bars in one plane. Good detailing should avoid spalling of the concrete in the neighbourhood of the splice and the width of the cracks should not exceed allowable values. Various types of splices are shown in Fig. 8.5.

8.2.2 Lap Splices (cl 8.7.2, EN)

The simplest, cheapest and most common type of splice is lapped as shown in Fig. 8.5(a). The force in one bar is transferred to the other by bond between the steel and the concrete. The design lap length (Eq. 8.10, EN),

$$l_o = \alpha_1 \alpha_2 \alpha_3 \alpha_4 \alpha_5 \alpha_6 l_{b,req} \geq l_{o,\,min}$$

The efficiency of splices depends on adequate cover and resistance to bursting of the concrete for deformed bars. Where cover is inadequate the situation is improved by transverse reinforcement, e.g. stirrups, which enclose the longitudinal reinforcement and increase the tensile bursting strength of the concrete.

8.2.3 Mechanical Splices (cl 8.7.1, EN)

The suitability of the coupling sleeve shown in Fig. 8.5(b) should be demonstrated. The sleeve is not in common use but is suitable for bars in compression provided that the sawn square cut ends are held in concentric contact. The sleeves advantage is that it saves on length of steel and reduces congestion of the bars. The disadvantages are increased cost and reduction of concrete cover. This type of sleeve may be used for bars in tension provided that it is demonstrated to be suitable.

8.2.4 Welded Splices (cls 3.2.5, 8.7.1, EN)

Welded splices are not often used on site because of the disadvantages of cost and supervision required. Welding is capable of producing a full strength connection and practical recommendations are:

(1) reinforcing steel must be suitable for welding (Pr EN 10080),
(2) welded joints should not occur at or near bends,

(3) where possible, joints in parallel bars of the principal tensile reinforcement should be staggered in the longitudinal direction,

(4) butt welded joints in compression resist 100 per cent of the design strength, but joints in tension should be designed to resist 80 per cent of the design strength.

Fillet welded lap connections, as shown in Fig. 8.5(d), are used where prefabricated cages of reinforcement, or mesh, are placed in a shutter and welded together in place. Alternatively, they may be used for the steel connections for precast reinforced concrete elements. The length and spacing of welds should be sufficient to transmit the design load in the bar (cl 8.6. EN).

8.3 CURTAILMENT OF REINFORCING BARS

8.3.1 Introduction

Curtailment of reinforcing bars is the stopping of some of the reinforcing bars in tension, or compression, where they are no longer needed for strength purposes. Curtailment can be applied to simply supported beams but it is generally not economic and the real advantages occur with repetitive multi-span beams and slabs. It can also be used to facilitate construction by curtailing bars to practical lengths which are easier to handle. However, the saving in steel is off-set by increased design calculations, detailing and site supervision.

8.3.2 Theoretical Curtailment

A part of a beam subject to a bending moment, which varies along is length, is shown in Fig. 8.6.

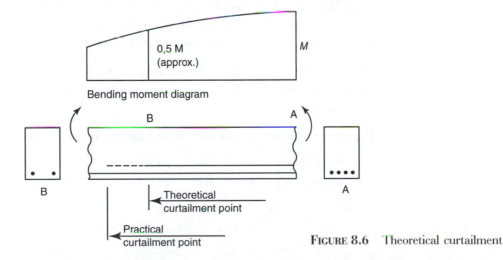

FIGURE 8.6   Theoretical curtailment

At section A, the bending moment is a maximum and four bars are required in tension. Further along the beam at section B, the bending moment has reduced to approximately half and only two bars are required. The bars are shown in Fig. 8.6 at different levels in the elevation for purposes of illustration. Theoretically two bars can be stopped, or curtailed, at this section. However, the bars must be anchored by further extension. The shear force at this section also introduces an axial stress in the bars and this must be taken into account (cl 6.2.3(7), EN).

8.3.3 Practical Curtailment (cl 9.2.1.3, EN)

A bar extends beyond the point at which it is theoretically no longer required for the following reasons:

(a) Continuing bars are at the design strength.
(b) Large cracks may appear at the curtailment section because of the abrupt change in section properties. This may reduce the shear strength of the member and therefore it is advisable to stagger the curtailment points in heavily reinforced members.
(c) To allow for inaccuracies in loading and theoretical analysis.
(d) To allow for inaccuracies in placing bars.
(e) To resist the shear force. The shear force at a section in bending introduces an additional tensile force in the longitudinal steel. For singly reinforced beams this is catered for by increasing the bond length by the effective depth (d) (cl 6.2.2, EN). For beams with shear reinforcement (cl 9.2.1.3(2), EN), the additional bond length is calculated from the additional tensile force in the longitudinal reinforcement which is related to the shear force (cl 6.2.3(7), EN),

$$\Delta F_{td} = 0{,}5 V_{Ed}(\cot \theta - \cot \alpha)$$

EXAMPLE 8.1 Curtailment of bars. Determine (a) the theoretical and (b) practical curtailment point for 2T20 bars for the cantilever beam shown in Fig. 8.7. Assume $b = 300$ mm, $d = 700$ mm, and C20/25 Grade concrete.

Stage (1) For a beam with shear reinforcement the theoretical curtailment point occurs where the design bending moment plus shear force can be resisted by 2T20 reinforcing bars.

Equating moments of resistance at a distance x from the end of the cantilever

$$M_r = M_{Ed}$$
$$A_s z(f_{yd} - \Delta F_{td}/A_s) = qx^2/2$$
$$A_s z(f_{yd} - 0{,}5qx/A_s) = qx^2/2$$

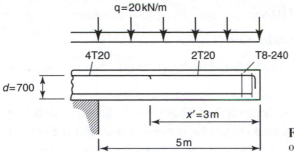

FIGURE 8.7 Example: curtailment of bars

Assuming that the steel stress is at the design strength, $z = 0{,}9d = 630$ mm, and $q = 20$ kN/m $= 20$ N/mm, the value of

$$x = -z/2 + ([z/2]^2 + 2A_s f_{yd} z/q]^{0,5}$$
$$= -630/2 + [(630/2)^2 + 2 \times 628 \times 500/1{,}15 \times 630/20]^{0,5} = 3{,}844E3 \text{ mm}$$

The theoretical curtailment point for 2T20 bars in 3,84 m from the end of the cantiliver.

Stage (2) Basic anchorage length (Eq. 8.3, EN)

$$l_{b,req} = (\phi/4)(\sigma_{sd}/f_{bd}) = 20/4 \times (500/1{,}15)/2{,}25 = 966 \text{ mm}$$

where the design bond stress for high bond bars (Eq. 8.2, EN)

$$f_{bd} = 2{,}25 n_1 n_2 f_{ctd} = 2{,}25 \times 1 \times 1 \times 1{,}5/1{,}5 = 2{,}25 \text{ MPa}$$

Design bond length with 30 mm cover (Eq. 8.4, EN)

$$l_{bd} = \alpha_1 \alpha_2 \alpha_3 \alpha_4 \alpha_5 \alpha_6 l_{b,req} = 1 \times 0{,}85 \times 966 = 821 \text{ mm}$$

where

$$\alpha_2 = 1 - 0{,}15 \times (c_d - \phi)/\phi = 1 - 0{,}15 \times (30 - 20)/20 = 0{,}85.$$

Check if greater than the minimum anchor length in tension (cl 8.4.4, EN)

$$l_{b,min} = 0{,}3 l_{b,req} = 0{,}3 \times 966 = 290; \text{ or } 10\phi = 10 \times 20 = 200; \text{ or } 100 \text{ mm}.$$

Stage (3) Distance of the practical curtailment point from the end of the cantilever

$$x' = x - l_{bd} = 3{,}84 - 0{,}821 = 3{,}02 \text{ m, Use } x' = 3 \text{ m}.$$

8.4 MEMBER CONNECTIONS

8.4.1 Introduction

Connections and joints in a structure may be classified as follows:

(a) Construction joints introduced between concrete pours, sections or levels to facilitate construction. These are particularly useful for in situ concrete work, e.g. foundations can be concreted first and the columns later. In many cases, these joints transmit forces but in other cases the requirements are that they should resist fire, corrosion, and be durable and sound proof.

(b) Movement joints are required to allow free expansion and contraction to occur in the structure due to temperature, shrinkage, creep, settlement, etc. and prevent a build-up of stresses (Alexander and Lawson, 1981).

(c) Structural connections are required to transmit forces from one member to another, or one part of a structure to another. It is these connections that require calculations.

It is possible for a structural connection to be subject to any combination of axial force, shear force and bending moment in relation to three perpendicular axes. Generally in design the situation is simplified to forces in one plane. The distribution of forces within the connection is generally complex and complete knowledge and understanding is not available. In practice, simple conservative assumptions are made in calculations as shown in the following examples.

The forces acting on a structural connection must be consistent with the structural analysis of the whole structure. Connections are assumed to be either 'rigid', or 'pinned', or 'on rollers'. In reality, such connections do not exist but some assumptions must be made when analysing the complete structure. The analysis produces moments and forces and the members and connections can then be designed.

8.4.2 Simple Beam Supports (cl 9.2.1.4, EN)

Beams which are described as simply supported are assumed in structural analysis to be pin-ended. For real structures, the reaction is distributed on the bearing and there is a frictional force acting between the bearing surfaces.

At the supports of simply supported beams there are special requirements to ensure that the reinforcing bars are anchored (Fig. 9.3, EN). The bearing area and bearing stresses also require detailed consideration. In general, the length of bearing in line with the span (a_1) should not be less than recommended minimum values (Table 10.2, EN) which range from 25 to 140 mm.

EXAMPLE 8.2   Direct beam support. Determine the bearing length for a simply supported floor beam shown in Fig. 8.8. Design forces at the support $V_{Ed} = 250$ kN

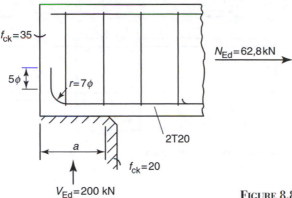

FIGURE 8.8   Example: direct beam support

and an axial force $N_{Ed} = 62,8\,\text{kN}$. The breadth of the beam $b_w = 300\,\text{mm}$, effective depth $d = 550\,\text{mm}$, concrete cover 60 mm. Concrete Grades C20/25 for the support and C35/50 for the beam.

Length of dry bearing (cl 10.9.5.2, EN),

$$a = V_{Ed}/(b\,0,4f_{cd}) = 250E3/(300 \times 0,4 \times 20/1,5) = 156,25\,\text{mm}.$$

Axial stress in the 2T20 tension bars in the beam at the support due to the tensile design force N_{Ed} and design shear force V_{Ed}

$$\sigma_{sd} = N_{Ed}/A_c + 0,5V_{Ed}(\cot\theta - \cot\alpha)/A_c$$
$$= 62,8E3/628 + 0,5 \times 200E3 \times 1/628 = 259\,\text{MPa}.$$

Basic anchorage length for a straight bar (Eq. 8.3, EN) for $\sigma_{sd} = 259\,\text{MPa}$

$$l_{b,req} = (\phi/4)(\sigma_{sd}/f_{bd}) = (20/4) \times (259/3,3) = 392\,\text{mm}$$

where the design bond stress for high bond bars (Eq. 8.2, EN)

$$f_{bd} = 2,25f_{ctk0,05}/\gamma_c = 2,25 \times 2,2/1,5 = 3,3\,\text{MPa}$$

and the value of $f_{ctk0,05}$ is obtained from Table 3.1, EN (Annex A2).

Design anchorage length (Eq. 8.4, EN) for $c_d = 3\phi = 3 \times 20 = 60\,\text{mm}$

$$l_{bd} = \alpha_1\alpha_2\alpha_3\alpha_4l_{b,req} = 0,7 \times 392 = 274\,\text{mm}$$

Check if greater than the minimum anchor length in tension

$$l_{b,min} = 0,3l_{b,req} = 0,3 \times 222 = 66,6;\ \text{or}\ 10\phi = 10 \times 20 = 200;\ \text{or}\ 100\,\text{mm}.$$

Length of support $= l_{bd} + \text{cover} + \text{chamfer} =$

$$274 + 60 + 15 = 349\,\text{mm, Use } 350\,\text{mm}.$$

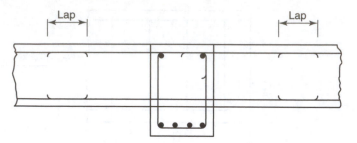

FIGURE 8.9   Beam-to-beam connection

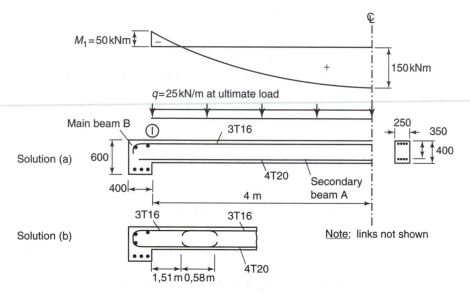

FIGURE 8.10   Example: beam-to-beam connection

Minimum length of bearing (Table 10.2, EN) for $\sigma_{Ed}/f_{cd} = 0,4$ for a floor, where $a_{1,min} = 30 < 350$ mm therefore satisfactory.

8.4.3 Beam-to-Beam Connections

Secondary beams of the same, or different, depths intersect main beams at right angles as shown in Fig. 8.9. This connection is assumed to be a 'rigid' connection in the structural analysis and therefore is designed to resist a bending moment, a shear force and sometimes an axial force.

Where secondary beams intersect the main beam from both sides the practical arrangement of steel is shown in Fig. 8.9. The lap lengths are the full design lengths. Alternative arrangements are shown in Figs 9.1 and 9.4, EN. Where only one secondary beam intersects the main beam the steel arrangement and design is more complicated (Fig. 8.10).

Similar arrangements can be used for the intersection of slabs and beams, except that the loop bars in the design example are arranged horizontally because of the reduced depth of slab.

EXAMPLE 8.3   Beam-to-beam connection. Design the connection between the secondary beam A and the main beam B shown in Fig. 8.10. The loading and bending moment diagram are shown above the diagram. Assume C30/35 Grade concrete.

Solution (a) (Fig. 8.10a)

Area of bars required to resist the hogging bending moment and shear force at point 1

$$A_s = M/(f_{yd}z) + \Delta F_{td}/f_{yd} = (1/f_{yd})(M/z + 0.5 V_{Ed})$$
$$= (1.15/500)[50E6/(0.9 \times 350) + 0.5 \times 100E3] = 480\,\text{mm}^2.$$

Use 3T16 (603 mm²), and extended into the top of the beam with a 90° bend.

Area of steel (603 mm²) is greater than the minimum (480 mm²) therefore the reduced axial steel stress

$$\sigma_{sd} = 480/603 \times 500/1.15 = 346\,\text{MPa}.$$

Basic anchorage length (Eq. 8.3, EN)

$$l_{b,req} = (\phi/4)(\sigma_{sd}/f_{bd}) = 16/4 \times 346/3 = 461\,\text{mm}$$

where the design bond stress for high bond bars (Eq. 8.2, EN)

$$f_{bd} = 2.25 f_{ctk0.05}/\gamma_c = 2.25 \times 2/1.5 = 3\,\text{MPa}.$$

Design anchorage length (Eq. 8.4, EN) if cover $c_d = 50$ mm

$$l_{bd} = \alpha_1\alpha_2\alpha_3\alpha_4 0.7 l_{b,req} = 0.7 \times 461 = 323\,\text{mm}$$

Design anchorage length + cover $= 323 + 50 = 373 < b$ (400) mm, acceptable.

Solution (b) (Fig 8.10 (b))

The essence of this variation is the insertion of 3T16 mm loop bars at the end of beam A as shown in Fig. 8.10(b).

Theoretical curtailment point of the 16 mm bars in the bottom of the secondary beam A at a distance x from the inside face of the main beam B.

Moments of resistance in kNm ($z = 0.9 \times 350 = 315$ mm),

$$M_r = A_s z(f_{yd} - \Delta F_{td}/A_s) = A_s z(f_{yd} - 0.5 V_{Ed}/A_s)$$
$$= A_s z[f_{yd} - 0.5/A_s(V - qx)] = A_s z f_{yd} - 0.5z(V - qx)]$$
$$= [603 \times 315 \times 500/1.15 - 0.5 \times 315 \times 100 + 0.5 \times 315 \times 25x]/1E6 \qquad \text{(i)}$$

$$M_{Ed} = R_1 x - M_1 - wx^2/2 = 100x - 50 - 25x^2/2 \tag{ii}$$

Equating (i) and (ii)

$$x^2 - 7{,}685x + 9{,}34 = 0; \quad \text{hence } x = 1{,}51\,\text{m}.$$

Basic required lap length for a fully stressed 16 mm diameter bar

$$l_{b,req} = (\phi/4)(\sigma_{sd}/f_{bd}) = 16/4 \times (500/1{,}15)/3 = 580\,\text{mm}$$

where the design bond stress for high bond bars (Eq. 8.2, EN)

$$f_{bd} = 2{,}25 f_{ctk0{,}05}/\gamma_c = 2{,}25 \times 2/1{,}5 = 3\,\text{MPa}.$$

Design lap length (Eq. 8.10, EN) assuming more than 30 per cent lapped in tension

$$l_o = \alpha_1 \alpha_2 \alpha_3 \alpha_4 \alpha_5 \alpha_6 l_{b,req} \geq 1 \times l_{b,req} = 580\,\text{mm}$$

which is greater than (Eq. 8.11, EN)

$$\begin{aligned} l_{o,min} &> (0{,}3\alpha_6 l_{b,req}; \ 15\phi; \ 200\,\text{mm}) \\ &= (0{,}3 \times 580; \ 15 \times 16; \ 200) = (174; \ 240; \ 200)\,\text{mm}. \end{aligned}$$

The arrangement of reinforcement is shown in Fig. 8.10(b).

8.4.4 Beam-to-Column Connections

Beams and columns, of the same or different widths, intersect at right angles as shown in Fig. 8.11. These connections are assumed to be 'rigid' in the theoretical structural analysis and therefore are designed to resist a combination of bending moment, shear force and axial force. The connections are most efficient when the intersecting members contain approximately 1 per cent reinforcement.

The two member connection shown in Fig. 8.11(a) occurs at the top of frame and may be subject to moments which open or close the connection. Loops are an adequate form of reinforcement when the moments are closing the joint, and additional diagonal reinforcement is required for opening moments arranged as shown.

The three member connection (Fig. 8.11(b)) may be reinforced with bends or with loops (Fig. 8.11(c)). Experiments (Taylor, 1974) have shown that the closing corner (Fig. 8.11(b)) is stronger than an opening corner.

The four member connection (Fig. 8.11(d)) is straightforward and reinforced with straight bars lapped at construction joints.

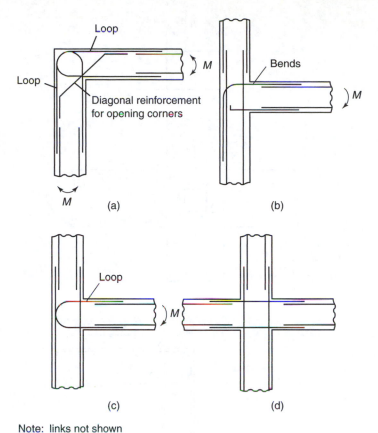

FIGURE 8.11   Beam-to-column connections

The three and four member connections may require checking for diagonal cracking in the joint block (Fig. 8.12) as follows. The shear stress $\tau = T/(bh)$ and axial stress $\sigma = N/(bh)$ are combined using a failure criterion (cl 12.6.3, EN). Where this requirement is not satisfied diagonal stirrups are placed to resist the tensile force, or alternatively the size of the joint block is increased.

EXAMPLE 8.4   Beam-to-column connection. Design the two member connection as shown in Fig. 8.13 which is subject to a moment of 70 kNm which may open or close the connection.

The beam and column are reinforced with 2T16 on both faces and lapped using 2T16 loop bars. Assume C25/30 Grade concrete.

Basic anchorage length based on maximum design strength (Eq. 8.3, EN)

$$l_{b,req} = (\phi/4)(\sigma_{sd}/f_{bd}) = (16/4) \times (500/1{,}15)/2{,}7 = 645 \text{ mm}.$$

where the design bond stress for high bond bars (Eq. 8.2, EN)

$$f_{bd} = 2{,}25 f_{ctk0{,}05}/\gamma_c = 2{,}25 \times 1{,}8/1{,}5 = 2{,}7 \text{ MPa}.$$

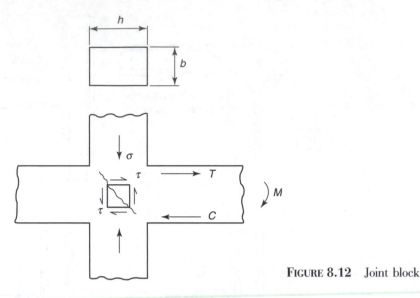

FIGURE 8.12   Joint block

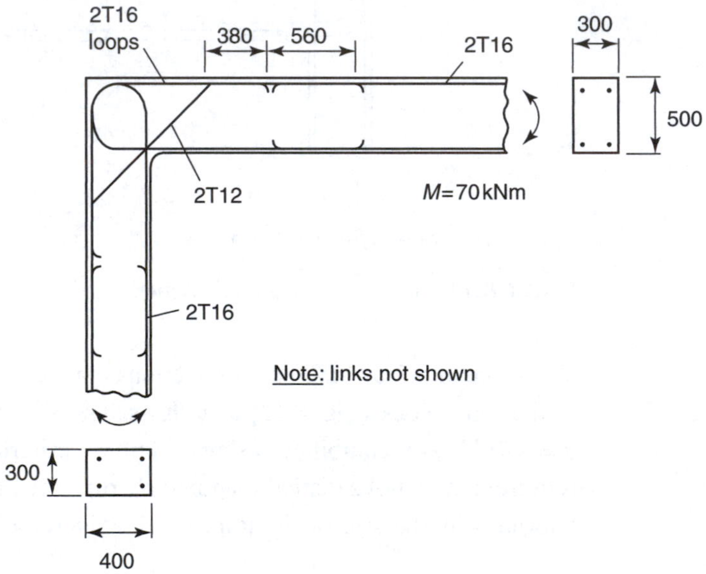

FIGURE 8.13   Example: beam-to-column connection

Design lap length for a straight bar, cover $c_d = 30$ mm (Eq. 8.10, EN)

$l_o = \alpha_1\alpha_2\alpha_3\alpha_4\alpha_5\alpha_6 l_{b,req} \geq 1 \times 0{,}87 \times 645 = 561$ mm, Use 560 mm.

$\alpha_1 = 1$ and $\alpha_3, \alpha_4, \alpha_5, \alpha_6$ not applicable

$\alpha_2 = 1 - 0{,}15(c_d - \phi)/\phi = 1 - 0{,}15(30 - 16)/16 = 0{,}87$.

Check if greater than the minimum lap length in tension (cl 8.7.3, EN)

$l_{b,min} = 0{,}3\alpha_6 l_{b,req} = 0{,}3 \times 1 \times 483 = 145$ mm; $15\phi = 15 \times 16 = 240$ mm;   200 mm.

The moment opens the connection therefore use diagonal bars. Area approximately 50 per cent of the main steel. Use 2T12.

Basic anchorage length based on maximum design strength (Eq. 8.3, EN)

$$l_{b,req} = (\phi/4)(\sigma_{sd}/f_{bd}) = (12/4) \times (500/1{,}15)/2{,}7 = 483\,\text{mm}$$

where the design bond stress for high bond bars (Eq. 8.2, EN)

$$f_{bd} = 2{,}25 f_{ctk0{,}05}/\gamma_c = 2{,}25 \times 1{,}8/1{,}5 = 2{,}7\,\text{MPa}.$$

Design lap length for a straight bar (Eq. 8.10, EN)

$$l_o = \alpha_1 \alpha_2 \alpha_3 \alpha_4 \alpha_5 \alpha_6 l_{b,req} \geq 1 \times 0{,}775 \times 483 = 374\,\text{mm, Use 380\,mm}$$
$$\alpha_1 = \alpha_3 = \alpha_4 = \alpha_5 = \alpha_6 = 1$$
$$\alpha_2 = 1 - 0{,}15\,(c_d - \phi)/\phi = 1 - 0{,}15(30 - 12)/12 = 0{,}775.$$

The column and beam should be checked for axial load and shear forces and links added. The arrangement of reinforcement to resist the moment is shown in Fig. 8.13 but the links are omitted for clarity.

8.4.5 Column-to-Base Connections

Columns generally intersect a base at right angles as shown in Fig. 8.14. This connection is assumed to be a 'rigid' connection in structural analysis and therefore is designed to resist a bending moment, a shear force and axial force.

The thickness of the base is related to the bond length required for the starter bars, or the shear resistance of the base. The starter bars are hooked behind the longitudinal bars in the base to reduce the design bond length. If the base is large and the thickness of the base has to be increased, consideration should be given to a pedestal base (Fig. 8.14). The lapped bars start approximately 75 mm above the top surface of the base to provide a construction joint and provide a 'kicker' to locate the column shuttering.

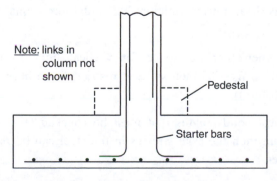

Note: links in column not shown

Pedestal

Starter bars

FIGURE 8.14 Column-to-base connection

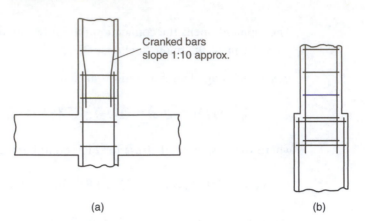

(a)

(b)

FIGURE 8.15   Column-to-column connections

8.4.6 Column-to-Column Connections

Column of the same, or different sizes, are connected in line a shown in Fig. 8.15. Connections are assumed to be 'rigid' in the structural analysis and therefore designed to resist a bending moment, shear force and an axial force. Calculations are required for the lap length. The percentage reinforcement at the lap may be high but should not be greater than 8 per cent.

8.4.7 Corbel Connection (Figs J6, 6.4(b) and 6.6, EN)

Corbels are brackets designed to support eccentric loads on columns (Fig. 8.16(a)). There are two possible types of failure as shown. Crack 1-2 is the one that is reinforced in design calculations. Crack 1–3 is possible if the load is close to the end of the bracket, but this is generally avoided because of limitations on bond length, and no calculations are required.

The bending forces acting about A are resisted by the main steel in the top of the corbel (Fig. 8.16(b)). This can take the form of horizontal or vertical links but must be well anchored on both sides of the crack 1-2. The tensile steel in the top of the corbel should be capable of resisting the bending moment due to the eccentricity of the load, plus shear, plus a minimum horizontal force equal to 20 per cent of the vertical load.

If $a_c/h_c > 0,5$ then there is space for vertical links to intercept crack 1-2 (Fig. 8.16(b)). If $a_c/h_c \leq 0,5$ then space is restricted and horizontal links are more effective (Fig. 8.16(c)).

There are critical bond lengths that must be checked (Fig. 8.16(d)). If there is insufficient room then the axial stresses in the steel can be reduced by increasing the area of steel.

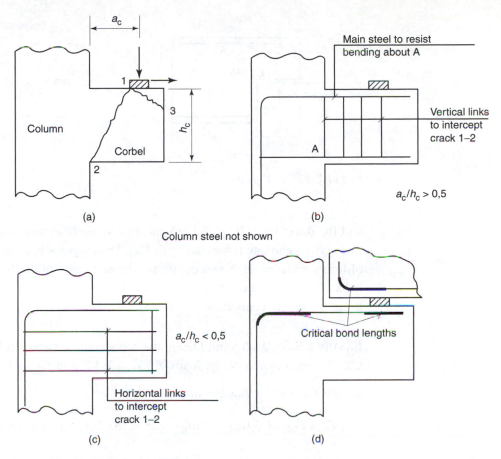

FIGURE 8.16   Corbel connections

Alternative shapes of corbels and arrangements of reinforcement are given in tests (Somerville and Taylor, 1972).

The shear resistance is calculated as for a singly reinforced concrete cantilever with an enhanced shear strength based on the effective depth/shear span ratio (cl 6.2.2(6), EN). Shear reinforcement is added to resist all of the shear. Typical calculations are shown in Example 7.10.

8.4.8 Half Joint (cl 10.9.4.6, EN)

This type of connection is useful when the construction depth is to be kept to a minimum, e.g. a bean supported on cantilever beams in line. The general shape and arrangement of reinforcement in a half joint is shown in Fig. 8.17.

The reinforcement is required to prevent the type of failure as shown in Fig. 8.17(a) (Reynolds, 1969). The horizontal link, or loop, resists the moment and shear of V_{Ed}

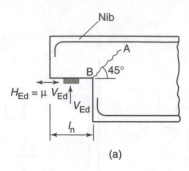
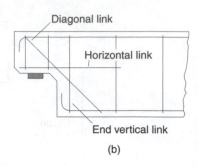

(a)   (b)

FIGURE 8.17   Half-joint

and the direct force H_{Ed}. The link must be securely anchored in the nib and beam. The vertical end link resists the force V_{Ed}. The diagonal link also resists the moment of forces about point A and duplicates the resistance of the horizontal and vertical links. The prime function of the diagonal link is to strengthen an opening joint as in beam to column connections.

EXAMPLE 8.5   Half joint. Design the connection as shown in Fig. 8.18(a). Assume C25/30 concrete and design loads of $V_{Ed} = 35$ kN and $H_{Ed} = 17,5$ kN.

Tensile force in the horizontal link

$$F_s = V_{Ed}e/d + H_{Ed} + 0,5V_{Ed} = 35 \times 165/250 + 17,5 + 0,5 \times 35 = 58,1 \text{ kN}.$$

Area of steel required in a horizontal link to resist F_s

$$A_s = F_s/f_{yd} = 58,1\text{E}3/(500/1,15) = 134 \text{ mm}^2$$

Use single horizontal link $2T12 = 226 \text{ mm}^2$

Reduced axial steel stress

$$\sigma_{sd} = 134/226 \times 500/1,15 = 258 \text{ MPa}.$$

Basic anchorage length for the 12 mm link (Eq. 8.3, EN)

$$l_{b,req} = (\phi/4)(\sigma_{sd}/f_{bd}) = (12/4) \times (258/2,7) = 287 \text{ mm}$$

where the design bond stress for high bond bars (Eq. 8.2, EN)

$$f_{bd} = 2,25f_{ctk0,05}/\gamma_c = 2,25 \times 1,8/1,5 = 2,7 \text{ MPa}.$$

Design bond length of the horizontal link into the nib (Eq. 8.4, EN)

$$l_{bd} = \alpha_1\alpha_2\alpha_3\alpha_4 l_{b,req} = 0,7 \times 287 = 201 \text{ mm}.$$

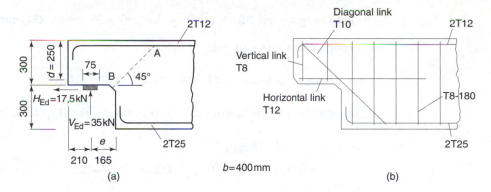

FIGURE 8.18   Example: half-joint

Minimum bond length in tension (cl 8.4.4, EN)

$$l_{b,min} = 0,3l_{b,req} = 0,3 \times 287 = 86\,mm; \quad 15\phi = 15 \times 12 = 180\,mm; \quad 200\,mm.$$

Bond length available for

nib $= e + 210 - cover = 165 + 210 - 50 = 325 > l_{bd}(201)\,mm$, acceptable.

Area of steel required in the end vertical single link to resist V_{Ed}

$$A_{sw} = V_{Ed}/(f_{yd}) = 35E3/(500/1,15) = 80,6\,mm^2$$

Use single end vertical T8 link $A_{sw} = 101\,mm^2$

Area of steel required in a diagonal 45° link, ignoring the horizontal and vertical links taking moments about A

$$A_{sw} = M_A/(f_{yd}z) = 18,9E6/(500/1,15 \times 0,9 \times 250 \times \sqrt{2}) = 137\,mm^2$$

Use single (2 legs) diagonal T10 link $A_{sw} = 157\,mm^2$

Basic anchorage length for the 10 mm diagonal link (Eq. 8.3, EN)

$$l_{b,req} = (\phi/4)(\sigma_{sd}/f_{bd}) = 10/4 \times (500/1,15)/2,7 = 402\,mm$$

where the design bond stress for high bond bars (Eq. 8.2, EN)

$$f_{bd} = 2,25f_{ctk0,05}/\gamma_c = 2,25 \times 1,8/1,5 = 2,7\,MPa.$$

Design bond length (Eq. 8.4, EN)

$$l_{bd} = \alpha_1\alpha_2\alpha_3\alpha_4 l_{b,req} = 0,7 \times 402 = 281\,mm.$$

Check if greater than the minimum bond length in tension (cl 8.4.4, EN)

$$l_{b,min} = 0,3l_{b,req} = 0,3 \times 402 = 121\,mm; \quad 10\phi = 10 \times 10 = 100\,mm; \quad 100\,mm.$$

Bond length available $= \sqrt{2}d = \sqrt{2} \times 250 = 354 > l_{bd}(281)$ mm, acceptable.

Design shear stress on nib

$$v_{Ed} = V_{Ed}/(b_w z) = 35E3/(400 \times 250) = 0,35 \text{ MPa}.$$

Design shear stress resistance of nib without links (Eq. 6.2a, EN)

$$v_{Rd,c} = C_{Rd,c}k(100\rho_l f_{ck})^{1/3} + 0,15\sigma_{cp}$$
$$= 0,18/1,5 \times 1,89(100 \times 0,0026 \times 25)^{1/3} - 0,15 \times 0,146 = 0,40 > v_{Ed}(0,35)\text{ MPa}$$

where

$$\sigma_{cp} = -H_{Ed}/A_c = -17,5E3/(400 \times 300) = -0,146 \text{ MPa}.$$

Minimum shear resistance

$$(v_{min} + k_1\sigma_{cp}) = 0,035k^{1,5}f_{ck}^{0,5} + k_1\sigma_{cp} = 0,035 \times 1,89^{1,5} \times 25^{0,5} - 0,15 \times 0,146$$
$$= 0,433 > V_{Rd,c}(0,40)\text{ MPa, add stirrups.}$$

Depth factor

$$k = 1 + (200/d)^{0,5} = 1 + (200/250)^{0,5} = 1,89$$

Longitudinal steel ratio

$$\rho_1 = 226/(400 \times 250) = 0,00226 < 0,02 \text{ (cl 6.4.4, EN), acceptable.}$$

Minimum area of tensile steel (cl 9.2.1.1, EN)

$$p_{min} = 0,26f_{ctm}/f_{yk} = 0,26 \times 2,6/500 = 0,00135 < \rho_1(0,00226), \text{ acceptable.}$$

Shear stress resistance $(\theta = 45°)$ with T8-180 links (Eq. 6.8, EN)

$$v_{Rd,s} = A_{sw}/s_l z f_{ywd} \cot\theta/(b_w d)$$
$$= 2 \times 50/180 \times 0.9 \times 250 \times 0,8 \times 500 \times 1/(400 \times 250) = 0,50 > v_{Ed}(0,35)\text{ MPa.}$$

Maximum shear stress resistance (Eq. 6.9, EN)

$$v_{Rd,max} = [\alpha_{cw}b_w z v_1 f_{cd}/(\cot\theta + \tan\theta)]/(b_w d)$$
$$= [1 \times 400 \times 0,9 \times 250 \times 0,6 \times 25/1,5/(1+1)]/(400 \times 250)$$
$$= 4,5 > v_{Ed}(0,35)\text{ MPa, acceptable.}$$

Maximum spacing of links (Eq. 9.6, EN)

$$s_{l,max} = 0,75d(1 + \cot\alpha) = 0,75 \times 250 = 187,5 > s_l(180)\text{ mm, acceptable.}$$

Shear reinforcement ratio (Eq. 9.4, EN)

$$\rho_w = A_{sw}/(s_1 b_w \sin\alpha) = 2 \times 50/(180 \times 400 \times 1) = 0{,}00139$$

Minimum shear reinforcement ratio (Eq. 9.5, EN)

$$\rho_{min} = 0{,}08 f_{ck}^{0,5}/f_{ywk} = 0{,}08 \times 25^{0,5}/500 = 0{,}0008 < \rho_w(0{,}00139), \text{ acceptable.}$$

Maximum effective area of shear reinforcement (Eq. 6.12, EN)

$$A_{sw,max} f_{ywd}/(b_w s_1) = 2 \times 50 \times 0{,}8 \times 500/(400 \times 180) = 0{,}555\,\text{MPa}$$
$$0{,}5\alpha_c v f_{cd} = 0{,}5 \times 1 \times 0{,}6 \times 25/1{,}5 = 5 > 0{,}555\,\text{MPa, acceptable.}$$

Compression reinforcement (Eq. 6.18, EN)

$$A_{sl} = 0{,}5 V_{Ed}(\cot\theta - \cot\alpha)/f_{yd}$$
$$= 0{,}5 \times 35E3 \times (1 - 0)/(500/1{,}15) = 40{,}3\,\text{mm}^2, \text{Use 2T12} = 226\,\text{mm}^2$$

REFERENCES AND FURTHER READING

Alexander, S.J. and Lawson, R.M. (1981) Movement design in buildings. *Cons. Ind. Res. and Info.*, Technical Note 107.

Paterson, W.S. and Ravenshill, K.R. (1981) Reinforcement connectors and anchorage, *Const. Ind. Res. And Info. Ass. Pub*, Report 92.

Reynolds, G.C. (1982) Bond strength of deformed bars in tension, *C. and C.A.*, Technical Report 548.

Reynolds G.C. (1969) The strength of half joints in reinforced concrete beams, *C. and C.A.*, Report 42.415.

Somerville, G. and Taylor, H.P.J. (1972) The influence of reinforcement detailing on the strength of concrete structures, *Struct. Eng.*, **50(1)**, 7–19.

Taylor, H.J.P. (1974) The behaviour of in-situ concrete beam to column joints, *C. and C.A*, Report 42.492, 188.

Williams, A. (1979) The bearing capacity of concrete loaded over a limited area, *C. and C.A.* Technical Report 526.

Pr EN 10080 (2006) Steel for the reinforcement of concrete – weldable reinforcing steel – General.

9 / Reinforced Concrete Columns

9.1 GENERAL DESCRIPTION

A reinforced column is a vertical member designed to resist an axial force and bending moments about the two principal axes. Most reinforced concrete columns are rectangular in section, but circular sections are not uncommon. This chapter covers the design of columns with a rectangular cross section where the larger dimension is not greater than four times the smaller dimension (cl 9.5.1, EN).

A typical arrangement of reinforcement for a column is shown in Fig. 9.1. Generally the reinforcement in the cross section is symmetrical. The vertical steel bars are placed near the outside faces to resist the bending moments which accompany the axial force. The column in Fig. 9.1, for example, is designed to resist bending moments about the Z-Z axis. Horizontal steel links at intervals along the length of the column prevent cleavage failure of the concrete, and to prevent buckling of the longitudinal steel bars as crushing of the concrete occurs at failure. Cover to the steel is decided from consideration of durability, bond and fire resistance.

Most columns are designed to resist an axial load and a bending moment about the major principal axis. However some columns, e.g. a corner column, are subject to biaxial bending and the resultant bending occurs about an axis skewed to the principal axes. In some cases, it is necessary to check the serviceability limit states of deflection and cracking.

9.1.1 Reinforcement (cls 9.5.2 and 9.5.3, EN)

There are limits for percentage steel and spacing of reinforcement to facilitate casting and to prevent premature failure. The minimum diameter of a longitudinal bar is 8 mm.

Minimum area of longitudinal steel to be effective is (cl 9.5.2, EN) $A_{s,min} = 0{,}10\, N_{Ed}/f_{yd}$ or $0{,}002A_c$ whichever is the greater.

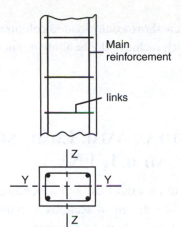

Main
reinforcement

links

FIGURE 9.1 Typical RC column

Maximum area of longitudinal steel

$A_{s,max} = 0{,}04A_c$ with twice this value at laps.

Transverse reinforcement, greater than 6 mm diameter, should be spaced at the least of the following distances ($s_{ci,max}$) (cl 9.5.3, EN)

(a) $20 \times$ minimum diameter of longitudinal bars,
(b) the lesser dimension of the column or
(c) 400 mm.

Additional detail requirements are given in cls 9.5.3 (4 to 6), EN.

9.2 THEORY FOR AXIALLY LOADED SHORT COLUMNS (cl 3.1.7, EN)

An axially loaded short column fails when the concrete crushes or the steel yields, or both. The ultimate axial load resistance

$$N = (\lambda\eta f_{ck}/\gamma_c)A_c + A_s(f_{yk}/\gamma_s) \tag{9.1}$$

where

A_c is the net cross-sectional area of the concrete,
A_s is the total area of the longitudinal reinforcement,
f_{ck} and f_{yk} are the characteristic strengths of the concrete and steel.
λ is the reduction factor for the depth of the rectangular stress block.
$\lambda = 0{,}8$ for $f_{ck} \leq 50$ MPa
$\lambda = 0{,}8 - (f_{ck} - 50)/400$ for $50 < f_{ck} \leq 90$ MPa
$\eta =$ is the strength factor for a rectangular stress block.
$\eta = 1{,}0$ for $f_{ck} \leq 50$ MPa
$\eta = 1{,}0 - (f_{ck} - 50)/200$ for $50 < f_{ck} \leq 90$ MPa.

In practice, however there are few short axially loaded columns, i.e. where $l_o/h < 12$ approximately, and a theory which includes the bending moment about a major axis is required.

9.3 THEORY FOR A COLUMN SUBJECT TO AN AXIAL LOAD AND BENDING MOMENT ABOUT ONE AXIS (cls 3.1.7 AND 6.1, EN)

Generally columns are subject to an axial load and a bending moment about one axis. The assumptions made in the design of sections in pure flexure can also be applied to column sections which are subject to a combination of flexure and axial force (ACI-ASCE Committee 441, 1966). Consider the section in Fig. 9.2 which is subject to an axial load and bending about the principal axis. Assuming

(a) a rectangular stress block,
(b) the neutral axis is within the section,
(c) both sets of reinforcement are in compression,
(d) compressive stress and strain are positive.

The section is subject to an axial force N acting along the centre line, and a bending moment M acting about the centroidal axis (Fig. 9.2). The compressive force on the concrete is $(\lambda \eta f_{ck}/\gamma_c)bx$. If f_1 and f_2 are the stresses (positive compression) in the reinforcement, then resolving forces horizontally

$$N = (\lambda \eta f_{ck}/\gamma_c)bx + f_1 A_1 + f_2 A_2 \tag{9.2}$$

Taking moments about the centre line,

$$M = (\lambda \eta f_{ck}/\gamma_c)bx(0{,}5h - \lambda/2x) + f_1 A_1(0{,}5h - d_1) - f_2 A_2(d - 0{,}5h) \tag{9.3}$$

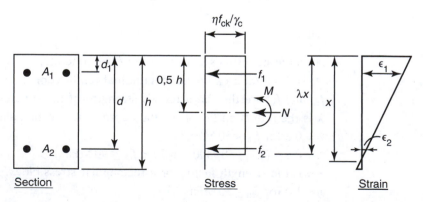

FIGURE 9.2   Distribution of stress and strain for a column section

In the case of a symmetrical section, total reinforcement (A_s), the equations can be simplified by putting $A_1 = A_2 = A_s/2$ and $d_1 = h - d$, as follows:

$$N = (\lambda \eta f_{ck}/\gamma_c)bx + (f_1 + f_2)A_s/2 \tag{9.4}$$

$$M = (\lambda \eta f_{ck}/\gamma_c)bx(0{,}5h - \lambda/2x) + (d - 0{,}5h)(f_2 - f_1)A_s/2 \tag{9.5}$$

In practice, the values of the axial force (N) and the bending moment (M) are known and the problem is to determine the required area of steel (A_s) for an assumed size of section and material strengths. The other unknowns in the equations are the depth x of the neutral axis and the steel stresses f_1 and f_2. The latter values can be expressed in terms of x by considering the strain diagram in Fig. 9.2. From similar triangles,

$$\varepsilon_1/(x - h + d) = 0{,}0035/x = \varepsilon_2/(x - d).$$

If the stress is below the design strength, its value in MPa may be obtained by multiplying the above strains by the modulus of elasticity of the steel (E_s). Hence,

$$f_1 = 0{,}0035 \times E_s(x - d + h)/x \tag{9.6}$$

$$f_2 = 0{,}0035 \times E_s(x - d)/x \tag{9.7}$$

When the stress is at the design strength, the right-hand sides of the above equations become $+$ or $-f_{yk}/\gamma_s$.

Eqs (9.4) to (9.7) can be solved to obtain the area of the steel (A_s) for particular values of the axial load (N) and bending moment (M), but the solution is difficult and requires a trial and error method. It is more convenient to use design graphs derived from these equations. A typical graph where $N/(bh f_{ck})$ is plotted against $M/(bh^2 f_{ck})$ for values of $\alpha = A_s f_{yk}/(bd f_{ck})$ is provided in Annex B3. It should be noted that the material factors for the concrete and steel are incorporated in the theory.

9.4 THEORY FOR COLUMNS SUBJECT TO AXIAL LOAD AND BIAXIAL BENDING (cl 5.8.9, EN)

Some columns are subject to an axial load and biaxial bending, e.g. a corner column. The exact theory (Bresler, 1960) is complicated but graphs can be produced to facilitate design. Alternatively, the solutions form an approximate quarter pear-shaped relationship between the axial load and the biaxial bending moments (Fig. 9.3). Horizontal sections of this diagram at each value of axial load can be represented by (Eq. 5.39, EN)

$$(M_{Edz}/M_{Rdz})^a + (M_{Edy}/M_{Rdy})^a \leq 1 \tag{9.8}$$

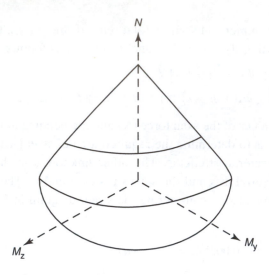

FIGURE 9.3 Biaxial bending of a column

where $a = 2$ for circular and elliptical sections. For rectangular sections

N_{Ed}/N_{Rd}	0,1	0,7	1,0
a	1,0	1,5	2,0

where $N_{Rd} = A_c f_{cd} + A_s f_{yd}$

9.5 TYPES OF FORCES ACTING ON A COLUMN (cl 5.8.9, EN)

A column is subject to an axial force and bending moments. These are broadly described as first order and second order forces and moments and it is important to distinguish them.

9.5.1 First Order Forces and Moments

The first order forces acting on a column are those obtained from the elastic or plastic analysis of the structure as shown in Chapter 4. In some cases, e.g. braced short columns with axial loading and bending moment, this is sufficient and design of the cross section can be carried out.

The first order forces in a column are the result of applied loads and sway in the structure, if it occurs. Some structures do not sway to any appreciable extent because of the existence of bracing or stiff elements, such as shear walls, as shown in Fig. 9.4(a).

First order bending moments may be induced into a column from;

(a) loading on the structure,
(b) column load not applied axially,

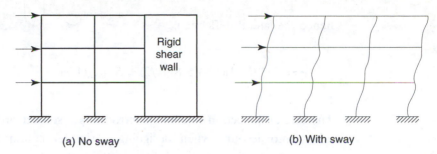

FIGURE 9.4 First order forces and moments

(c) column not perfectly straight and

(d) sway of the column.

9.5.2 Second Order Forces and Moments

In addition to first order forces and moments, there are second order forces and moments. If a column is slender second order forces arise from the lateral deflection from buckling.

The strength of a slender column is reduced by the transverse deflections (MacGregor *et al.*, 1970 and Cranston, 1972). If a column is very slender the deflections become very large and failure is by buckling before the concrete crushes or the steel yields.

9.6 BASIC THEORY FOR SECOND ORDER FORCES FROM BUCKLING

The expression given in the European Code (cl 5.8.8.2, EN) is based on theory derived from the fundamental bending equation $M/l = f/y = E/r$ and related to research by Cranston. The variation in M/EI over the height of a pin-ended column is shown in Fig. 9.5(b). Assuming a parabolic bending moment diagram and applying the area moment theorem (Croxton and Martin, 1990) between one end and mid-height

$$x(\delta y/\delta x) - y = Ax/(EI)$$

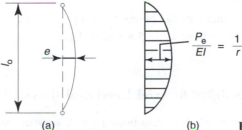

FIGURE 9.5 Second order eccentricity for a pin-ended strut

hence the lateral deflection

$$e = (2/3)(l_o/2)(1/r)(5/8)(l_o/2) = 5/48l_o^2(1/r) \qquad (9.9)$$

The lateral deflection (e), which introduces a second order moment Ne, must be taken into account when designing a slender column. The factor of 5/48 is approximately the factor of $1/c = 1/10$ in the European Code. The curvature $1/r = f/(Ey) = \varepsilon/(h/2)$ are values given in the code (cl 5.8.8.2, EN). An additional factor is introduced to allow for non-elastic behaviour such as creep and cracking.

9.6.1 Methods for Accounting for Second Order Forces

If second order effects are taken into account then there is a choice of methods.

(a) General method: This includes first and second order effects and is based on non-linear analysis, including geometric non-linearity i.e. second order effects, cracking and creep (cl 5.8.5, EN).
(b) Second order analysis based on nominal stiffness (cl 5.8.7, EN). Nominal values of flexural stiffness are used, taking into account material non-linearity, cracking and creep.
(c) Method based on nominal curvature (cl 5.8.8.1, EN). Suitable for isolated members with constant normal force and effective lengths. The method gives a second order moment (M_2) based on deflection, which is related to effective length and an estimated maximum curvature. The design moment $M_{Ed} = M_{OEd} + M_2$, where M_{OEd} is the first order moment and M_2 is the second order moment.

9.6.2 Conditions for Ignoring Second Order Effects

If a column is short and the axial load and the first order moment are small then second order effects can be ignored if, the

(a) second order moments are less than 10 per cent of the corresponding first order effects (cl 5.8.2(6), EN),
(b) slenderness of a member (λ) is less than a prescribed value (cl 5.8.3.1(1), EN).
(c) total vertical load is limited (cl 5.8.3.3(1), EN).

9.6.3 Columns Subject to Axial Load and Biaxial Bending (5.8.9, EN)

Some columns are subject to biaxial bending e.g. a corner column. For a complete analysis the general method may be used. This includes first and second order moments and is based on non-linear analysis, including geometric non-linearity i.e. second order effects, cracking and creep (cl 5.8.5, EN).

If ordinary elastic or plastic analysis is used and if the slenderness ratios are small and the bending moments are not large, then no further check is necessary, provided that (cl 5.8.9(3), EN) the slenderness ratios

$$\lambda_y/\lambda_z \le 2 \text{ and } \lambda_z/\lambda_y \le 2$$

and the eccentricity ratios (axes defined in Fig. 9.1)

$$(e_y/h)/(e_z/b) \le 0{,}2 \text{ or } (e_z/b)/(e_y/h) \le 0{,}2$$

If these conditions are not met then the methods described previously for uniaxial bending with second order effects may be applied about each axis (cl 5.8.9(4), EN). In the absence of design methods or graphs Eq. 5.39, EN may be used to determine the reinforcement.

9.6.4 Effective Length of a Column (cl 5.8.3.2, EN)

Included in Eq. (9.9) is the effective length (l_o). This modifies the actual length of the strut to allow for the rotation and translation at the ends as shown (Fig. 9.6). The basic value of the effective length equal to unity is the Euler value for a pin-ended strut. Values for other end conditions vary.

Two methods for determining the effective length in the European Code:

(1) Estimate of the effective length (l_o), (Fig. 9.6) and Fig. 5.7, EN;
(2) Calculate effective lengths using Eqs 5.15 (braced) and 5.16 (unbraced), EN for members in regular frames. These values are based on research for columns in multi-storey buildings (Wood, 1974).

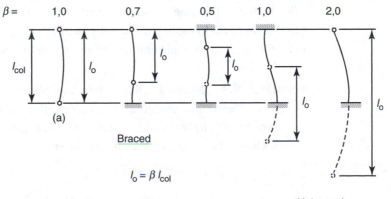

FIGURE 9.6 Effective height of an ideal column

9.6.5 Slenderness Ratio of a Column (cl 5.8.3.2, EN)

A better measure of the susceptibility to buckling is the slenderness ratio which includes section properties of the column. The slenderness ratio

$$\lambda = \text{(effective length)}/\text{(radius of gyration)} = l_o/i = l_o l(I/A)^{0.5} \tag{9.10}$$

For rectangular sections, cross-sectional area $b \times h$, bending about the major axis, ignoring the steel reinforcement, the radius of gyration

$$i = (I/A)^{0.5} = (bh^3/12bh)^{0.5} = h/(2 \times 3^{0.5}) \tag{9.11}$$

9.7 SUMMARY OF THE DESIGN METHOD FOR COLUMNS SUBJECT TO AXIAL LOAD AND BENDING

Assuming that an elastic or plastic analysis of the structure has been carried out which gives only primary moments.

(1) Assume a size of column and material strengths for concrete and steel.
(2) Determine the first order moment (M_{OEd}) about the major axis based on the plastic or elastic analysis, of part, or whole of the structure.
(3) Check if first order moment about the major axis and eccentricity can be ignored.
(4) Assume area of longitudinal steel and calculate the second order buckling moment about the major axis $M_2 = N_{Ed}e_2$.
(5) Check if second order moment about the major axis can be ignored.
(6) Calculate the design moment about the major axis $M_{Ed} = M_{OEd} + M_2$.
(7) Confirm, or revise, the assumed area of reinforcement from a column graph (Annex B3) which relates $N_{Ed}/(bh\,f_{ck})$ and $M_{Ed}/(bh^2\,f_{ck})$.
(8) If biaxial bending, repeat bending calculations and checks about minor axis to determine the design moment.
(9) If biaxial bending, check whether combined actions should be considered.
(10) If no biaxial graph, use biaxial equation $(M_{Edz}/M_{Rdz})^a + (M_{Edy}/M_{Rdy})^a \leq 1$.
(11) Check limits for maximum and minimum reinforcement.

EXAMPLE 9.1   Isolated column bending about the major axis in a non-sway structure. Determine the steel reinforcement required in column in a non-sway structure subject to first order actions of 1750 kN axial load and a bending moment of 100 kNm. The height of the column is 3 m.

Following the steps in Section 9.7.

Step (1) Assume section 300×300 mm, Grade C50/60 concrete and $f_{yk} = 500$ MPa.

Step (2) Given the first order bending moment $= 100$ kNm.

Step (3) Check if the first order bending moment > minimum (cl 6.1(4), EN)

$$N_{Ed}h/30 = 1750 \times 0,3/30 = 17,5 < M_1 = 100\,\text{kNm, Use } 100\,\text{kNm.}$$

Check if first order eccentricity greater than minimum (cl 6.1(4), EN)

$$e_1 = M_{Ed}/N_{Ed} = 100\text{E}3/1750 = 57 > \text{min. } (20) > h/30(10)\,\text{mm. Use } e_1 = 57\,\text{mm.}$$

Step (4) Assume 4T16 (804 mm^2). Apply method (Eq. 5.33, EN) based on nominal curvature to determine the additional buckling second order moment

$$M_2 = N_{Ed}e_2.$$

Calculate the following factors to obtain a value for the second order buckling eccentricity (e_2)

Curvature (Eq. 5.34, EN)

$$
\begin{aligned}
(1/r) &= K_r K_\phi (1/r_0) = K_r K_\phi (f_{yd}/E_s)/(0,45d) \\
&= 0,745 \times 1,37 \times (500/1,15)/200\text{E}3/(0,45 \times 0,85 \times 300) = 19,4\text{E}{-}6 \\
K_r &= (n_u - n)/(n_u - n_{bal}) \text{ (Eq. 5.36, EN)} \\
&= (1,117 - 0,583)/(1,117 - 0,4) = 0,745 < 1 \\
n_u &= 1 + \omega = 1 + 0,117 = 1,117 \\
n &= N_{Ed}/(A_c f_{cd}) = 1750\text{E}3/(300 \times 300 \times 50/1,5) = 0,583 \\
\omega &= A_s f_{yd}/(A_c f_{cd}) = 804 \times (500/1,15)/(300 \times 300 \times 50/1,5) = 0,117
\end{aligned}
$$

Effective creep factor (Eq. 5.37, EN)

$$
\begin{aligned}
K_\phi &= 1 + \beta\phi_{ef} = 1 + 0,427 \times 0,867 = 1,37 > 1, \text{ satisfactory.} \\
\phi_{ef} &= \phi_{(\alpha,to)}M_{OEqp}/M_{OEd} = 1,3 \times 2/3 = 0,867 \quad \text{(cl 5.19, EN)} \\
\phi_{(\alpha,to)} &= 1,3 \quad \text{(cl 3.1.4(2), EN)} \\
\beta &= 0,35 + f_{ck}/200 - \lambda/150 = 0,35 + 50/200 - 26/150 = 0,427
\end{aligned}
$$

Slenderness ratio (cl 5.8.3.1, EN)

$$\lambda = l_0/i = 0,75 \times 3\text{E}3/[300/(2 \times 3^{0,5}] = 26$$

Second order buckling deflection of the column (cl 5.8.8.2(4), EN)

$$e_2 = [(1/r)l_0^2/c] = 19,4\text{E-}6 \times (0,75 \times 3\text{E}3)^2/10 = 9,82\,\text{mm.}$$

Second order moment (Eq. 5.33, EN)

$$M_2 = N_{Ed}e_2 = 1750 \times 9,82 \times 1\text{E-}3 = 17,2\,\text{kNm.}$$

Step (5) Check if the second order moment can be ignored.
Method (a) (cl 5.8.2(6), EN)

$M_2/M_{OEd} = 17{,}2/100 \times 100 = 17.2\% > 10\%$, therefore second order moment cannot be ignored.

Method (b) (cl 5.8.3.1, EN)

$$\lambda = 26$$
$$A = 1/(1+0{,}2\phi_{ef}) = 1/(1+0{,}2 \times 0{,}867) = 0{,}852$$
$$B = (1+2\omega)^{0{,}5} = (1+2\times 0{,}117)^{0{,}5} = 1{,}11$$
$$C = 1{,}7 - r_m = 1{,}7 - M_{01}/M_{02} = 1{,}7 - 0{,}7 = 1$$
$$n = N_{Ed}/(A_c f_{cd}) = 1750E3/(300 \times 300 \times 50/1{,}5) = 0{,}583$$
$$\lambda_{lim} = 20ABC/n^{0{,}5} = 20 \times 0{,}852 \times 1{,}11 \times 1/0{,}583^{0{,}5} = 24{,}8$$

$\lambda(26) < \lambda_{lim}(24{,}8)$, therefore second order effects cannot be ignored. Method(c) (cl 5.8.3.3, EN) is not applicable to this structure.

Step (6) Design values.

$$N_{Ed} = 1750\,\text{kN, and } M_{Ed} = M_1 + M_2 = 100 + 17{,}2 = 117{,}2\,\text{kNm.}$$

Step (7) Reinforcement is determined from the design graph (Annex B3)

$$M_{Ed}/(hb^2 f_{ck}) = 117{,}2E6/(300 \times 300^2 \times 50) = 0{,}0868$$
$$N_{Ed}/(bhf_{ck}) = 1750E3/(300 \times 300 \times 50) = 0{,}389$$
$$d/h = (300 - 30 - 16/2)/300 = 0{,}873.$$

From graph, for $d/h = 0{,}85$ and $f_{ck} = 50\,\text{MPa}$

$\alpha = A_{sc} f_y/(bhf_{ck}) = 0{,}5$ and hence

$A_{sc,req} = 0{,}05 \times 300^2 \times 50/500 = 450 < (4T16)804\,\text{mm}^2$, acceptable.

Use 4T16 longitudinal corner bars $(804\,\text{mm}^2)$.

Steps (8) to (10) Ignore because no biaxial bending.

Step (11) Check limits for reinforcement.

Maximum reinforcement ratio (cl 9.5.2(3), EN)

$A_s = 804 < A_{sc,max} = 0{,}04 \times 300^2 = 3600\,\text{mm}^2$, satisfactory

Minimum reinforcement ratio (cl 9.5.2.(2), EN)

$$A_{s,min} = 0{,}10 N_{Ed}/f_{yd} = 0{,}10 \times 1750E3/(500/1{,}15)$$
$$= 402{,}5 < (4T16)804\,\text{mm}^2, \text{ satisfactory}$$

or

$$0,002 A_c = 0,002 \times 300^2 = 180 < (4T16)804 \, \text{mm}^2, \text{ satisfactory.}$$

Transverse reinforcement (cl 9.5.3(3), EN)

$\phi/4 = 16/4 = 4 < 6$, use 6 mm.

Spacing of transverse reinforcement (cl 9.5.3(3), EN) the lesser value of $s_{cl,tmax} = 20\phi = 20 \times 16 = 320 \, \text{mm}$; or $b = 300 \, \text{mm}$; or 400 mm, Use $s = 300 \, \text{mm}$

EXAMPLE 9.2 Column subject to biaxial bending. A column in the bottom storey of a building sways in bending about the z-z axis but shear walls prevent sway about the y-y axis (Fig. 9.7).

Following the steps in section 9.7.

Step (1) Assume section $400 \times 800 \, \text{mm}$, C50/60 Grade concrete and $f_{yk} = 500 \, \text{MPa}$.

Step (2) From an elastic analysis the axial load is 2000 kN and the first order moments at the ends of the column are as shown (Fig. 9.7). This analysis does not give second order bending moments.

Step (3) Check if the first order bending moment > minimum (cl 6.1(4), EN)

$$N_{Ed} h/30 = 2000 \times 0,8/30 = 53,3 < M_1 = 150 \, \text{kNm, Use 150 kNm.}$$

Check if first order eccentricity is greater than minimum (cl 6.1(4), EN)

$e_1 = M_{Ed}/N_{Ed} = 150\text{E}3/2000 = 75 > \text{min. } (20) > h/30(800/30 = 26,7) \, \text{mm}$, Use $e_1 = 75 \, \text{mm}$

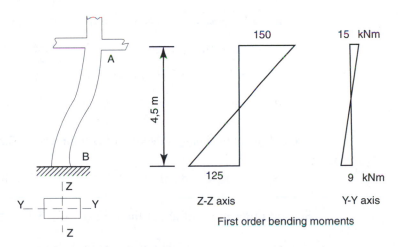

FIGURE 9.7 Example: column in biaxial bending

Step (4) Assume 8T25 ($3927\,\text{mm}^2$). Determine the second order bending ($M_{2z}=N_{Ed}e_{2z}$) about the z-z axis based on nominal curvature (Eq. 5.33, EN). Calculate the following to obtain a value for the second order deflection of the column (e_{2z}).

Estimated effective length of column bending about the z-z axis (Fig. 9.4)

$$l_o = 1{,}5 \times 4{,}5 = 6{,}75\,\text{m}$$
$$\lambda_z = l_o/i_z = 6{,}75\text{E}3/[800/(2 \times 3^{0{,}5})] = 29{,}2.$$

Curvature (Eq. 5.34, EN)

$$(1/r) = K_r K_\phi (1/r_o) = K_r K_\phi (f_{yd}/E_s)/(0{,}45d)$$
$$= 1 \times 1{,}25 \times (500/1{,}15)/200\text{E}3/(0{,}45 \times 0{,}85 \times 800) = 8{,}88\text{E-6}$$

which necessitates calculating the following factors (Eq. 5.36, EN)

$$K_r = (n_u - n)/(n_u - n_{bal}) = (1{,}16 - 0{,}188)/(1{,}16 - 0{,}4)$$
$$= 1{,}28 > 1, \text{ use } 1 \text{ (Eq. 5.36, EN)}$$
$$n_u = 1 + \omega = 1 + 0{,}16 = 1{,}16$$
$$n = N_{Ed}/(A_c f_{cd}) = 2000\text{E}3/(400 \times 800 \times 50/1{,}5) = 0{,}188$$

assuming 8T25 ($3927\,\text{mm}^2$)

$$\omega = A_s f_{yd}/(A_c f_{cd}) = 3927 \times (500/1{,}15)/(400 \times 800 \times 50/1{,}5) = 0{,}16$$

Effective creep factor (Eq. 5.37, EN)

$$K_\phi = 1 + \beta\phi_{ef} = 1 + 0{,}405 \times 0{,}867 = 1{,}35 > 1, \text{ satisfactory.}$$
$$\phi_{ef} = \phi_{(\alpha,to)} M_{OEqp}/M_{OEd} = 1{,}3 \times 2/3 = 0{,}867 \quad \text{(Eq. 5.19, EN)}$$
$$\phi_{(\alpha,to)} = 1{,}3 \quad \text{(cl 3.1.4(2), EN)}$$
$$\beta = 0{,}35 + f_{ck}/200 - \lambda/150 = 0{,}35 + 50/200 - 29{,}2/150 = 0{,}405$$

Second order buckling deflection of column (cl 5.8.8.2(4), EN)

$$e_{2z} = (1/r)l_o^2/c = 8{,}88\text{E-6} \times (1{,}5 \times 4{,}5\text{E}3)^2/10 = 40{,}5\,\text{mm}.$$

Second order moment about the z-z axis (Eq. 5.33, EN)

$$M_{2z} = N_{Ed}e_{2z} = 2000 \times 40{,}5 \times 1\text{E-3} = 81\,\text{kNm}.$$

Step (5) Check if second order effects can be ignored.

Method (a) Second order effects can be ignored if they are less than 10 per cent of the corresponding first order effects (cl 5.8.2(6), EN). This necessitates calculating the second order effects. Consider method (b) and (c) below as alternatives.

Second order moments have been calculated and $M_{2z}/M_{OEd} = 81/150 \times 100 = 54\% > 10\%$, unacceptable.

Method (b) As an alternative second order effects may be ignored if the slenderness $\lambda_z < \lambda_{lim}$ (cl 5.8.3.1, EN)

$$\lambda = l_o/i_z = 1,5 \times 4,5E3/[800/(2 \times 3^{0,5})] = 29,2$$
$$A = 1/(1 + 0,2\phi_{ef}) = 1/(1 + 0,2 \times 0,867) = 0,852$$
$$B = (1 + 2\omega)^{0,5} = (1 + 2 \times 0,16)^{0,5} = 1,15$$

assuming 8T25 (3927 mm^2)

$$\omega = A_s f_{yd}/(A_c f_{cd}) = 3927 \times 500/1,15/(400 \times 800 \times 50/1,5) = 0,16$$
$$C = 1,7 - r_m = 1,7 - M_{01}/M_{02} = 1,7 - 125/150 = 0,867$$
$$n = N_{Ed}/(A_c f_{cd}) = 2000E3/(400 \times 800 \times 50/1,5) = 0,188$$
$$\lambda_{lim} = 20ABC/n^{0,5} = 20 \times 0,852 \times 1,15 \times 0,867/0,188^{0,5} = 39,2$$

$\lambda(29,2) < \lambda_{lim}$ (39, 2), acceptable.

Method (c) Ignoring global second order effects in buildings (cl 5.8.3.3, EN). This is not applicable to this structure.

The second order effects could be ignored but calculations continue.

Step (6) Design actions about the z-z axis are

$$N_{Ed} = 2000 \text{ kN and } M_{Edz} = 150 + 81 = 231 \text{ kNm.}$$

Step (7) Considering these bending actions about the z-z axis only with three of the eight bars in tension

$$A_{sc} f_y/(b\,d f_{ck}) = (3/8 \times 3927) \times 500/(400 \times 0,85 \times 800 \times 50) = 0,0541$$

From graph (Annex B1) for a doubly reinforced beam in bending

$$M_{Rdz}/(b\,d^2 f_{ck}) = 0,046 \text{ and hence}$$
$$M_{Rdz} = 0,046 \times 400(0,85 \times 800)^2 \times 50/1E6 = 426 \text{ kNm.}$$

Step (8) Consider axial load and bending about the y-y axis and repeat steps (3) to (7).

Step (3) repeat. Check if the first order bending moment > minimum (cl 6.1(4), EN)

$$N_{Ed}h/30 = 2000 \times 0,45/30 = 30 > M_1 = 15 \text{ kNm, Use } M_1 = 30 \text{ kNm.}$$

Check if first order eccentricity is greater than minimum (cl 6.1(4), EN)

$$e_1 = M_{Ed}/N_{Ed} = 15E3/2000 = 7,5 < h/30(400/30 = 13,3) < \text{min. } (20)\,\text{mm, Use } e_1$$
$$= 20\,\text{mm and } M_1 = 20 \times 2E3 = 40\,\text{kNm.}$$

Estimated effective length of column bending about the y-y axis (Fig. 9.4) $l_o = 0,75 \times 4,5 = 3,375$ m.

Slenderness ratio

$$\lambda_y = l_o/i_y = 3,375E3/[400/(2 \times 3^{0,5})] = 29,2.$$

Step (4) repeat. Determine the second order bending ($M_2 = N_{Ed}e_{2y}$) about the y-y axis based on nominal curvature (Eq. 5.33, EN). The following quantities need to be calculated to obtain a value for the deflection of the column (e_{2y})

Curvature (Eq. 5.34, EN)

$$(1/r) = K_r K_\phi(1/r_o) = K_r K_\phi(f_{yd}/E_s)/(0,45d)$$
$$= 1 \times 1,25 \times (500/1,15)/200E3/(0,45 \times 0,85 \times 400) = 17,8E - 6$$

which includes factors (Eq. 5.36, EN)

$$K_r = (n_u - n)/(n_u - n_{bal}) = (1,16 - 0,188)/(1,16 - 0,4) = 1,28 > 1, \text{ Use } 1$$
$$n_u = 1 + \omega = 1 + 0,16 = 1,16$$
$$n = N_{Ed}/(A_c f_{cd}) = 2000E3/(400 \times 800 \times 50/1,5) = 0,188$$
$$\omega = A_s f_{yd}/(A_c f_{cd}) = 3927 \times (500/1,15)/(400 \times 800 \times 50/1,5) = 0,16$$

Effective creep factor (Eq. 5.37, EN)

$$K_\phi = 1 + \beta\phi_{ef} = 1 + 0,405 \times 0,867 = 1,35 > 1, \text{ satisfactory.}$$

which includes

$$\phi_{ef} = \phi_{(\alpha,to)} M_{OEqp}/M_{OEd} = 1,3 \times 2/3 = 0,867 \quad \text{(Eq. 5.19, EN)}$$
$$\phi_{(\alpha,to)} = 1,3 \text{ (cl 3.1.4(2), EN)}$$
$$\beta = 0,35 + f_{ck}/200 - \lambda/150 = 0,35 + 50/200 - 29,2/150 = 0,405$$

Buckling deflection of strut (cl 5.8.8.2(4), EN)

$$e_{2y} = (1/r)l_o^2/c = 17,8\text{E-}6 \times (0,75 \times 4,5\text{E3})^2/10 = 20,3\,\text{mm.}$$

Second order moment about the y-y axis (Eq. 5.33, EN)

$$M_{2y} = N_{Ed}e_{2y} = 2000 \times 20,3 \times 1\text{E-}3 = 40,6\,\text{kNm.}$$

Step (6) repeat. Design actions about the y-y axis are

$$N_{Ed} = 2000 \, kN \text{ and } M_{Edy} = 40 + 40{,}6 = 80{,}6 \, kNm.$$

Step (7) repeat. Considering these design bending actions acting about the y-y axis only with three of the eight bars in tension

$$A_s f_y/(b \, d \, f_{ck}) = (3/8) \times 3927 \times 500/(800 \times 0{,}85 \times 400 \times 50) = 0{,}0541$$

From graph (Annex B1) for a doubly reinforced concrete beam in bending $M_{Rdy}/(b \, d^2 \, f_{ck}) = 0{,}046$ from which

$$M_{Rdy} = 0{,}046 \times 800(0{,}85 \times 400)^2 \times 50/1E6 = 213 > 80{,}6 \, kNm, \text{ satisfactory.}$$

Step (9) Check whether combined actions about both axes should be considered. Check relative slenderness ratios and eccentricities (cl 5.8.9(3), EN).
Relative slenderness ratios

$$\lambda_y/\lambda_z = \lambda_z/\lambda_y = 29{,}2/29{,}2 = 1 < 2, \text{ satisfactory}$$

Relative eccentricities

$$(e_y/h)/(e_z/b) = [(231E3/2E3)/800)]/[55{,}6E3/2E3)/400)] = 2{,}07 > 0{,}2, \text{ unacceptable.}$$

or

$$(e_z/b)/(e_y/h) = [(55{,}6E3/2E3)/400)]/[231E3/2E3)/800)]$$
$$= 0{,}481 > 0{,}2, \text{ unacceptable.}$$

Further calculations are necessary because only two out of four conditions are satisfied.

Step (10) Combined bending actions about the z-z and y-y axes (Eq. 5.39, EN).

$$N_{Rd} = A_c f_{cd} + A_s f_{yd} = (400 \times 800 \times 50/1{,}5 + 3927 \times 500/1{,}15)/1E3 = 12374 \, kN$$
$$N_{Ed} = 2000 \, kN$$
$$N_{Ed}/N_{Rd} = 2000/12374 = 0{,}162$$

From table exponent (cl 5.8.9, EN)

$$a = 1 + 0{,}5(0{,}162 - 0{,}1)/0{,}6 = 1{,}05$$

Combining bending actions

$$(M_{Edz}/M_{Rdz})^a + (M_{Edy}/M_{Rdy})^a \leq 1$$
$$(231/426)^{1{,}05} + (80{,}6/213)^{1{,}05} = 0{,}886 < 1, \text{ acceptable}$$

Use 8T25 (3927 mm^2) longitudinal bars.

Step (11) Check limits for reinforcement
Check maximum reinforcement ratio (cl 9.5.2(3), EN)

$$A_s = 3927 < A_{s,\text{max}} = 0{,}04 \times 400 \times 800 = 12800 \, \text{mm}^2, \text{ satisfactory}$$

Check minimum reinforcement ratio (cl 9.5.2(2), EN)

$$A_{s,\text{min}} = 0{,}10 N_{\text{Ed}}/f_{\text{yd}} = 0{,}10 \times 2000\text{E}3/(500/1{,}15) = 460 < 3927 \, \text{mm}^2, \text{ acceptable}$$

or

$$0{,}002 A_c = 0{,}002 \times 400 \times 800 = 640 < 3927 \, \text{mm}^2, \text{ satisfactory}.$$

Transverse reinforcement (cl 9.5.3(3), EN)

$$\phi/4 = 25/4 = 6{,}25 > 6, \text{ use } 8 \, \text{mm}.$$

Spacing of transverse reinforcement (cl 9.5.3(3), EN) the lesser value of $s_{\text{cl,tmax}} = 20\phi = 20 \times 25 = 500 \, \text{mm}$; or $b = 400 \, \text{mm}$; or $400 \, \text{mm}$; use $s = 400 \, \text{mm}$.

REFERENCES AND FURTHER READING

ACI-ASCE Committee 441(1966). Reinforced concrete columns. *ACI Pub. SP-13*.

Bresler B. (1960). Design criteria for reinforced concrete columns under axial load and bending. *Proc. ACI*, 57, No. 5, Nov.

Cranston, W.B. (1972). Analysis and design of reinforced concrete columns. *C and C. Assoc.* Research Report No. 20.

Croxton, P.C.L. and Martin, L.H. (1990). *Solving Problems in Structures*. Longman Scientific and Technical.

Furlong, R.W. (1961). Ultimate strength of square columns under biaxially eccentric loads. *Proc. ACI*, 57, No. 9, March.

MacGregor, J.D., Breen, J.E. and Pfrang, E.O. (1970) Design of slender concrete columns. *Proc. ACI*, 67, No. 1, January.

Pannel, F.N. (1963). Failure surfaces for members in compression and biaxial bending. *ACI Jnl.*, Proc. Vol. 60, No. 1.

Warner, R.F. (1969). Biaxial Moment thrust curvature relations. *Jnl. Struct. Div., ASCE*, Vol. 95, No. ST5.

Wood, R.H. (1974). Effective lengths of columns in multi-storey buildings. *Struct. Eng.*, Vol. 52, No. 7, July.

This chapter covers design of slabs and provides an introduction to both Yield Line and Hillerborg Strip Methods. The definition of a slab is when the panel width is substantially greater than 5 times the overall slab thickness (cl 5.3.1 (4)). This distinction is unimportant for flexural design except that it is customary to consider unit widths in the calculations but becomes potentially critical for shear especially punching shear around columns for flat slabs.

10.1 TYPES OF SLAB

The various types of slab are illustrated in Fig. 10.1. The simplest type of slab is solid and of uniform thickness. This type of slab is potentially uneconomic in terms of efficient usage of materials owing to the inherent high ratio of permanent to variable actions. It is however relatively simple to construct as it has a flat soffit.

For large areas of slab, it is necessary to consider ways in which the self weight of the slab can be reduced without compromising its load carrying capacity. This reduction can be achieved in one of the two ways. The first is to place voids in the soffit of slab and thus creating a regular pattern of ribs in both directions. This type of slab is known as a waffle or coffered slab. The portions round the columns are left solid to counteract the effect of punching shear and to enhance resistance to hogging moments. Such slabs are often left with exposed soffits. The second method is to reduce the thickness of slab over most of the span where the shear is low but to keep an increased thickness at the support. Such a system is called a flat slab with drop-heads around the columns.

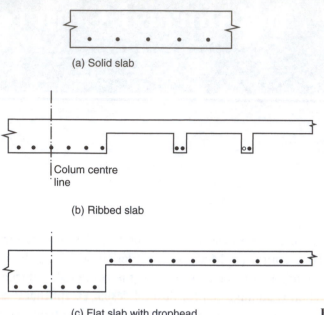

(a) Solid slab

Colum centre
line

(b) Ribbed slab

(c) Flat slab with drophead

FIGURE 10.1   Types of slab

10.2 DESIGN PHILOSOPHIES

10.2.1 Ultimate Limit States (cl 2.5.3.5.2)

A series of methods are available to determine the internal stress resultants:

(1) The Principles in Cl 5.5 indicate that linear elastic analysis with or without redistribution of moments, subject to limits on the neutral axis depths and the ratio between lengths of adjacent spans must lie between 0,5 and 2 if no check is to be made on rotation capacity.

(2) Section 5.6 indicates Upper or lower bound plastic methods may be used subject to the conditions that x_u is limited to 0,25 for concrete strength classes less than C50/60 or 0,15 above this value, the reinforcement is Class B or C, and the ratio of moments at intermediate supports to the span moments should lie between 0,5 and 2, and

(3) Non-linear numerical approaches. In practice the most common methods are the Johansen Yield Line approach which corresponds to an Upper Bound plastic method, the Hillerborg Strip to a Lower Bound plastic method and Finite Element Analysis (FEA). A linear elastic analysis may either be based on longitudinal and transverse frames or FEA loaded at ultimate limit state and the design then carried out on the resulting moments and shears. Non-linear analysis will not be considered.

Beeby (2004) indicates that currently high yield reinforcement of sizes below 16 mm may be Class A owing to their method of manufacture. Since smaller bar sizes than this are more likely to be used in slabs, it is essential where plastic methods are used for the analysis that Class B or C reinforcement is specified and that this is checked on site before the reinforcement is fixed or concreting has commenced. It is suggested that even for the Hillerborg Strip Method, Class B or C should specified.

10.2.2 Serviceability Limit States

As deemed to satisfy, clauses are normally used to design against excessive deflection or cracking a full analysis is rarely necessary.

10.3 PLASTIC METHODS OF ANALYSIS

For a plastic method of analysis to give the true collapse load the following three criteria must, by the uniqueness theorem, be satisfied:

- Equilibrium
  The internal forces, or stress resultants, must be in equilibrium with the applied loading.

- Yield
  The moments in any part of the structure must be less than or equal to the ultimate, or yield, moment at that point in the structure.

- Mechanism Formation
  In order that partial or total collapse can occur, a mechanism must occur.

For a frame structure comprising linear elements it is relatively easy to determine a situation which satisfies all the three criteria. For slabs it is less easy, and therefore acceptable methods which satisfy only two of the criteria have been developed. Each of these methods must satisfy equilibrium. It is NOT permissible to generate solutions which are not in equilibrium with the applied loading.

10.3.1 Lower Bound Solution

Here a set of moments are found which are in equilibrium with the applied loading and satisfy the yield criterion. Such a set of moments will not cause collapse to occur. An example of such a solution is the result of an elastic analysis under ultimate loads with no redistribution. A design method was formulated by Hillerborg and is known as the Hillerborg Strip Method (Wood and Armer, 1968).

10.3.2 Upper Bound Solution

A set of moments is found that are in equilibrium and form mechanisms but which may violate the yield criterion. This type of solution was studied extensively by

Johansen and gave rise to the Johansen Yield Line Method (Johansen, 1972). Each of these methods will be considered in turn.

10.4 JOHANSEN YIELD LINE METHOD

It has been observed in tests that when slabs are loaded to failure, large cracks occur normal to the directions of maximum moments. These cracks are observed only on the tension zones of the slab and are caused by the reinforcement yielding and large ductile rotations occurring (similar to those in under-reinforced beams). The sections between the yield lines undergo rigid body rotation (or displacement) and thus the concept of a failure mechanism can be postulated. Although the yield line approach is an upper bound solution in that a yield criterion is not checked, the results from such an analysis are satisfactory since the analysis ignores both the effects of strain hardening in the reinforcement as the reinforced concrete slab is assumed to behave as a rigid, elastic, perfectly plastic material and any beneficial effect generated by membrane action caused by in-plane forces in the slab. Both these phenomena increase the load carrying capacity beyond the level calculated, assuming the loading is carried by flexure to the extent that the calculated load carrying capacity is below the actual. It is essential that all potential mechanisms are checked especially where point loads occur or where a sagging yield line intersects the angle between two hogging yield lines.

10.4.1 Methodology

A possible yield mechanism is proposed and the load that may be carried is determined in terms of the moment capacity of the slab, or the moment field required is calculated in terms of a predetermined loading. It is often necessary to characterize the yield line pattern by a series of unknown dimensions. Symmetry and the existence of point loads may fix some of the yield line patterns. The resultant algebraic expressions involving the loading, moment field and unknown dimensions may be solved by either explicit solutions in relatively simple cases or by numerical techniques or spreadsheets where an explicit solution is not possible or difficult.

There are two methods of determining the inter-relations between the loading, moments and yield line patterns: Virtual work approach or Nodal force approach.

10.4.2 Virtual Work Approach

In this method, the equilibrium equation is set up by imposing virtual deformations to the yield lines and determining the (virtual) work done by the loads and the moments.

10.4.3 Nodal Force Approach

In this method, nodal forces due to internal shears are calculated at the points where the yield lines intersect, other than under point loads or at supports. Equilibrium equations are then formulated for each portion of the slab, the unknown dimensions characterizing the yield lines eliminated and the resultant relationship between the loads and moments determined.

Attention was drawn earlier to the situation where a sagging yield line intersects the angle between two hogging yield lines. In this situation the yield criterion is automatically violated and thus alternatives need to be sought. It should be noted that such alternatives often involve the use of curved rather than straight yield lines. The situation involving the intersection of hogging and sagging yield lines may however often produce a solution which is still acceptable in terms of the needs of design.

10.4.4 Postulation of the Layout of Yield Line Mechanisms

A basic series of rules were laid out by Johansen (1972) for situations not involving the formation of curved yield lines:

  (i) Yield lines are normally straight and form axes of rotation.
 (ii) Yield lines must terminate at a boundary unless parallel to a support.
(iii) Axes of rotation coincide with simply supported edges, cut free edges and pass over column supports.
 (iv) Axes of rotation of adjacent regions must either meet at a point or be parallel.
  (v) Hogging yield lines will form along encastré supports.

The examples shown in Fig. 10.2 illustrate the application of these rules.

Note, in all the examples on yield line and Hillerborg strip, the conventional notation of m with no subscripts has been retained for the moment per unit width.

Reinforcement is generally provided in an orthogonal pattern (unless the slab is a very odd shape) and thus a yield line is likely to intersect the reinforcement at an angle (Fig. 10.3).

If the reinforcement in a particular direction has a moment of resistance m per unit width normal to the direction of the reinforcement, the moment across the yield line m_{n1} can be obtained by resolving forces,

$$m_{n1} L = [m_{Rd}(L \sin \theta)] \sin \theta \qquad (10.1)$$

or

$$m_{n1} = m_{Rd} \sin^2 \theta \qquad (10.2)$$

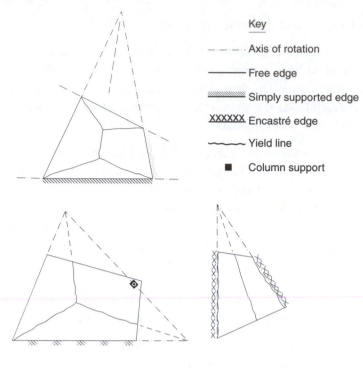

Key

— · — · — Axis of rotation

———— Free edge

▨▨▨▨▨ Simply supported edge

XXXXXX Encastré edge

～～～ Yield line

■ Column support

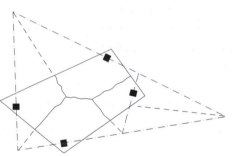

FIGURE 10.2 Some typical yield line patterns

If there is an additional reinforcement orthogonal to the first with a moment of resistance μm_{Rd}, then the contribution of this reinforcement to the moment across the yield line m_{n2} is

$$m_{n2} = \mu m_{Rd} \sin^2(90 - \theta) \tag{10.3}$$

Total moment across the yield line m_n is given by the sum of the contributions m_{n1} and m_{n2} from Eqs (10.2) and (10.3), or

$$m_n = m_{Rd} \sin^2 \theta + \mu m_{Rd} \sin^2(90 - \theta) \tag{10.4}$$

If as is usual, the reinforcement is isotropic ($\mu = 1$), then

$$m_n = m_{Rd} \tag{10.5}$$

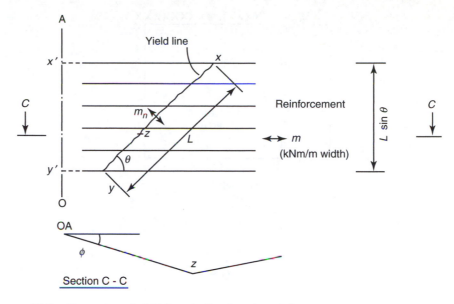

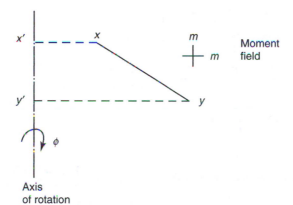

FIGURE 10.3 Formation of yield lines inclined to the reinforcement

FIGURE 10.4 Work done by yield lines

To determine the work done by the yield line consider the yield line xy in Fig. 10.4. The work done may be stated as the moment times the rotation times the projected length of the yield line on the axis of rotation, so for yield line xy, the work done is $m \times (\text{length } x'y') \times \phi$. The alternative approach is to use the moment normal to the yield line given by Eqs (10.4) or (10.5) and the work done calculated as the normal moment times the length of the yield line times the total rotational angle.

To determine the angles of rotation, it is a general practice to impose (virtual) displacements of unity on the yield line undergoing the maximum displacement.

The work done by the loads is simply calculated as load times (virtual) displacement for a point load and the total load times the (virtual) displacement of the centroid of the load for either uniform line loads (loads per unit length) or uniformly distributed loads (loads per unit area). For uniformly distributed loads, the work done is also

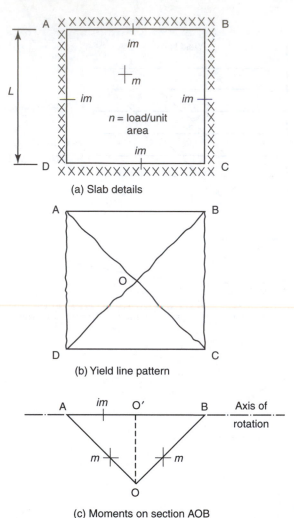

(a) Slab details

(b) Yield line pattern

(c) Moments on section AOB

FIGURE 10.5 Yield line collapse for an encastré square slab

given by the load times the volume between the original slab and its displaced shape. For non-uniform loads, it will be necessary to carry out integration to determine the work done.

A series of examples will now be undertaken. In all cases an encastré slab will be analysed since the solution for a simply supported slab can be obtained by setting the moment parameter i defining the encastré moment equal to zero.

EXAMPLE 10.1 Square slab. Consider a square slab of side L carrying a UDL at ultimate limit state of n per unit area (Fig. 10.5(a)). The slab has isotropic orthogonal reinforcement giving sagging moments of resistance m per unit run and hogging im.

The only possible configuration using simple straight yield lines is that shown in Fig. 10.5(b). Owing to symmetry on a quarter of the slab need be considered.

Using virtual work and giving the centre of the slab O a displacement of unity, the internal work done by the moments can be determined as follows:

Sagging yield Line:

Yield Line AO:
moment per unit run: m
Projected length of AO: $AO' = L/2$
Rotation: $\phi = 1/OO' = 1/(L/2) = 2/L$

$$W.D. = m\frac{L}{2}\frac{1}{L/2} = m \tag{10.6}$$

Yield Line OB:
This is identical to yield line AO, so work done $= m_{Rd}$
Thus total work done by sagging yield line

$$W.D. = 2m \tag{10.7}$$

Hogging yield line:
moment: im
Projected length of AB: $AB = L$
Rotation: $\phi = 1/(L/2) = 2/L$

$$W.D. = imL\frac{1}{L/2} = 2im \tag{10.8}$$

Total work done by hogging and sagging yield lines from Eqs (10.7) and (10.8) is given by

$$W.D. = 2m + 2im = 2(i+1)m \tag{10.9}$$

Work done by loads:
Displaced volume between original and deflected slab for one-quarter of the slab is $(1/4)(1 \times L^2/3) = L^2/12$.

$$W.D. = n\frac{L^2}{12} \tag{10.10}$$

Equating internal and external work done from Eqs (10.9) and (10.10) gives

$$2(i+1)m = \frac{nL^2}{12} \tag{10.11}$$

or

$$m = \frac{nL^2}{24(i+1)} \tag{10.12}$$

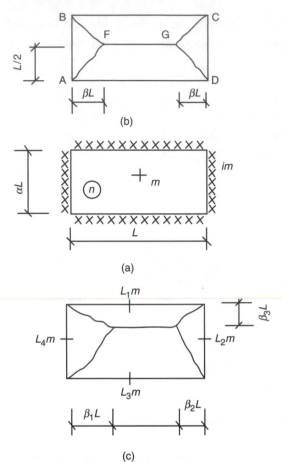

FIGURE 10.6 Yield line collapse for an encastré rectangular slab

Note that this is a classic case where the yield criterion is violated at the corner (Jones and Wood, 1967). Alternative solutions are given in Example 10.9.

EXAMPLE 10.2 Rectangular slab encastré on all edges under a UDL of n per unit area. The geometry and collapse mechanism are given in Fig. 10.6(a) and (b). The yield line position is characterized by the length βL and the yield line FG is displaced by unity vertically downwards. Consider half the slab i.e. region ABFGD.

Work done by loads:
Volume displaced (for half the slab).
Volume under the pyramid in region AFB,

$$V_I = \frac{1}{3}(\alpha L \beta L) \times 1 \tag{10.13}$$

Volume under half the region between F and G

$$V_{II} = \frac{1}{2}\left((1 - 2\beta)L \frac{\alpha L}{2}\right) \times 1 \tag{10.14}$$

The total volume displaced is given by

$$V = V_{\mathrm{I}} + V_{\mathrm{II}} = \frac{(3 - 2\beta)\alpha L^2}{12} \tag{10.15}$$

The work done is given by $n \times V$, or

$$W.D. = n\frac{(3 - 2\beta)\alpha L^2}{12} \tag{10.16}$$

Work done by moments:
Region ABF (I):
The length of the hogging yield line is AB $(=\alpha L)$, and the projected length of the sagging yield line on the axis of rotation AB is also αL, so total work done is given by

$$W.D._{\mathrm{I}} = im\alpha L\frac{1}{\beta L} + m\alpha L\frac{1}{\beta L} = m(1 + i)\frac{\alpha}{\beta} \tag{10.17}$$

Region AFGD (II):
The length of the hogging yield line is L as is the projected length of the sagging yield line, so work done is given by

$$W.D._{\mathrm{II}} = imL\frac{1}{\alpha L/2} + mL\frac{1}{\alpha L/2} = 2m(1 + i)\frac{1}{\alpha} \tag{10.18}$$

Total work done by moments is given by summing Eqs (10.17) and (10.18), or

$$W.D. = m(1 + i)\left(\frac{\alpha}{\beta} + \frac{2}{\alpha}\right) \tag{10.19}$$

Equating work done by the moments from Eq. (10.19) and the load from either Eq. (10.14) or Eq. (10.15) gives

$$m(1 + i)\left(\frac{\alpha}{\beta} + \frac{2}{\alpha}\right) = \frac{n(3 - 2\beta)\alpha L^2}{12} \tag{10.20}$$

or

$$\frac{12m(1 + i)}{n\alpha L^2} = \frac{3 - 2\beta}{\alpha/\beta + 2/\alpha} \tag{10.21}$$

The right-hand side of Eq. (10.21) still contains β.

There are two approaches that can be used to eliminate the effect of β. The first is to substitute numerical values of β until the right-hand side of Eq. (10.21) attains a stationary value (maximum or minimum). If the curve given by the right-hand side is plotted, the curve is relatively flat and slight inaccuracies in the value of β are unimportant. Note this is only recommended if numerical values of all geometric parameters (or moment ratios i, if these vary on some of the edges) are given.

The alterative is to proceed to find an algebraic solution by differentiating the right-hand side. There is an easy way to do this. From the basic quotient rule

$$\frac{d}{dx}\left(\frac{u}{v}\right) = \frac{v(du/dx) - u(dv/dx)}{v^2} \tag{10.22}$$

As a stationary value of Eq. (10.22) is required

$$\frac{d}{dx}\left(\frac{u}{v}\right) = 0 \tag{10.23}$$

or

$$\frac{u}{v} = \frac{du/dx}{dv/dx} \tag{10.24}$$

The value of the variable determined from Eq. (10.24) may be back-substituted into either side of Eq. (10.24) to determine m (or n).

So

$$\frac{3 - 2\beta}{(\alpha/\beta) + (2/\alpha)} = \frac{-2}{(-\alpha/\beta^2)} = \frac{2\beta^2}{\alpha} \tag{10.25}$$

Multiplying out the extreme left- and right-hand sides of Eq. (10.25) gives

$$3\alpha^2 - 4\beta\alpha^2 - 4\beta^2 = 0 \tag{10.26}$$

or, solving

$$\beta = \frac{-\alpha^2}{2} \pm \frac{\alpha}{2}\sqrt{\alpha^2 + 3} \tag{10.27}$$

Note, the positive root must be taken as $\beta > 0$.
Back-substituting the value of β into Eq. (10.21) (or the right-hand side of Eq. (10.25)) gives

$$m = \frac{n(\alpha L)^2}{24(1 + i)}\left[\sqrt{\alpha^2 + 3} - \alpha\right]^2 \tag{10.28}$$

Where the support (hogging) moments have different moment ratios i then the symmetry of the problem no longer exists and up to three different parameters are then required to characterize the yield line pattern even though the central yield line remains parallel to the supports (Fig. 10.6(c)). Although in this case an algebraic solution is possible, it will need the work equation to be (partially) differentiated with respect to each of the variables and the resultant equations solved simultaneously to obtain optimal values. This type of problem is best solved by numerical techniques (or spreadsheet).

EXAMPLE 10.3 Rectangular slab supported on three sides. A rectangular slab of sides L and αL ($\alpha \leq 1,0$) is supported on the long side L and the two shorter sides αL (Fig. 10.7(a)). There are two possible collapse mechanisms (Fig. 10.7(b) and (c)).

Mechanism I:
Work done by loads (EF is displaced by unity):
Volume under displaced surface is given by

$$V = \frac{1}{3}L\beta L + \frac{1}{2}L(\alpha - \beta)L \tag{10.29}$$

Work done is given by nV, or

$$W.D. = nV = \frac{n L^2}{6}(3\alpha - \beta) \tag{10.30}$$

Work done by moments:
Region I (Axis BC):

$$W.D._I = m L \frac{1}{\beta L} = \frac{m}{\beta} \tag{10.31}$$

Region II (Axis AB or CD):

$$W.D._{II} = m\alpha L \frac{1}{L/2} = 2m\alpha \tag{10.32}$$

The total work done for Region I and both regions II is given by

$$W.D. = W.D._I + W.D._{II} = m\left(\frac{1}{\beta} + 4\alpha\right) \tag{10.33}$$

Equating work done from Eqs (10.30) and (10.33) gives

$$\frac{6m}{n L^2} = \frac{3\alpha - \beta}{(1/\beta) + 4\alpha} \tag{10.34}$$

Differentiating Eq. (10.34) with respect to β gives

$$\frac{3\alpha - \beta}{(1/\beta) + 4\alpha} = \frac{-1}{-(1/\beta^2)} = \beta^2 \tag{10.35}$$

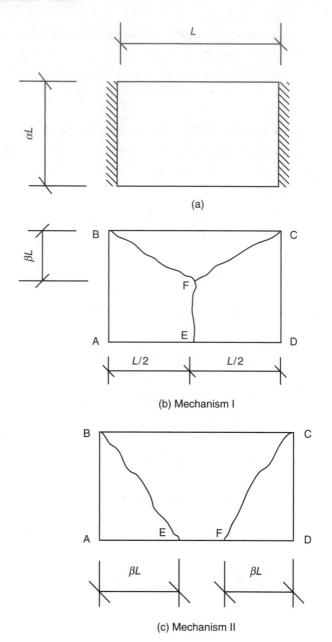

FIGURE 10.7 Yield line collapse for a rectangular slab simply supported on three sides

or

$$4\alpha\beta^2 + 2\beta - 3\alpha = 0 \tag{10.36}$$

or

$$\beta = \frac{-1 \pm \sqrt{1 + 12\alpha^2}}{4\alpha} \tag{10.37}$$

The plus sign must be taken in Eq. (10.37) as $\beta > 0$. Note, for $\alpha \leq 0.5$, $\beta \geq \alpha$.

Back-substituting β into Eq. (10.34) for $\alpha \geq 0,5$ gives

$$m = \frac{nL^2}{24}\left[\sqrt{3 + \frac{1}{4\alpha^2}} - \frac{1}{2\alpha}\right]^2 \tag{10.38}$$

For $\alpha \leq 0,5$ set $\alpha = \beta$, and Eq. (10.34) then gives

$$m = \frac{nL^2}{3}\frac{\alpha^2}{1 + 4\alpha^2} \tag{10.39}$$

Mechanism II:

Work done by loads:
Region I (ABE or CDF):

$$W.D._{\text{I}} = \frac{1}{3}n\frac{\alpha L}{2}\beta L \tag{10.40}$$

Region II:

$$W.D._{\text{II}} = 2\left(\frac{1}{3}n\frac{\alpha L}{2}\beta L\right) + \frac{1}{2}n\alpha L(1 - 2\beta)L \tag{10.41}$$

Total work done for Region II and both Regions I is given by

$$W.D. = 2W.D._{\text{I}} + W.D._{\text{II}} = \frac{n\alpha L^2}{6}(3 - 2\beta) \tag{10.42}$$

Work done by moments:
Region I (Axis AB or CD):

$$W.D._{\text{I}} = m\alpha L\frac{1}{\beta L} \tag{10.43}$$

Region II (Axis BC):

$$W.D._{\text{II}} = 2\left(m\beta L\frac{1}{\alpha L}\right) \tag{10.44}$$

Total work done is given by

$$W.D. = 2W.D._{\text{I}} + W.D._{\text{II}} = 2m\left(\frac{\beta}{\alpha} + \frac{\alpha}{\beta}\right) \tag{10.45}$$

Equating work done from Eqs (10.42) and (10.45) gives

$$\frac{12m}{n\alpha L^2} = \frac{3 - 2\beta}{(\beta/\alpha) + (\alpha/\beta)} \tag{10.46}$$

Differentiate Eq. (10.46) with respect to β to give

$$\frac{3 - 2\beta}{(\beta/\alpha) + (\alpha/\beta)} = \frac{-2}{(1/\alpha) - (\alpha/\beta^2)} \tag{10.47}$$

or

$$\frac{3}{\alpha}\beta^2 + 4\alpha\beta - 3\alpha = 0 \tag{10.48}$$

Solving and taking the positive sign gives

$$\beta = \frac{2\alpha}{3}\left(\sqrt{\alpha^2 + \frac{9}{4}} - \alpha\right) \tag{10.49}$$

Note that from Eq. (10.49) if $\alpha > 0{,}867$, then β is greater than $0{,}5$ which is inadmissible. In this case the value of β must then be set equal to its geometrical maximum, i.e. $0{,}5$.

If $\alpha < 0{,}867$, then back-substituting β into Eq. (10.46) gives

$$m = \frac{nL^2}{12}\left[\sqrt{\alpha^2 + \frac{9}{4}} - \alpha\right] \tag{10.50}$$

For the case when $\alpha > 0{,}867$,

$$m = \frac{n\alpha L^2}{3(1 + 4\alpha^2)} \tag{10.51}$$

EXAMPLE 10.4   Triangular slab simply supported on two edges under a UDL of n. The slab geometry is given in Fig. 10.8(a), with the collapse mechanism and associated geometry in Fig. 10.8(b). The yield line position is characterized by the angle α. For a general triangular shape any attempt to use the distance BX or CX as the variable will produce horrendous algebra. An isosceles or equilateral triangle can be solved either by use of an angle or a distance on the unsupported edge.

Work done by loads:
Region I (Triangle ABX):

$$W.D._{\mathrm{I}} = \frac{1}{3}n \text{ Area } ABX \tag{10.52}$$

Region II (Triangle ACX):

$$W.D._{\mathrm{II}} = \frac{1}{3}n \text{ Area } ACX \tag{10.53}$$

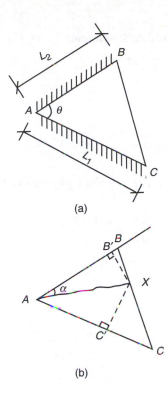

(a)

(b)

FIGURE 10.8 Yield line collapse for a triangular slab simply supported on two sides

Total work done is given by

$$W.D. = W.D._{I} + W.D._{II}$$

$$= \frac{1}{3}n(\text{Area } ABX + \text{Area } ACX)$$

$$= \frac{1}{3}n \text{ Area } ABC = \frac{1}{3}n\frac{1}{2}L_1 L_2 \sin\theta \qquad (10.54)$$

Work done by Yield line:
Region I:
Projected length AB'; rotation $1/B'X$

$$W.D._{I} = mA'B\frac{1}{B'X} = m\frac{A'B}{B'X} = m\cot\alpha \qquad (10.55)$$

Region II:
Projected length AC'; rotation $1/C'X$

$$W.D._{II} = mA'C\frac{1}{C'X} = m\frac{A'C}{C'X} = m\cot(\theta - \alpha) \qquad (10.56)$$

Total work done is given by,

$$W.D. = W.D._{I} + W.D._{II} = m(\cot\alpha + \cot(\theta - \alpha)) \qquad (10.57)$$

Equating Work Done by the loads and moments from Eqs (10.54) and (10.57), respectively gives

$$\frac{n L_1 L_2 \sin \theta}{6m} = \cot \alpha + \cot(\theta - \alpha)$$

$$= \cot \alpha + \frac{\cot \alpha \cot \theta + 1}{\cot \alpha - \cot \theta} \tag{10.58}$$

Differentiate the right-hand side of Eq. (10.58) with respect to $\cot \alpha$ to give

$$0 = 1 + \frac{(\cot \alpha - \cot \theta) \cot \theta - (\cot \theta \cot \alpha + 1)(1)}{(\cot \alpha - \cot \theta)^2} \tag{10.59}$$

or

$$(\cot \alpha - \cot \theta)^2 = 1 + \cot^2 \theta \tag{10.60}$$

or

$$\cot^2 \alpha - 2 \cot \alpha \cot \theta - 1 = 0 \tag{10.61}$$

or

$$\cot \alpha = \cot \theta + \sqrt{\cot^2 \theta + 1} = \frac{1 + \cos \theta}{\sin \theta}$$

$$= \frac{1 + (2 \cos^2(\theta/2) - 1)}{2 \sin(\theta/2) \cos(\theta/2)} = \cot \frac{\theta}{2} \tag{10.62}$$

or

$$\alpha = \frac{\theta}{2} \tag{10.63}$$

Thus substituting Eq. (10.63) into Eq. (10.58) gives

$$\frac{n L_1 L_2 \sin \theta}{6m} = 2 \cot \frac{\theta}{2} \tag{10.64}$$

or

$$m = \frac{n L_1 L_2}{6} \sin^2 \frac{\theta}{2} \tag{10.65}$$

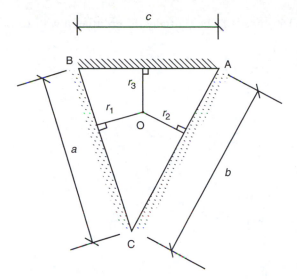

FIGURE 10.9 Yield line collapse for a triangular slab simply supported on all sides

EXAMPLE 10.5 Triangular slab simply supported on all three sides. Consider a triangular slab ABC (sides a, b and c) under a UDL of n simply supported on all three sides with a sagging moment of resistance m (Fig. 10.9). The point O is characterized by the lengths r_1, r_2 and r_3.

If the area of the triangle is Δ, then the work done by the load is $\Delta n/3$. The work done by the moment on each of the slab segments is m times the side length divided by the relevant value of r, thus

$$\frac{n}{m}\frac{\Delta}{3} = \frac{a}{r_1} + \frac{b}{r_2} + \frac{c}{r_3} \tag{10.66}$$

Once the values of any two of the three dimensions r_i are selected, the third is fixed. Thus the values of r_i are not independant. However the area of the triangle is also given as

$$\Delta = \frac{a r_1 + b r_2 + c r_3}{2} \tag{10.67}$$

or

$$r_3 = \frac{2\Delta}{c} - \frac{a}{c}r_1 - \frac{b}{c}r_2 \tag{10.68}$$

Thus Eq. (10.66) becomes

$$\frac{n}{m}\frac{\Delta}{3} = \frac{a}{r_1} + \frac{b}{r_2} + \frac{c^2}{2\Delta - a r_1 - b r_2} \tag{10.69}$$

Equation (10.69) needs to be partially differentiated with respect to r_1 and r_2, thus

$$\frac{-a}{r_1^2} - \frac{(-a)c^2}{(2\Delta - ar_1 - br_2)^2} = 0 \tag{10.70}$$

or

$$\frac{1}{r_1^2} = \frac{c^2}{(2\Delta - ar_1 - br_2)^2} \tag{10.71}$$

and, similarly

$$\frac{-b}{r_2^2} - \frac{(-b)c^2}{(2\Delta - ar_1 - br_2)^2} = 0 \tag{10.72}$$

or

$$\frac{1}{r_2^2} = \frac{c^2}{(2\Delta - ar_1 - br_2)^2} \tag{10.73}$$

The right-hand sides of Eqs (10.71) and (10.73) are identical, thus

$$r_1 = r_2 = r \tag{10.74}$$

From either Eq. (10.71) or (10.73),

$$\frac{1}{r^2} = \frac{c^2}{(2\Delta - ar - br)^2} \tag{10.75}$$

or

$$cr = 2\Delta - ar - br \tag{10.76}$$

or from Eq. (10.68),

$$r_3 = r \tag{10.77}$$

Thus, the point O is the centre of the inscribed circle of the triangle. Thus

$$\frac{n}{3m} \frac{(a+b+c)r}{2} = \frac{a+b+c}{r} \tag{10.78}$$

or

$$n = \frac{6m}{r^2} \tag{10.79}$$

To determine r it should be noted that the area of a triangle Δ is given by

$$\Delta = \sqrt{s(s-a)(s-b)(s-c)} = sr \tag{10.80}$$

or

$$r = \sqrt{\frac{(s-a)(s-b)(s-c)}{s}} \qquad (10.81)$$

where s is the semi-perimeter given by

$$s = \frac{a+b+c}{2} \qquad (10.82)$$

EXAMPLE 10.6   Circular slab under a UDL. Consider a circular slab of radius R with encastré hogging moments resistance im and sagging moments of resistance m (Fig. 10.10(a)). The failure pattern under such loading is also shown in Fig. 10.10(a), and an enlargement of one of the sectors of the slab in Fig. 10.10(b).

Work done by the moments:

(a) Sagging:

$$W.D. = m(R\delta\theta)\frac{1}{R} = m\delta\theta \qquad (10.83)$$

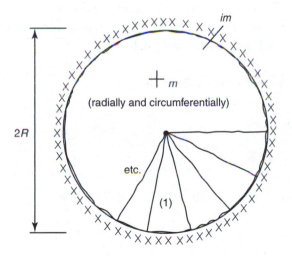

(a) Failure mode of complete slab

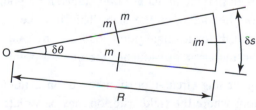

(b) Enlargement of element (1)

FIGURE 10.10   Yield line collapse for a circular slab

(b) Hogging:

$$W.D. = im(R\delta\theta)\frac{1}{R} = im\delta\theta \tag{10.84}$$

From Eqs (10.83) and (10.84) the total work done by the moments is

$$W.D. = m\delta\theta + im\delta\theta = (1+i)m\delta\theta \tag{10.85}$$

Work done by loads:
The sector of a circle may be treated as a triangle as $\delta\theta$ is small.

$$W.D. = \frac{1}{3}n\left(\frac{1}{2}R^2\,\delta\theta\right) = \frac{nR^2\,\delta\theta}{6} \tag{10.86}$$

Equating Work done from Eqs (10.85) and (10.86) gives

$$m = \frac{nR^2}{6(1+i)} \tag{10.87}$$

EXAMPLE 10.7 Circular slab with a central point load P. Total Work done by the moments is obtained by summing the incremental work done over a single sector given by Eq. (10.85) over the whole slab

$$W.D. = \sum m(1+i)\delta\theta = 2\pi m(1+i) \tag{10.88}$$

Work done by load:

$$W.D. = P \times 1 \tag{10.89}$$

Equating work done from Eqs (10.88) and (10.89) gives

$$m = \frac{P}{2\pi(1+i)} \tag{10.90}$$

It should be noted that the moment of resistance for a point load on a circular slab is independent of the radius of the slab. This has the implication that a possible collapse mechanism for a single point load on any shape slab is circular with the radius such that the circle just touches the nearest point of the slab to the load.

Yield lines comprising circular portions occur in a number of possible instances, namely situations where the yield criterion may be violated with straight yield lines, around point supports (or columns) and under any instances of point loads.

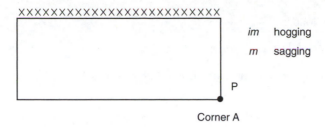

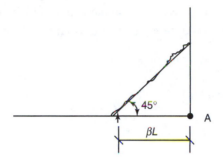

(a) Basic layout

(b) Corner yield line

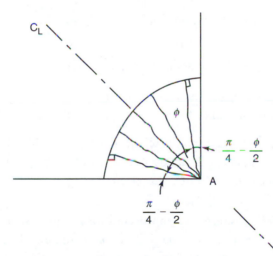

(c) Part fan

FIGURE 10.11 Yield line collapse for a point load at the corner of a cantilever slab

EXAMPLE 10.8 Point load at the corner of a cantilever slab (Fig. 10.11). The overall collapse mechanism involving a hogging yield line at the encastré support must be checked. This needs only simple statics and will not be considered further.

There are two further collapse mechanisms that must be considered.
The first is a straight yield line which requires top or hogging steel (Fig. 10.11(b)).

W.D. by moments:

$$W.D. = im\sqrt{2}\,\beta L\,\frac{\sqrt{2}}{\beta L} = 2im \qquad (10.91)$$

W.D. by loads:

$$W.D. = P \times 1 \tag{10.92}$$

So equating work done from Eqs (10.91) and (10.92) gives

$$m = \frac{P}{2i} \tag{10.93}$$

An alternative pattern is where the yield line is in part circular and in part two straight lines (Fig. 10.11(c)). Note a complete quarter circle cannot occur as this would imply a moment could be generated normal to a free edge.

The work equation can be written as

$$P = (1 + i)m\phi + 2im \tan\left(\frac{\pi}{4} - \frac{\phi}{2}\right) \tag{10.94}$$

Differentiating Eq. (10.94) with respect to ϕ gives

$$\phi = \frac{1}{\sqrt{i}} \tag{10.95}$$

or, substituting Eq. (10.95) into Eq. (10.94) gives

$$m = \frac{P}{(i + 1)((\pi/2) - 2\arctan(1/\sqrt{i})) + (2/\sqrt{i})} \tag{10.96}$$

Note that if the sagging and hogging yield moments are equal ($i = 1$), then the second mechanism cannot form and reduces to the straight line mechanism above.

With the above mechanisms in mind, and the fact that the classical straight line solution for the square slab can be shown to violate the yield criterion at the corners, alternative mechanisms can be suggested.

EXAMPLE 10.9   Alternative approaches for a square slab. There are two possible additional mechanisms for the failure of a square slab. To keep the algebra reasonably simple, the hogging and sagging yield lines have equal moments of resistance m, i.e. $i = 1$. The classic solutions for equal hogging and sagging reinforcement are for a uniformly distributed load n, $m = nL^2/48$, and for a central point load P, $m = P/16$.

The first solution involves partial collapse with straight lines leaving the corners intact (known as corner levers), and the second involves fans.

Mechanism 1 (UDL):

This is illustrated in Fig. 10.12(a). Only one quarter of the slab need be considered. First a little geometry:

$$XY = \sqrt{2}x \tag{10.97}$$

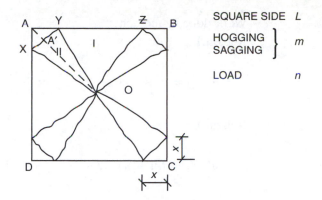

(a) Mechanism I – Corner lever

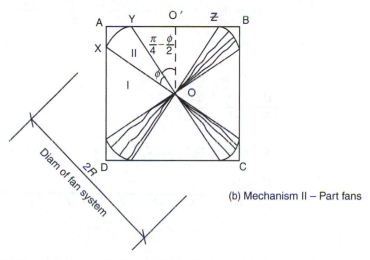

(b) Mechanism II – Part fans

FIGURE 10.12   Alternative yield line collapse for an encastré square slab

$$A'O = AO - A'A = \frac{L}{2}\sqrt{2} - \frac{x}{\sqrt{2}} = \frac{L-x}{\sqrt{2}} \qquad (10.98)$$

Work done by loads:
Region I:

$$W.D._I = n\left(\frac{1}{2}\frac{L}{2}(L-2x)\frac{1}{3}\right) = \frac{nL(L-2x)}{12} \qquad (10.99)$$

Region II:

$$W.D._{II} = n\left(\frac{1}{2}\frac{L-x}{\sqrt{2}}\sqrt{2}x\frac{1}{3}\right) = \frac{nx(L-x)}{6} \qquad (10.100)$$

The total work done is given by Eqs (10.99) and (10.100), i.e.

$$W.D. = W.D._I + W.D._{II} = \frac{n}{12}(L^2 - 2x^2) \qquad (10.101)$$

Work done by moments:
Region I:

$$W.D._{\mathrm{I}} = (m+m)(L-2x)\frac{1}{L/2} = 4m\frac{L-2x}{L} \tag{10.102}$$

Region II:

$$W.D._{\mathrm{II}} = (m+m)\sqrt{2}x\frac{1}{(L-x)/\sqrt{2}} = 4m\frac{x}{L-x} \tag{10.103}$$

Total work done is given by Eqs (10.102) and (10.103), i.e.

$$W.D. = W.D._{\mathrm{I}} + W.D._{\mathrm{II}} = \frac{4m}{L}\frac{L^2 - 2xL + 2x^2}{L-x} \tag{10.104}$$

Equating work done by the loads from Eq. (10.101) and moments from Eq. (10.104) gives

$$\frac{nL}{48m} = \frac{L^2 - 2xL + 2x^2}{(L-x)(L^2 - 2x^2)} \tag{10.105}$$

Differentiate the right-hand side of Eq. (10.105) with respect to x,

$$\frac{L^2 - 2xL + 2x^2}{(L-x)(L^2 - 2x^2)} = \frac{4x - 2L}{(L-x)(-4x) + (L^2 - 2x^2)(-1)}$$
$$= \frac{4x - 2L}{6x^2 - 4xL - L^2} \tag{10.106}$$

or, simplifying

$$4x^4 - 8Lx^3 + 12L^2x^2 - 8L^3x + L^4 = 0 \tag{10.107}$$

Solving Eq. (10.107) gives $x/L = 0{,}1594;\ 0{,}8576;\ 0{,}5182 \pm 1{,}249i$.

The last three roots are inadmissible either because they are outside the limits of $0 \leq x \leq 0{,}5$ or are complex. The only acceptable solution is $x/L = 0{,}1594$.

Substituting this value into Eq. (10.105) gives

$$m = \frac{nL^2}{48 \times 0{,}917} = \frac{nL^2}{44{,}02} \tag{10.108}$$

i.e. 9 per cent more reinforcement is required over the traditional solution.

Mechanism II:
This is illustrated in Fig. 10.12(b). Again only one quarter of the slab need be considered.

Again a little geometry:

$$YZ = 2O'Y = 2\frac{L}{2}\tan\left(\frac{\pi}{4} - \frac{\phi}{2}\right) = L\tan\left(\frac{\pi}{4} - \frac{\phi}{2}\right) \tag{10.109}$$

$$R = \frac{L}{2\cos[(\pi/4) - (\phi/2)]} \tag{10.110}$$

The fans are treated as if each contained a large number of sectors subtending small angles at the centroid with each incremental sector displacing by 1/3 (as in the complete circular collapse).

Work done by loads:
Region I:

$$
\begin{aligned}
W.D._{\mathrm{I}} &= n\left(\frac{1}{2}\frac{L}{2}L\tan\left(\frac{\pi}{4} - \frac{\phi}{2}\right)\frac{1}{3}\right)\\
&= \frac{nL^2}{12}\tan\left(\frac{\pi}{4} - \frac{\phi}{2}\right)
\end{aligned}
\tag{10.111}
$$

Region II:

$$W.D._{\mathrm{II}} = n\left(\frac{R^2}{2}\phi\frac{1}{3}\right) = \frac{nL^2}{12}\frac{\phi/2}{\cos^2[(\pi/4) - (\phi/2)]} \tag{10.112}$$

The total work done is given by Eqs (10.111) and (10.112), i.e.

$$
\begin{aligned}
W.D. &= W.D._{\mathrm{I}} + W.D._{\mathrm{II}}\\
&= \frac{nL^2}{12}\left[\frac{\phi/2}{\cos^2[(\pi/4) - (\phi/2)]} + \tan\left(\frac{\pi}{4} - \frac{\phi}{2}\right)\right]
\end{aligned}
\tag{10.113}
$$

Work done by moments:
Region I:

$$W.D._{\mathrm{I}} = (m + m)L\tan\left(\frac{\pi}{4} - \frac{\phi}{2}\right)\frac{1}{L/2} = 4m\tan\left(\frac{\pi}{4} - \frac{\phi}{2}\right) \tag{10.114}$$

Region II:

$$W.D._{\mathrm{II}} = (m + m)\phi \tag{10.115}$$

Total work done is given by Eqs (10.114) and (10.115), i.e.

$$W.D. = W.D._{\mathrm{I}} + W.D._{\mathrm{II}} = 4m\left[\frac{\phi}{2} + \tan\left(\frac{\pi}{4} - \frac{\phi}{2}\right)\right] \tag{10.116}$$

Equating work done by the loads from Eq. (10.113) and moments from Eq. (10.116) gives

$$\frac{n L^2}{48m} = \frac{(\phi/2) + \tan[(\pi/4) - (\phi/2)]}{(\phi/2)/\cos^2[(\pi/4) - (\phi/2)] + \tan[(\pi/4) - (\phi/2)]} \tag{10.117}$$

The RHS of Eq. (10.117) was optimized using a spread-sheet to give a value of 0,9055 with $\phi = 0,488$ rad (28°) thus from Eq. (10.117),

$$m = \frac{n L^2}{48 \times 0,9055} = \frac{n L^2}{43,46} \tag{10.118}$$

i.e. 10 per cent more reinforcement is required over the traditional solution.

Thus, it is observed that the classical straight yield line pattern overestimates the carrying capacity of a square slab by around 10 per cent. In practice this error may not be significant as detailing generally overprovides reinforcement (and must not underprovide!). It does however indicate that care needs to be taken. It is also to be noted that the correct solution is greater than either the straight yield line or the case when the circular yield line forms the complete pattern, i.e. when $\varphi = \pi/2$ and $R = L/2$. In the latter case,

$$m = \frac{n R^2}{6(i + 1)} = \frac{n[(L/2)]^2}{6(1 + 1)} = \frac{n L^2}{48} \tag{10.119}$$

EXAMPLE 10.10   Central Point load, P. The mechanisms take exactly the same form as in Example 10.9, hence the work done by the moments is unchanged.

In both mechanisms I and II, for one quarter of the slab, the work done by the load is $(P/4) \times 1$.

Mechanism I:
The work done by the moments is given by Eq. (10.104), thus the complete work equation may be written as

$$\frac{P}{4} = \frac{4m}{L} \frac{L^2 - 2xL + 2x^2}{L - x} \tag{10.120}$$

or

$$\frac{PL}{16m} = \frac{L^2 - 2xL + 2x^2}{L - x} \tag{10.121}$$

Differentiate the right-hand side of Eq. (10.121) with respect to x to give

$$\frac{L^2 - 2xL + 2x^2}{L - x} = \frac{4x - 2L}{-1} = 2L - 4x \tag{10.122}$$

Simplifying gives

$$2x^2 - 4xL + L^2 = 0 \tag{10.123}$$

or

$$x = \frac{L}{\sqrt{2}}\left(\sqrt{2} - 1\right) \tag{10.124}$$

Back-substituting gives

$$m = \frac{P}{32(\sqrt{2} - 1)} = \frac{P}{13,25} \tag{10.125}$$

This is around 17 per cent lower than the classical solution.

Mechanism II:
The work done by the moments is given by Eq. (10.116), thus the complete work equation is given by

$$\frac{P}{4} = 4m\left[\frac{\phi}{2} + \tan\left(\frac{\pi}{4} - \frac{\phi}{2}\right)\right] \tag{10.126}$$

or

$$\frac{P}{16m} = \frac{\phi}{2} + \tan\left(\frac{\pi}{4} - \frac{\phi}{2}\right) \tag{10.127}$$

Differentiate the right-hand side of Eq. (10.127) with respect to ϕ to give

$$\frac{1}{2} + \left(\frac{-1}{2}\right)\frac{1}{\cos^2[(\pi/4) - (\phi/2)]} = 0 \tag{10.128}$$

This gives $\phi = \pi/2$, i.e. the fan becomes a complete circle with m given by

$$m = \frac{P}{4\pi} = \frac{P}{12,57} \tag{10.129}$$

and is around 21 per cent lower than the classical solution. The results are summarized in Table 10.1 for both the UDL and point load.

In conclusion, it may be observed that from the complete set of results substantially larger errors are produced by not considering fans (or alternative straight line patterns) for loading due to point loads (or column reactions) than for uniformly distributed loads.

TABLE 10.1 Comparison of failure moments of square slabs.

Type of failure	UDL	% Error	Point load	% Error
Classical	1/48	—	1/16	—
Corner lever	1/44,02	9	1/13,25	21
Part Fans	1/43,46	10	1/12,57	27
Complete circle	1/48	—	1/12,57	27
	$\times nL^2$		$\times P$	

Note: The percentage error is determined using the classical solution as the basis.

EXAMPLE 10.11   Slabs on column supports. Consider a square slab supported at its corners by columns carrying a UDL (Fig. 10.13(a)).

Mechanism I:

The immediate response for a simple solution to the problem is to adopt that is shown in Fig. 10.13(b). It can be readily shown that this gives

$$m = \frac{nL^2}{8} \tag{10.130}$$

Mechanism II:

A possible alternative is that shown in Fig. 10.13(c) with diagonal yield lines. The virtual work equation for this case may be written as,

$$4m(\sqrt{2}\beta L)\left(\frac{\sqrt{2}}{\beta L}\right) = n\left[L^2 - 4\frac{(\beta L)^2}{2}\right] \times 1 + 4n\frac{(\beta L)^2}{2} \times \frac{2}{3} \tag{10.131}$$

where the first term on the right-hand side is due to the centre of the slab deflecting by unity and the second term is due to the four corner triangles, the centres of gravity of which deflect by 2/3.

The work equation reduces to

$$m = \frac{nL^2}{8}\left(1 - \frac{2\beta^2}{3}\right) \tag{10.132}$$

The value of m tends to a maximum when β tends to zero. This maximum value of m is again $nL^2/8$.

An elastic solution to the problem of a slab supported on corner columns indicates that there will be hogging and twisting moments over the columns. Neither yield solution gives any indication of the existence of any such moments, nor any mobilization of any top steel. It is thus necessary to consider the possibility of the formation of fans. Although it has already been demonstrated that a complete quarter circle yield line pattern cannot physically occur (Example 10.8), for this exercise it will be assumed possible (Fig. 10.13(d)). It should also be noted that if this

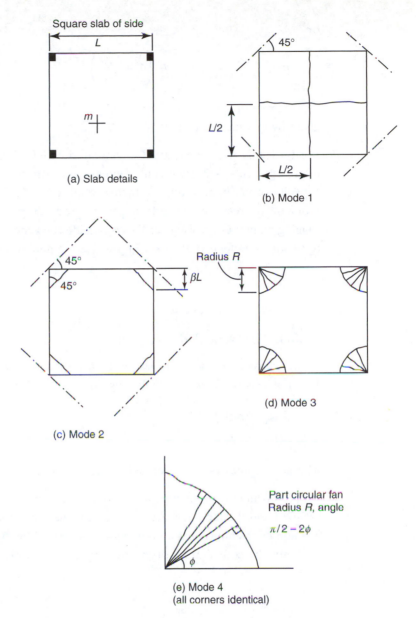

(a) Slab details

(b) Mode 1

(c) Mode 2

(d) Mode 3

(e) Mode 4
(all corners identical)

FIGURE 10.13 Collapse mechanisms for a square slab on corner columns

slab were part of a multi-bay flat slab, circular yield lines around the column would be possible, and would need considering.

The centroid of a quarter circle is at a distance of $4\sqrt{2}R/3\pi$ from the centre. Thus, if the perimeter of the quarter circle undergoes a virtual displacement of unity, the centroid is displaced by $(1/R) \times (4\sqrt{2}R/3\pi) = 4\sqrt{2}/3\pi$. Thus, the equilibrium equation can be written as,

$$2m(1+i)\pi R\left(\frac{1}{R}\right) = n(L^2 - \pi R^2) \times 1 + n\pi R^2\left(\frac{4\sqrt{2}}{3\pi}\right) \qquad (10.133)$$

or

$$m = \frac{n}{2\pi(1+i)}\left[L^2 - \pi R^2\left(1 - \frac{4\sqrt{2}}{3\pi}\right)\right]$$ (10.134)

The maximum value of m is obtained when R tends to zero. The value of m is then given by $m = nL^2/2\pi$ $(1+i)$.

For $i = 1$, this gives m as $0,08nL^2$, compared with $0,125nL^2$ for a straight yield lines. This represents a 36 per cent decrease in carrying capacity! Should a carrying capacity of $nL^2/8$ be required, then a value of $i = 0,27$ is needed, i.e. the hogging moment of resistance needs to be around 25 per cent of the value of the sagging moment of resistance. This is outside the generally accepted ratio of hogging to sagging moments of 0,5 to 3. In practice the minimum value of 0,5 would be used.

10.5 HILLERBORG STRIP METHOD

10.5.1 Theoretical Background

Using classical elastic analysis of plates (Timoshenko and Woinowsky-Krieger, 1959), it may be shown the following governing equation holds for the moments,

$$\frac{\partial^2 M_{xx}}{\partial x^2} + \frac{\partial^2 M_{yy}}{\partial y^2} + \frac{\partial^2 M_{xy}}{\partial x \partial y} + \frac{\partial^2 M_{yx}}{\partial y \partial x} = -n$$ (10.135)

where the force system is defined in Fig. 10.14.

An elastic analysis is justified as this is a lower bound solution in that equilibrium is satisfied and the moments are less than the yield moments. A further simplification is made by assuming that all the loading is carried by the flexural moments M_{xx} and M_{yy} with the contribution of the twisting moments M_{xy} and M_{yx} set equal to zero.

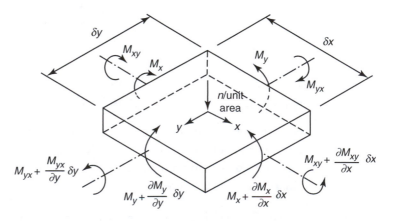

FIGURE 10.14   Forces acting on an incremental volume

Thus Eq. (10.135) becomes

$$\frac{\partial^2 M_{xx}}{\partial x^2} + \frac{\partial^2 M_{yy}}{\partial y^2} = -n \tag{10.136}$$

The load n may be considered to be partially resisted by xx axis flexure and partially by yy axis flexure with a distribution coefficient α introduced such that

$$\frac{\partial^2 M_{xx}}{\partial x^2} = -\alpha n \tag{10.137}$$

and

$$\frac{\partial^2 M_{yy}}{\partial y^2} = -(1-\alpha)n \tag{10.138}$$

Equations (10.137) and (10.138) will be recognized as the equations governing flexure of simple beam elements. Thus, a numerical approach to utilizing elastic plate bending theory can be set up by dividing the slab into a series of strips in the x and y directions and distributing the loading between these strips. It is normal practice to use only two values of α, namely zero or unity with the resultant load distribution being in either the x or the y directions. The exact pattern of load distribution is at the designer's discretion but clearly it is more economic to allow the majority of the loading where possible to be taken on the shorter span. At slab corners it is generally acceptable to use load dispersion lines at 45°. The number of load carrying strips should not be excessive as this may produce detailing problems with continually changing bar spacing. Too small a number however will lead to inefficient use of reinforcement. For background information on the Hillerborg Strip Method and for advanced applications, reference should be made to Wood and Armer (1968). Note the Hillerborg Strip Method does not impose compatibility of deformations between individual strips.

EXAMPLE 10.12 Application of Hillerborg Strip Method. Consider the simply supported rectangular slab shown in Fig. 10.15, taking load distribution lines at 45° from the corners. Fig. 10.15(b) shows a possible distribution of load carrying strips.

Strip 1 (Fig. 10.15(c)):
The whole slab is under a UDL of n, hence

$$m_{\text{Sd},1} = \frac{n L_y^2}{8} \tag{10.139}$$

Strip 2 (Fig. 10.15(d)):
The loading here comprises two triangular shaped load patches at the ends of the slab. From Moy (1996), it may be shown that for this case

$$m_{\text{RSd}} = Kn \frac{(L_1 + L_2)^2}{8} \tag{10.140}$$

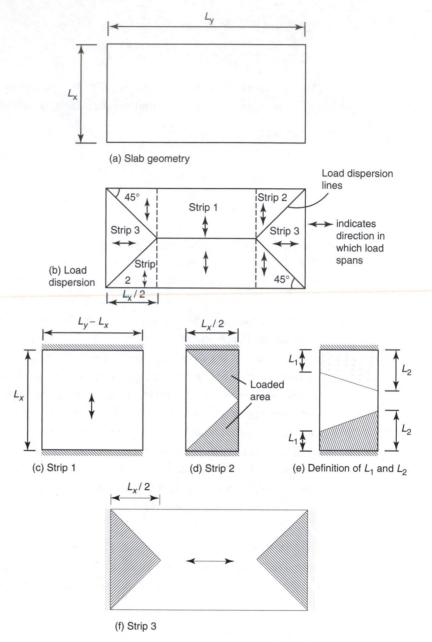

FIGURE 10.15 Hillerborg Strip Method

where in the general case (Fig. 10.15(e)) K is given by

$$K = \frac{4}{3}\left(1 - \frac{1}{2 + (L_1/L_2) + (L_2/L_1)}\right)$$ (10.141)

Strip 3 (Fig. 10.15(f)):
This can be treated in a similar fashion to Strip 2.

10.6 SLAB DESIGN AND DETAILING

10.6.1. Ratio of Hogging and Sagging Moments of Resistance

The ratio of hogging to sagging moments of resistance i may according to cl 5.6.2 (2) be taken between values of 0,5 and 2,0. It is suggested however that to avoid problems with the serviceability limit states of deflection and cracking that the likely distribution of moments corresponding to an elastic analysis be borne in mind and that the value of i be taken between 1 and 2.

10.6.2 Design Considerations

The maximum value of x_u/d should be 0,25 for concretes of strengths less than or equal to C50, above C50 the limit is 0,15 (cl 5.6.2 (2)).

10.6.3 Detailing Requirements

In general these are the same as beams except that:

(1) The maximum spacing of bars $s_{max,slabs}$ is for the main reinforcement $3h \leq 400$ mm (cl 9.3.1.1 (3)), and for secondary reinforcement $3,5h \leq 450$ mm. Provided these rules are complied with and $h \leq 200$ mm further measures (or checks) on crack widths need not be made (cl 7.3.3 (1)).

(2) At least half the sagging reinforcement should continue to the support and be fully anchored (cl 9.3.1.2 (1)). Where the slab has partial fixity not taken into account in the analysis, then the hogging reinforcement should be capable of resisting at least 25 per cent of the maximum moment in the adjacent span and should extend for a length of 0,2 times the adjacent span (cl 9.3.1.2 (2)).

(3) At a free edge, reinforcement should be detailed as in Fig. 10.16 to allow correct placement and to resist any implicit twisting moments.

(4) Curtailment. This is in accordance with beams. Where, however, a full fire safety engineering approach is adopted; it will be necessary to check the anchorage

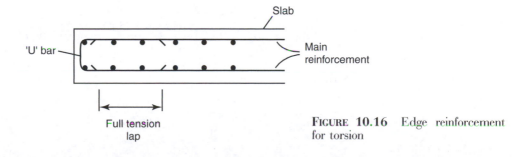

FIGURE 10.16 Edge reinforcement for torsion

capabilities of the top reinforcement owing the gradual shift of any points of contra-flexure towards the centre of the span as the moments redistribute when the sagging moment of resistance gradually drops to zero.

EXAMPLE 10.13 Design of a rectangular simply supported slab. The slab has dimensions of 6 m by 4,5 m and is 200 mm thick. It carries a variable action of 5 kPa. The concrete is grade C25/30 (specific weight 25 kN/m^3) and the steel grade is 500. Total factored actions on the system:

$$n_{Sd} = 1,5 \times 5 + 1,35 \times (0,200 \times 25) = 14,25 \text{ kPa} \tag{10.142}$$

(a) Yield line analysis:
  The solution for a rectangular slab is given in Example 10.2.
  In this case, $i = 0$, $L = 6m$, $\alpha L = 4,5$ (hence $\alpha = 0,75$), giving

$$m_{Sd} = \frac{14,25 \times 4,5^2}{24} \left(\sqrt{0,75^2 + 3} - 0,75 \right)^2$$
$$= 15,56 \text{ kNm/m} \tag{10.143}$$

(b) Hillerborg Strip:
  Use load dispersion lines at 45° from the corners to give the strip pattern shown in Fig. 10.17(a). This gives the short side divided into 3 strips and the long side in 5.

  The moment m_{Sd} is determined from Eq. (10.140). From Eq. (10.141) it should be noted that K takes a value of 4/3 if either L_1 or L_2 equals zero.

  Strip 1:
  In this case $L_1 = 1,125$ m and $L_2 = 0$, so

$$m_{Sd,1} = \frac{4}{3} \times 14,25 \frac{1,125^2}{8} = 3,01 \text{ kNm/m} \tag{10.144}$$

  Strip 2:

$$L_1 = 1,125 \text{ m}, L_2 = 2,25 \text{ m}, \text{ so } K = 1,037, \text{ and}$$

$$m_{Sd,2} = 1,037 \times 14,25 \frac{(1,125 + 2,25)^2}{8} = 21,0 \text{ kNm/m} \tag{10.145}$$

  Strip 3:

$$m_{Sd,3} = 14,25 \times \frac{4,5^2}{8} = 36,1 \text{ kNm/m} \tag{10.146}$$

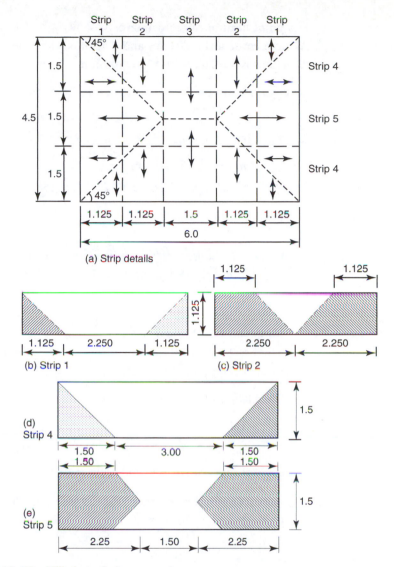

FIGURE 10.17 Hillerborg Strip approach

Strip 4:

$$L_1 = 0, L_2 = 1{,}5\,\text{m}, K = 4/3$$

$$m_{\text{Sd},4} = \frac{4}{3} \times 14{,}25\frac{1{,}5^2}{8} = 5{,}34\,\text{kNm/m} \tag{10.147}$$

Strip 5:

In order to calculate K it should be noted that the strip is that symmetric about its centre line, thus

$$L_1 = 2{,}25\,\text{m}, L_2 = 1{,}50, K = 1{,}013:$$

$$m_{\text{Sd},4} = 1{,}013 \times 14{,}25\frac{(1{,}5 + 2{,}25)^2}{8} = 25{,}4\,\text{kNm/m} \tag{10.148}$$

Design of flexural reinforcement:

Assume cover to be 20 mm, and D12 bars, thus the mean effective depth for both directions is $d = 200 - 20 - 12 = 168$ mm.

Strictly d should be calculated for each direction, however using a mean value will not induce significant errors.

(a) Yield line:

$$m_{Sd} = 15,56 \text{ kNm/m}$$
$$M/bd^2f_{ck} = 15,564 \times 10^6/(1000 \times 168^2 \times 25) = 0,0221$$

From Eq. (6.22)

$$A_s f_{yk}/dbf_{ck} = 0,652 - \sqrt{0,425 - (1,5M/bd^2f_{ck})} = 0,026$$
$$x_u/d = 1,918A_s f_{yk}/dbf_{ck} = 1,918 \times 0,026 = 0,05 < 0,25$$
$$A_s = 0,026 \times 1000 \times 168 \times 25/500 = 218 \text{ mm}^2/\text{m}.$$

Minimum longitudinal steel:

The greater of $0,0013b_td$ or $0,26f_{ctm}b_td/f_{yk}$ where b_t is the mean width of the tension zone (cl 9.2.1.1 (1)).

Calculation of $A_{s,min}$:

$$0,0013b_td = 0,0013 \times 1000 \times 168 = 218 \text{ mm}^2/\text{m}.$$
$$f_{ctm} = 0,3f_{ck}^{(2/3)} = 0,3 \times 25^{(2/3)} = 2,56 \text{ MPa}$$
$$0,26f_{ctm}b_td/f_{yk} = 0,26 \times 2,56 \times 1000 \times 168/500 = 224 \text{ mm}^2$$

Actual required is less than $A_{s,min}$. Maximum spacing is lesser of $3h$ or 400 mm. Critical value is 400 mm.

Deflection check:

$$\rho = 0,001\sqrt{f_{ck}} = 0,001\sqrt{25} = 5,0 \times 10^{-3}$$

$$\rho = A_s/bd = 262/(1000 \times 168) = 1,56 \times 10^{-3}. \text{ Thus } \rho < \rho_0$$

Thus allowable l/d is given by

$$\frac{l}{d} = K\left[11 + 1,5\sqrt{f_{ck}}\frac{\rho_0}{\rho} + 3,2\sqrt{f_{ck}}\left(\frac{\rho_0}{\rho} - 1\right)^{3/2}\right]$$

or

$$\frac{l}{d} = 1,0\left[11 + 1,5\sqrt{25}\frac{0,00500}{0,00156} + 3,2\sqrt{25}\left(\frac{0,00500}{0,00156} - 1\right)^{3/2}\right]$$

$$= 87.43$$

This value is high, Table 7.4N (EN 1992-1-1) appears to suggest a limiting value of 20. The actual l/d is $4500/168 = 26,8$, thus implying the reinforcement needs increasing by $26,8/20 = 1,34$ to $1,34 \times 218 = 292 \, \text{mm}^2/\text{m}$.

Fix D10 at 300 ctrs [377 mm²/m].

(b) Hillerborg Strip:
Moment corresponding to minimum reinforcement:
From above, $A_{s,min} = 224 \, \text{mm}^2/\text{m}$

$$A_{s,min} f_{yk}/bd f_{ck} = 224 \times 500/(1000 \times 168 \times 25) = 0,0267$$

From Eq. (6.22),

$$M_{min}bd^2 f_{ck} = 0,667 \times 0,0267(1,303 - 0,0267) = 0,0227$$
$$M_{Rd,min} = 0,0227 \times 1000 \times 168^2 \times 25 \times 10^{-6} = 16,0 \, \text{kNm/m}.$$

The moments in Strips 1 and 4 are both less than the moment corresponding to minimum reinforcement.

Strip 2:

$$m_{Sd} = 21,0 \, \text{kNm/m}, M_{Sd}/bd^2 f_{ck} = 0,0298, A_s f_{yk}/bd f_{ck} = 0,0353$$
$$(x_u/d = 0,068), A_s = 297 \, \text{mm}^2/\text{m} \, [\text{D10 at 250, 314 mm}^2/\text{m}]$$

Strip 3:

$$m_{Sd} = 36,1 \, \text{kNm/m}, M_{Sd}/bd^2 f_{ck} = 0,0512, A_s f_{yk}/bd f_{ck} = 0,0619$$
$$(x_u/d = 0,119), A_s = 520 \, \text{mm}^2/\text{m} \, [\text{D10 at 125, 628 mm}^2/\text{m}]$$

Strip 5:

$$M_{Sd} = 25,4 \, \text{kNm/m}, M_{Sd}/bd^2 f_{ck} = 0,0360, A_s f_{yk}/bd f_{ck} = 0,0429$$
$$(x_u/d = 0,082), A_s = 360 \, \text{mm}^2/\text{m} \, [\text{D10 at 200, 393 mm}^2/\text{m}]$$

Strip 3 will control deflection. $\rho = A_s/bd = 520/1000 \times 168 = 0,00310$. $\rho_0 = 0,005$ (from above).
For a simply supported slab $K=1$ (Table 7.4N, EN 1992-1-2).

As $\rho < \rho_0$, basic L/d ratio is given by

$$\frac{L}{d} = K\left[11 + 1,5\sqrt{f_{ck}}\frac{\rho_0}{\rho} + 3,2\sqrt{f_{ck}}\left(\frac{\rho_0}{\rho} - 1\right)^{3/2}\right]$$
$$\frac{L}{d} = 1,0\left[11 + 1,5\sqrt{25}\frac{0,005}{0,00310} + 3,2\sqrt{25}\left(\frac{0,005}{0,00310} - 1\right)^{3/2}\right]$$
$$= 26,8$$

TABLE 10.2   Comparison of reinforcement required in Example 10.13.

Yield line:

$2 \times 377 \times 4{,}5 \times 6 \times 10^{-6} = 0{,}0203 \, \text{m}^3$

Hillerborg Strip:

Strip 1: $2 \times 1{,}125 \times 4{,}5 \times 262 \times 10^{-6} = 0{,}0027$
Strip 2: $2 \times 1{,}125 \times 4{,}5 \times 314 \times 10^{-6} = 0{,}0032$
Strip 3: $1{,}5 \times 4{,}5 \times 628 \times 10^{-6} = 0{,}0042$
Strip 4: $2 \times 1{,}5 \times 6 \times 262 \times 10^{-6} = 0{,}0047$
Strip 5: $2 \times 1{,}5 \times 6 \times 393 \times 10^{-6} = 0{,}0070$

Total $= 0{,}0198 \, \text{m}^3$

Factor for service stress:

$$310/\sigma_s = (500/f_{yk}) \, (A_{s,prov}/A_{s,req})$$

or

$$310/\sigma_s = (500/500) \, (628/520) = 1{,}21$$

Thus allowable $L/d = 26{,}8 \times 1{,}21 = 32{,}4$
Actual $L/d = 4500/168 = 26{,}8$. This is satisfactory.

A comparison of amounts of reinforcement by volume is carried out in Table 10.2. There is virtually no difference between the two solutions. Note, however that the distribution of reinforcement is significantly different as under the Hillerborg Strip Method, parts of the slab require only minimum with a resultant redistribution to the parts of the slab with high bending moments.

EXAMPLE 10.14   Encastré slab. Determine the design moments in a slab all of whose edges are encastré. The dimensions and loading actions are those of Example 10.13. The value of the ratio of hogging to sagging moments is taken as 1,5.

(a) Yield line:
     The design moment from Example 10.13 is divided by $1 + i = 1 + 1{,}5 = 2{,}5$ (see Example 10.2), so $m_{Sd} = 14{,}64/2{,}5 = 5{,}86 \, \text{kNm/m}$ (sagging) and $i \times m_{Sd} = 1{,}5 \times 14{,}64/2{,}5 = 8{,}78 \, \text{kNm/m}$ (hogging).

(b) Hillerborg Strip Method:
     In this case the free bending moments can be resisted by a sagging moment of resistance of m and a hogging moment of resistance of im:

     Strip 1:

$$(1 + i)m_{Sd,1} = 2{,}83 \, \text{kNm/m};$$

     so, $m_{Sd,1} = 1{,}13 \, \text{kNm}$ and $im_{Sd,1} = 1{,}70 \, \text{kNm/m}$

Strip 2:

$$(1+i)m_{Sd,2} = 19{,}80 \text{ kNm/m};$$

so, $m_{Sd,2} = 7{,}92 \text{ kNm/m}$ and $im_{Sd,2} = 11{,}88 \text{ kNm/m}$

Strip 3:

$$(1+i)m_{Sd,3} = 33{,}94 \text{ kNm/m};$$

so, $m_{Sd,3} = 13{,}58 \text{ kNm/m}$ and $im_{Sd,3} = 20{,}6 \text{ kNm/m}$

Strip 4:

$$(1+i)m_{Sd,4} = 5{,}03 \text{ kNm/m};$$

so, $m_{Sd,4} = 2{,}01 \text{ kNm/m}$ and $im_{Sd,4} = 3{,}02 \text{ kNm/m}$

Strip 5:

$$(1+i)m_{Sd,5} = 23{,}88 \text{ kNm/m};$$

so, $m_{Sd,5} = 9{,}55 \text{ kNm/m}$ and $im_{Sd,5} = 14{,}33 \text{ kNm/m}$.
The reinforcement may now be designed in the usual manner.

EXAMPLE 10.15 Slab with mixed supports. A slab is restrained on one long side and one short side, but is simply supported on the others. The remainder of the data is as Example 10.13.

(a) Yield line:

Without proceeding through the full solution, a convenient standard solution exists for an isotropic slab with mixed supports (Jones and Wood, 1967) (Fig. 10.18), so m_{Sd} is given by

$$m = \frac{na_r^2}{24}\left[\sqrt{3+\left(\frac{a_r}{b_r}\right)^2} - \frac{a_r}{b_r}\right]^2 \qquad (10.149)$$

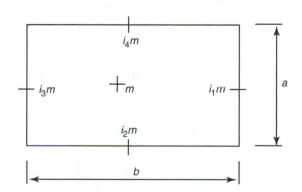

FIGURE 10.18 Notation for Eq. (10.149)

where

$$a_r = \frac{2a}{\sqrt{1+i_1} + \sqrt{1+i_4}}$$

(10.150)

and

$$b_r = \frac{2b}{\sqrt{1+i_1} + \sqrt{1+i_3}}$$

(10.151)

Note that for Eq. (10.149) to be valid $b_r/a_r > 1{,}0$.

From Eq. (10.150),

$$a_r = 2 \times 4{,}5/(\sqrt{(1+1{,}5)} + 1) = 3{,}487 \, \text{m}$$

From Eq. (10.151),

$$b_r = 2 \times 6{,}0/(\sqrt{(1+1{,}5)} + 1) = 4{,}649 \, \text{m} \ (b_r > a_r)$$
$$a_r/b_r = 3{,}487/4{,}649 = 0{,}75$$

From Eq. (10.149),

$$m_{Sd} = \frac{13{,}41 \times 3{,}487^2}{24} \left[\sqrt{3 + 0{,}75^2} - 0{,}75 \right]^2$$
$$= 8{,}79 \, \text{kNm/m}$$

(10.152)

$$im_{Sd} = 1{,}5 \times 8{,}79 = 13{,}19 \, \text{kNm/m}.$$

(b) Hillerborg Strip Method:

For each strip, the moment at one end equals zero, the moment at the other equals im. The free bending moment in any strip is given by Eq. (10.140), thus assuming the maximum sagging moment occurs at the centre of a strip, the sagging moment is given by the value from Eq. (10.140) less $im/2$. This is not strictly correct as, of course, the maximum sagging moment occurs at the point of zero shear. The above assumption will produce answers for m within 5 per cent. The values of free bending moments are those in Example 10.12.

Strip 1:
Free bending moment $= 2{,}83 \, \text{kNm/m}$

so

$$m_{Sd,1} + 0{,}5(1{,}5m_{Sd,1}) = 2{,}83$$

(10.153)

or,

$$m_{Sd,1} = 1{,}62 \, \text{kNm/m} \quad \text{and} \quad im_{Sd,1} = 2{,}43 \, \text{kNm/m}.$$

Strip 2:
Free bending moment $= 19,8\,\text{kNm/m}$

so

$$m_{\text{Sd},2} + 0,5(1,5m_{\text{Sd},2}) = 19,8 \qquad\qquad (10.154)$$

or

$m_{\text{Sd},2} = 11,3\,\text{kNm/m}$ and $im_{\text{Sd},2} = 17,0\,\text{kNm/m}$.

Strip 3:
Free bending moment $= 33,94\,\text{kNm/m}$

so

$$m_{\text{Sd},3} + 0,5(1,5m_{\text{Sd},3}) = 33,94 \qquad\qquad (10.155)$$

or

$m_{\text{Sd},3} = 19,4\,\text{kNm/m}$ and $im_{\text{Sd},3} = 29,1\,\text{kNm/m}$.

For strip 3, the exact solution for m_{Sd} is given by

$$m_{\text{Sd}} = nL^2 \frac{i + 2 - 2\sqrt{i+1}}{2i^2} \qquad\qquad (10.156)$$

or

$$m_{\text{Sd}} = 13,41 \times 4,5^2 \frac{1,5 + 2 - 2\sqrt{1,5+1}}{2 \times 1,5^2} = 20,38\,\text{kNm/m} \qquad\qquad (10.157)$$

and $im_{\text{Sd},5} = 1,5 \times 20,38 = 30,57\,\text{kNm/m}$.
The approximate solutions are within 5 per cent.

Strip 4:
Free bending moment $= 5,03\,\text{kNm/m}$

so

$$m_{\text{Sd},4} + 0,5(1,5m_{\text{Sd},4}) = 5,03 \qquad\qquad (10.158)$$

or

$m_{\text{Sd},4} = 2,88\,\text{kNm/m}$ and $im_{\text{Sd},4} = 4,31\,\text{kNm/m}$.

Strip 5:
Free bending moment $= 23,9\,\text{kNm/m}$

so

$$m_{\text{Sd},5} + 0,5(1,5m_{\text{Sd},5}) = 23,9 \qquad\qquad (10.159)$$

or,

$m_{\text{Sd},5} = 13,7\,\text{kNm/m}$ and $im_{\text{Sd},5} = 20,5\,\text{kNm/m}$.

10.7 Beam and Slab Assemblies

Yield line theory by itself gives no indication on how much load is transferred from the slab to a monolithic supporting beam. The Hillerborg Strip Method allows the reactions from the strips to be directly taken as reactions on any monolithic beam support system.

The geometric parameters determined whilst undertaking the yield line collapse for the slab generally do not give an indication of the load distribution to the support beams. For yield line methods, therefore, an analysis has to be undertaken involving the beams and slab.

10.7.1 Yield Line Method for Beam and Slab Assemblies

Where the moment capacities of the beam system and the slab are independent, the following procedure is adopted:

(a) The collapse load for the slab is determined using the methods previously illustrated assuming the beams form rigid supports to the slab. For single bay systems or free edges, the beam is considered as a simple support. For continuous edges, the effect of adjacent bays may be taken into account.
(b) The collapse load for the complete system is considered with the slab moment of resistance taken as that determined in the previous step.

EXAMPLE 10.16   Beam and slab. Consider a square slab of side L carrying a UDL of n per unit area supported on edge beams and the whole assembly on corner columns (Fig. 10.19(a)). The slab has a (sagging) moment of resistance of m per unit width and the beam has a local moment of resistance M.

Stage 1: Collapse of slab alone.
This has already been covered in Example 10.1 and $m = nL^2/24$.

Stage 2: Collapse of beam and slab assembly.
There are four possible failure modes to be considered:
Mode 1 (Fig. 10.19(b)):
The equilibrium equation using virtual work for this case is given by

$$\frac{nL^2}{2}\frac{L}{4} = 2M + mL \tag{10.160}$$

Defining a normalized beam failure moment $\bar{m}$ as $\bar{m} = M/mL$, Eq. (10.160) becomes

$$nL^2 = 8m(1 + \bar{m}) \tag{10.161}$$

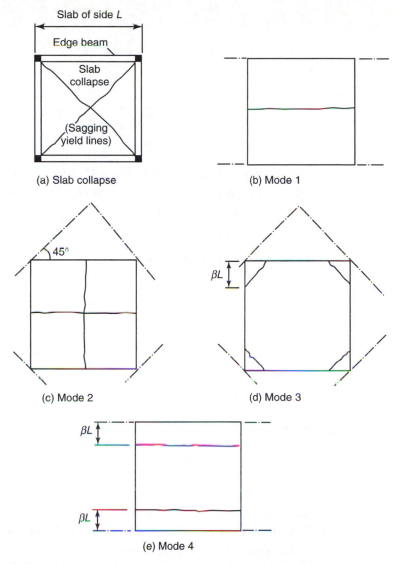

FIGURE 10.19 Collapse mechanisms for a square slab with edge beams

If the slab collapse moment m is set equal to $nL^2/24$, Eq. (10.161) gives $\bar{m} = 1,00$, i.e. the beam has constant strength of $M = mL$.

Mode 2 (Fig. 10.19(c)):
This produces the same result as Mode 1. It is left to the reader to verify this.

Mode 3 (Fig. 10.19(d)):
The equilibrium equation for this mode is

$$nL^2 - 4n\frac{(\beta L)^2}{3} \times \frac{1}{3} = \frac{8M}{\beta L} + 4m\sqrt{2}\beta L\frac{\sqrt{2}}{\beta L} \tag{10.162}$$

or

$$\frac{nL^2}{24m} = \frac{1 + (\bar{m}/\beta)}{3 - 2\beta^2}$$

(10.163)

or with the slab collapse load as $nL^2/24$,

$$\bar{m} = 2\beta(1 - \beta^2)$$

(10.164)

For $\beta = 0$, $m = 0$, and $\beta = 0,5$, $m = 0,75$. The reason why this mode gives a lower collapse moment for the beam than Modes 1 or 2 is that with a collapse moment of $nL^2/24$ the slab is 'overstrong'.

Mode 4 (Fig. 10.19(e)):
The equilibrium equation for this mode is

$$nL^2(1 - 2\beta) + 2n\beta L^2\frac{1}{2} = 2(mL + 2M)L\frac{1}{\beta L}$$

(10.165)

or

$$\frac{nL^2}{m} = \frac{2(1 + 2\bar{m})}{\beta(1 - \beta)}$$

(10.166)

Setting $m = nL^2/24$ gives

$$\bar{m} = 6\beta(1 - \beta) - 0,5$$

(10.167)

At $\beta = 0$, Eq. (10.167) gives $m = 0,50$ and at $\beta = 0,5$, $m = 1,00$. The existence of the need for an apparent hogging moment near the column is again due to the slab being overstrong for a moment of $nL^2/24$ and is also due to the neglect of the possibility demonstrated by an elastic analysis of the existence of hogging moments in the slab close to the support.

The values of m for Modes 1, 3 and 4 are plotted in Fig. 10.20. This figure shows that although the assumption of a uniform moment is safe for the sagging moment of resistance of the support beams, it would be prudent to consider that some allowance should be made for possible hogging moments at the supports.

For the situation where the beam and slab assembly is multi-bay, consideration needs to be given to both the collapse of a single bay and to the whole system (Fig. 10.21).

Equally the hogging moments across a beam have to be the same for the slabs spanning from either side. It is not sufficient to use the same values of i (the ratio of hogging to sagging moment), it is necessary to use the same absolute value of the hogging moment. The easiest way to ensure this is to analyse the slabs with the most free edges (i.e. slabs A in Fig. 10.21), then gradually work toward slabs with no free

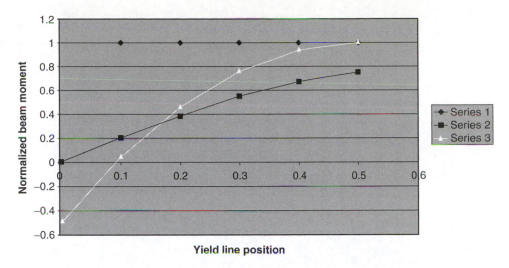

FIGURE 10.20   Variation of normalized beam moment with yield line position

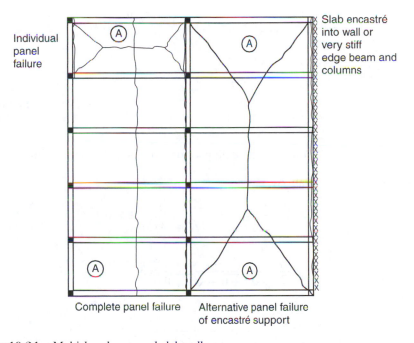

FIGURE 10.21   Multi-bay beam and slab collapse

edges using the hogging moments already determined at the slab boundary and then applying these as known moments when carrying out the analysis either by yield line or Hillerborg Strip. This is demonstrated in the next example.

EXAMPLE 10.17   Two bay continuous slab. Consider the two bay slab shown in Fig. 10.22 both loaded with a load of n with the common side continuous. The factor i is initially considered as 4/3.

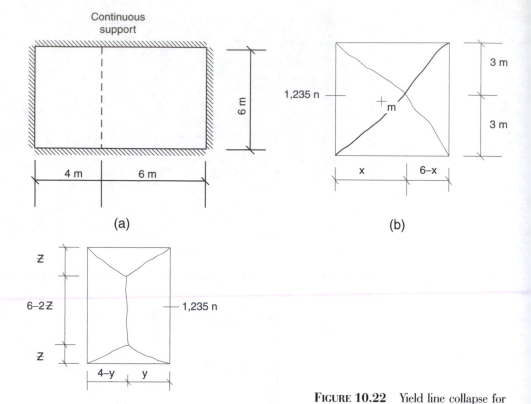

FIGURE 10.22 Yield line collapse for two-bay slab

Consider the collapse load of each slab separately using Eq. (10.149) to (10.151),

Slab A:

$$\frac{a_r}{2} = \frac{4}{\sqrt{1+0} + \sqrt{1+(4/3)}} = 1{,}583 \tag{10.168}$$

$$\frac{b_r}{2} = \frac{6}{\sqrt{1+0} + \sqrt{1+0}} = 3 \tag{10.169}$$

and

$$
\begin{aligned}
m &= \frac{n a_r^2}{24}\left[\sqrt{3 + \left(\frac{a_r}{b_r}\right)^2} - \frac{a_r}{b_r}\right]^2 \\[2mm]
&= \frac{n(2 \times 1{,}583)^2}{24}\left[\sqrt{3 + \left(\frac{1{,}583}{3}\right)^2} - \frac{1{,}583}{3}\right]^2 \\[2mm]
&= 0{,}688\,n
\end{aligned}
\tag{10.170}
$$

The hogging moment over the support is then given by

$$(4/3) \times 0{,}688n = 0{,}917n$$

Slab B:

$$\frac{a_r}{2} = \frac{6}{\sqrt{1+0} + \sqrt{1+(4/3)}} = 2{,}374 \tag{10.171}$$

$$\frac{b_r}{2} = \frac{6}{\sqrt{1+0} + \sqrt{1+0}} = 3 \tag{10.172}$$

and

$$m = \frac{na_r^2}{24}\left[\sqrt{3 + \left(\frac{a_r}{b_r}\right)^2} - \frac{a_r}{b_r}\right]^2$$

$$= \frac{n(2 \times 2{,}374)^2}{24}\left[\sqrt{3 + \left(\frac{2{,}374}{3}\right)^2} - \frac{2{,}374}{3}\right]^2$$

$$= 1{,}164\,n \tag{10.173}$$

The hogging moment over the support is then given by

$$(4/3) \times 1{,}164n = 1{,}552n$$

There are three alternative approaches to handling the apparent mismatch of hogging moments:

- Use the value from slab A and re-analyse slab B with a known hogging moment,
- Use the value from slab B and re-analyse slab A with a known hogging moment,
- Use a value lying between those determined and re-analyse both slabs.

In this example, the third approach will be adopted, use a hogging moment of resistance equal to the mean of the two values determined, i.e. 1,235n.

Slab B:
Consider the yield line pattern in Fig. 10.22(b). This may not be completely correct but it will give an answer close to the actual value.

Work equation:

$$n\frac{6^2}{3} = \frac{6 \times 1{,}235\,n + 6m}{x} + \frac{6m}{6-x} + \frac{2 \times 6m}{3} \tag{10.174}$$

where the left-hand term is the work done by the load and the right-hand term the work done by the sagging moment m and the hogging moment 1,235n.

Re-arranging Eq. (10.176) gives

$$\frac{n}{m} = \frac{(6/6x - x^2) + (4/6)}{2 - (1{,}235/x)} \tag{10.175}$$

Differentiate Eq. (10.175) to give

$$\frac{(6/6x - x^2) + (4/6)}{2 - (1{,}235/x)} = \frac{-6[(6 - 2x)/(6x - x^2)^2]}{1{,}235/x^2} \tag{10.176}$$

After re-arranging Eq. (10.176) gives

$$(0{,}823x^2 - 33{,}88x + 108{,}85)x^2 = 0 \tag{10.177}$$

The roots of this equation are 0 (twice), 37,65 and 3,512. The only applicable root is $x = 3{,}512$.

Back-substituting gives $m = 1{,}218n$. The final ratio of hogging to sagging moment is $1{,}235n/1{,}218n = 1{,}02$ which is acceptable.

Slab A (Fig. 10.22(c)):
Displaced volume is given by

$$Vol = \frac{2z \times 4}{3} + \frac{4}{2}(6 - 2z) = \frac{4}{3}(9 - z) \tag{10.178}$$

work equation is given by

$$\frac{4n}{3}(9 - z) = \frac{6}{y}(1{,}235\,n + m) + \frac{6m}{4y} + \frac{2x4m}{z} \tag{10.179}$$

or

$$\frac{n}{m} = \frac{(1/y) + (1/4 - y) + (4/3z)}{(2/9)(9 - z) - (1{,}235/y)} \tag{10.180}$$

Although Eq. (10.180) could be partially differentiated with respect to y and z, the resulting equations cannot be solved analytically, so a spreadsheet was used to optimize the value of n/m. The results are given in Table 10.3. The least value of n/m of 1,633 is given when $z = 2$ and $y = 2{,}5$.

It will also be observed that a slight error in y or z will not significantly affect the value of n/m. The final ratio of hogging to sagging moment is $1{,}235n/1{,}633n = 0{,}756$ which is acceptable, although on the low side (this could give problems with deflection).

TABLE 10.3   Optimization of Eq. (10.180).

y	2	2,25	2,5	2,75	3
z					
1,5	1,800	1,704	1,668	1,686	1,771
1,75	1,773	1,674	1,637	1,657	1,747
2,0	1,777	1,671	1,633	1,654	1,748
2,25	1,804	1,691	1,649	1,671	1,770
3	1,854	1,730	1,683	1,705	1,807

10.7.2 Load Dispersion on to Supporting Beams

An advantage of using the Hillerborg Strip approach is that the loads carried by the supporting beams may be directly determined from the reactions from the loading on the individual strips. Yield line analysis, unless combined beam and slab mechanisms are considered, does not provide the reactions on to supporting systems. The final yield line pattern does not necessarily provide the load distribution onto supporting systems, although clearly in some cases it does, e.g. simply supported square slab.

The need for calculation of beam support loads is often performed assuming the existence as in the Hillerborg Strip Method of lines indicating load dispersion. It is common to assume corner lines at 45°. This is probably not unreasonable for a slab with the same edge condition all round the slab, but it may need reassessing where unequal edge conditions occur.

10.7.3 Shear in One- and Two-way Spanning Slabs

10.7.3.1 Shear Effects from Distributed Loads

The shear at the periphery of a slab is determined from the loading carried to that support. The shear resistance is determined in the usual manner as for beams, except that should problems arise with a deficiency in shear capacity, which is rare, then it is made up by increasing the percentage of reinforcement effective at the edge being considered. Shear links are never used in normal one or two spanning slabs.

10.7.3.2 Concentrated Loads

The only loads that need to be considered under this heading are where high concentrated loads are physically applied to a slab. There is no general need to check the optional concentrated variable load. Where high concentrated loads are applied then punching shear needs to be checked around the load. If high concentrated loads are applied, it may be prudent to consider either a beam and slab assembly or using the Hillerborg Strip approach to design whereby 'strong' strips can be employed to resist flexure and shear from such loads.

10.8 FLAT SLABS

These will not be covered in any detail, as their flexural design generally follows the principles above. The problems occur if drop heads are used as failure needs considering both for the drop head and the slab surrounding the drop head. Additionally, problems may arise with punching shear. These can be circumvented by using drop heads and proprietary shear reinforcement rather than traditional links as these are often very difficult to detail and place within thin slabs. For further details see Reagan (1981), Oliviera, Regan and Melo (2004), Whittle (1994) and Chana and Clapson (1992). Additional information is given in (informative) Annex I. This suggests that,

(a) For an elastic analysis the stiffness of the slab should be based on the full width of the panels for vertical loading, but 40 per cent for horizontal loading in order to reflect the increased flexibility of beam–slab joints in this type of construction.

(b) The slab should be divided into column and middle strips (as Fig. 10.23), and the moments apportioned between the middle and column strips as in Table 10.3 (Table I.1 of EN 1992-1-1). The values that have been adopted in conventional British practice are also included in square parentheses in Table 10.4. Where there are column drop heads, the values of the widths of the strips may be adjusted appropriately.

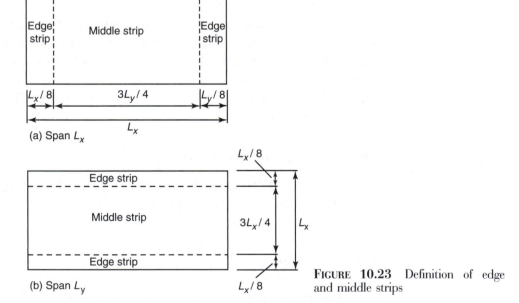

FIGURE 10.23 Definition of edge and middle strips

TABLE 10.4 Bending moment distribution in flat slabs.

	Negative moments	*Positive moments*
Column Strip	60–80 [75]	50–70 [55]
Middle Strip	40–20 [25]	50–30 [45]

Note: In all cases the values selected for the column and middle strips should sum to 100.

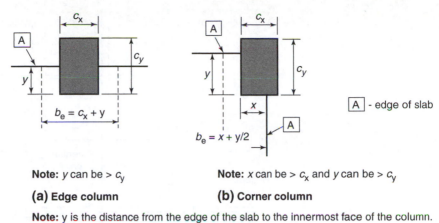

Note: y can be > c_y

Note: x can be > c_x and y can be > c_y

(a) Edge column

(b) Corner column

Note: y is the distance from the edge of the slab to the innermost face of the column.

FIGURE 10.24   Definition of effective width b_e

(c) The moment transfer at external columns should be limited unless there are perimeter beams designed to take torsion. The value of the moment transfer should be limited to $0{,}17b_e d^2 f_{ck}$ (where b_e is defined in Fig. 10.24).

10.9 WAFFLE SLABS

The design of such slabs is generally similar to flat slabs (Whittle, 1994) but the following may be noted:

(a) Owing to the presence of the ribs deflection is likely to be more critical than that for flat slabs and the ribs should be allowed for;

(b) It is recommended that the section at midspan should be used to calculate the flexural stiffness and taken as constant over the complete span for equivalent frame analyses unless the solid section at the column has a dimension greater than one-third the shorter panel dimension in which case the effect of the solid may be included;

(c) It is further recommended that the solid portion should extend at least 2,5 times the slab effective depth from the face of the column;

(d) Should critical shear perimeters extend outside the solid region into the ribbed area, then the shear should be divided equally between all the ribs. Also at the corners of a solid section whence two ribs frame then a potential 45° shear plane should be considered (Fig. 10.25);

(e) Should shear links be required in the ribs, although they should be avoided if at all possible, then they should extend for a distance along the rib at least equal to the effective depth after the theoretical cut-off point.

Cl 5.3.6 indicates that waffle or ribbed slabs need not be treated as discrete elements in the analysis if

• Rib spacing is less than 1500 mm,
• Depth of the rib below the flange does not exceed 4 times its width,

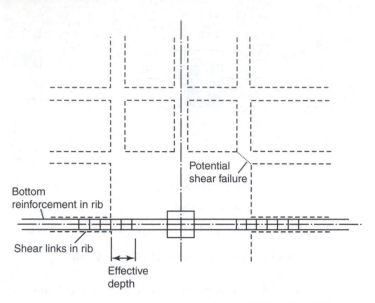

FIGURE 10.25   Shear detailing in waffle slabs

- The depth of the flange is at least 50 mm or 1/10 the clear distance between the ribs (the minimum flange thickness may be reduced to 40 mm where permanent blocks are introduced between the ribs),
- Transverse rib spacing is at least 10 times the overall depth of the slab between ribs.

REFERENCES AND FURTHER READING

Beeby, A. (2004) Why do we need ductility in reinforced concrete structures? *Concrete*, **38(5)**, 27–9.

Chana, P. and Clapson, J. (1992) Innovative shearhoop system for flat slab construction. *Concrete*, **26(1)**, 21–24.

Johansen, K.W. (1972) Yield Line Formulae for Slabs, *Cement and Concrete Association*.

Jones, I.L. and Wood, R.H. (1967) *Yield-Line Analysis of Slabs*. Thames and Hudson/Chatto and Windus.

Moy, S.S.J. (1996) *Plastic Methods for Steel and Concrete Structures* (2nd Edition), MacMillan.

Oliviera, D.R.C., Regan, P.E. and Melo, G.S.S.A. (2004) Punching resistance of RC slabs with rectangular columns. *Magazine of Concrete Research*, **56(3)**, 123–138.

Reagan, P.E. (1981) *Behaviour of reinforced concrete slabs*. Construction Industry Research and Information Association. Report No 89.

Timoshenko, S.P. and Woinowsky-Kreiger, S. (1959) *Theory of Plates and Shells*, McGraw Hill.

Whittle, R.T. (1994) *Design of reinforced concrete slabs to BS8110 (revised edition)*. Construction Industry Research and Information Association. Report No 110.

Wood, R.H. and Armer, G.S.T. (1968) The theory of the strip method for design of slabs. *Proceedings of the Institution of Civil Engineers*, **41**, 285–311.

Chapter 11 / Foundations and Retaining Walls

This chapter details design considerations for various types of foundations and retaining walls. Retaining walls are required where it is necessary for part of a structure to be below ground level, to retain earthworks in a cutting for a road, say, where due to space full stable slopes are impossible, or may form part of the design of other parts of structures such as bridge abutments. The function of a foundation is to transmit the actions from a structure to the underlying ground in such a way that the bearing strata taking the actions are not over-stressed and that undue deformations occur neither in the structure nor in the ground. The section on foundation design only considers some of the most common cases. For further information reference should be made to Henry (1986).

11.1 TYPES OF FOUNDATION

The type of foundation used in any given case is determined from consideration of the functions of the foundation, load carrying capacity and minimal deformation, and the type of load bearing strata available in a given case. Unfortunately, it is often not possible to determine the details of ground conditions at the start of a design when the new structure is to be built on the site of existing buildings. This means that the design of the foundations is often not possible until the latter stages of the design process. This may cause problems as the structure layout will already have determined and that a more complex foundation detail will then be needed. It is absolutely vital that in all cases a full geotechnical survey be carried out not only to determine the bearing pressures at the level likely to be needed for the foundation but also below this level where there is a potential likelihood of weak or soft strata which will cause problems with long term settlement. A properly carried out geotechnical survey is absolutely essential for any major project, since any long term problems with the foundations may cause severe structural distress. The cost of carrying out an appropriate geotechnical survey on any project is only a very small

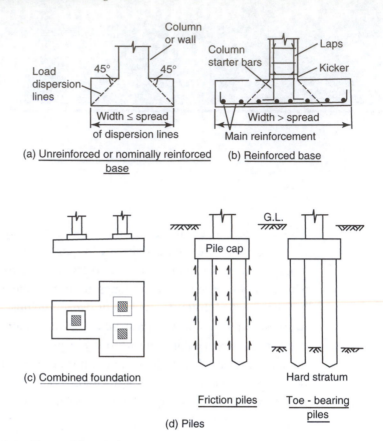

FIGURE 11.1   Types of foundation

proportion of the total cost, whereas remedial works to a structure and its environs, if such a survey is either imperfectly carried out or worse, omitted, may cost many times the original cost of the new structure.

Typical examples of commonly found structural foundations are illustrated in Fig. 11.1 and are briefly described as follows.

11.1.1 Simple Unreinforced Strip or Pad Foundation (Fig. 11.1a)

This type of foundation is restricted to the case where the loading is relatively light with reasonably high bearing pressures. The 45° load dispersion lines must intersect the sides of the foundation whose width in the case of strip footings is generally governed by the available widths of excavator buckets. For strip footings carrying cavity walls it is essential to remember that it is the internal leaf that is the more heavily loaded and that the centre line of the foundation must coincide with the resultant load in the two leaves of the wall. It may also be prudent to consider a nominal eccentricity of 50 to 100 mm for this resultant for consideration of non-uniform bearing pressures due to construction tolerances. Also for long strip footings nominal mesh reinforcement is often provided.

11.1.2 Pad Foundation (Fig. 11.1(b))

The principal use for a pad foundation is to resist the reaction from a single concrete column or steel stanchion. Since the reactions are likely to be large then the area of a pad foundation will be such that the 45° lines will fall within the base. This means that the area of the base will need to be reinforced in order to carry such reactions.

11.1.3 Combined Foundations and Raft Foundations (Fig. 11.1(c))

Combined foundations are used where either the column spacing is such that single pads would overlap or the bearing capacity is low. To avoid differential movement it is essential that the bearing pressure is uniform over the base. A raft foundation is simply a combined foundation that extends over the whole of the structure allowing the structure to behave as an entity. Because of the high loads in the lowest level of columns in such structures a raft foundation will often be several metres thick in order to avoid shear failure between the column face and the slab forming the raft.

11.1.4 Piles (Fig. 11.1(d))

These are required where any suitable bearing stratum is at too greater depth for economic use of any above types owing to the potentially high costs of excavation. Piles can be divided into two categories: end bearing piles where the load is transferred to a suitable stratum by bearing on the end of the pile and friction piles where skin friction between the shaft of the pile and clay strata is mobilized to transfer the loading. A combination of end bearing and skin friction is also possible.

Piles may also be classified by their method of construction, whether driven or bored. The possible method of construction must be taken into account when considering load carrying capacity as bored piles can rarely mobilize skin friction. Pile design is largely based on experience and empirical methods and will not be considered further. The reader is referred to Tomlinson (1994) for further information.

11.2 Basis of Design

The actual bearing pressure distribution below a foundation is complex as it depends both upon the stiffness of the foundation (and the structure above) and the physical properties and deformation characteristics of the strata below the foundation. Whilst it is possible to determine a distribution using the elastic properties of the ground, it is not generally considered necessary. The common simplification is to assume a linear distribution of bearing pressures (EN 1992-1-1 cl G.1.2). The bearing pressures or other strength parameters, used in a design should be cautious estimates based on tests or experience (EN 1997-1 cl 2.4.5.2(3)P). All designs must have a Geotechnical Design Report as an Annex to the main calculations (EN 1997-1, cl 2.8). This report must contain details of the assumptions made in the foundation design together with a summary of the relevant actions and design parameters. It need not necessarily be a complex report if the foundation design is simple or straightforward.

EN 1997-1 cl 2.4.7 recognizes five potential limit states:

EQU: This is concerned with equilibrium of the structure or ground considered as a rigid body where strength of the materials are insignificant in providing resistance. An example of where this limit state applies is overturning of a retaining wall, it does not apply to normal foundations.

STR: This is concerned with internal failure or excessive deformation within the structural materials.

GEO: This is concerned with excessive deformation or failure within the soil mass.

UPL and HYD: These are concerned with failure due to uplift or hydraulic heave, respectively.

To size foundations, the limit state STR or GEO is used. UPL and HYD will not be further considered in this text.

The partial safety factors to be used on actions are given in Table 11.1 (from Table A.1 and A.3 of EN 1997-1).

For soil parameters the values of the partial safety factors are given in Table 11.2 (from Table A.2 and A.4 of EN 1997-1).

Table 11.3 gives the partial safety factors to be applied to resistances for spread foundations and earth retaining structures for STR and GEO (from Tables A.5 and A.13 of EN 1997-1).

EN 1997-1 outlines the possible Design Approaches for STR and GEO that may be used. These are summarized in Table 11.4.

In all cases partial safety factors are applied to the actions rather than the effects of the actions. The eventual National Annex to EN 1997-1 may specify the use of one or more of these Design Approaches (possibly by prohibiting the use of some). As the National Annex had not been produced at the time of writing the text, the author has felt it expedient to consider all three, thus making design examples slightly laborious.

TABLE 11.1 Partial safety factors on actions (Table A1 and A3 of EN 1997-1).

Action	EQU		STR & GEO SET		
				A1	A2
Permanent					
Unfavourable	$\gamma_{G;dst}$	1,1	γ_G	1,35	1,0
Favourable	$\gamma_{G;stb}$	0,9	γ_G	1,0	1,0
Variable					
Unfavourable	$\gamma_{Q;dst}$	1,5	γ_Q	1,5	1,3
Favourable	$\gamma_{Q;stb}$	0	γ_Q	0	0

TABLE 11.2 Partial safety factors on soil parameters (Table A4 of EN 1997-1).

Angle of shearing	EQU	STR & GEO SET	
		M1	M2
Resistance[1] ($\gamma_{\phi'}$)	1,25	1,0	1,25
Effective cohesion ($\gamma_{c'}$)	1,25	1,0	1,25
Undrained shear strength (γ_{cu})	1,4	1,0	1,4
Unconfined strength (γ_{qu})	1,4	1,0	1,4
Weight density (γ_γ)	1,0	1,0	1,0

Note 1: For the angle of shearing resistance, the partial safety factor is applied to $\tan\phi'$.

TABLE 11.3 Partial safety factors (γ_R) for resistances (Table A13 of EN 1997-1).

	R1	R2	R3
Bearing ($\gamma_{R;v}$)	1,0	1,4	1,0
Sliding ($\gamma_{R;h}$)	1,0	1,1	1,0
Earth Resistance ($\gamma_{R;e}$)	1,0	1,4	1,0

TABLE 11.4 Design approaches.

Design approach	Action set	Soil set	Resistance set
ONE			
Combination 1	A1	M1	R1
Combination 2	A2	M2	R1
TWO	A1	M1	R2
THREE	A1 or A2	M2	R3

Note: In Design Approach 3, the factors A1 are applied to structural actions and A2 to geotechnical actions.

11.3 BEARING PRESSURES UNDER FOUNDATIONS

Under no circumstances can tension occur between the ground and the underside of the base. Further the bearing pressure distribution should be sensibly constant for the combination of permanent and variable actions (excluding wind or other accidental actions). Ideally the ratio of maximum to minimum bearing pressures should not exceed around 1,5 to 2. For cases involving transient load effects such as wind or other accidental actions this ratio can increase to around 2,5. For retaining walls a slightly higher value is permissible as some movement must occur in order to mobilize active or passive pressures (See Section 11.6.1).

Under no account should partial bearing be allowed as this will cause undue rotation of the base and a possible deleterious redistribution of internal forces in the structure above the foundation. It is generally accepted that a linear distribution of bearing pressures may be taken. The moment applied to the base of the foundation giving rise to the non-linear component of the bearing stress distribution must include the effect of any horizontal reaction applied at the top face of the foundation. Thus the total moment $M_{Sd,y.net}$ is given by

$$M_{Sd,y.net} = M_{Sd,y} + H_{Sd,z}h \tag{11.1}$$

where $M_{Sd,y}$ is the moment about the y axis from the structural analysis, $H_{Sd,z}$ is the corresponding horizontal reaction and h is the overall depth of the foundation. It should be noted that the effect of $H_{Sd,z}$ may oppose the effect of $M_{Sd,y}$, in which case it should probably be neglected.

11.3.1 Uniaxial Moments (Fig. 11.2)

Consider a base of dimensions length L and width B ($B \leq L$) with actions N_{Sd} and $M_{Sd,y.net}$. The maximum bearing pressure σ_1 is given by

$$\sigma_1 = \frac{N_{Sd}}{A} + \frac{M_{Sd,y.net}}{W_{el}} \tag{11.2}$$

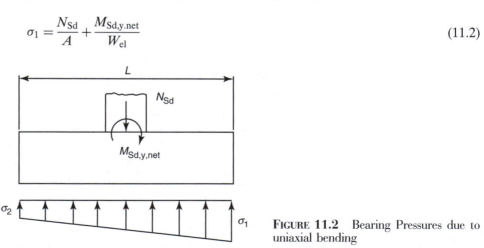

FIGURE 11.2 Bearing Pressures due to uniaxial bending

where A is the area of the base (BL) and W_{el} is the elastic section modulus of the base about the YY axis $(BL^2/6)$. The minimum bearing pressure σ_2 is given by

$$\sigma_2 = \frac{N_{Sd}}{A} - \frac{M_{Sd,y.net}}{W_{el}} \tag{11.3}$$

The pressure can be taken as uniform across the width of the base.

11.3.2 Biaxial Bending (Fig. 11.3)

The actions are N_{Sd}, $M_{Sd,y.net}$ and $M_{Sd,z.net}$ and the bearing pressures at each corner σ_1 to σ_4 are given by

$$\sigma_1 = \frac{N_{Sd}}{A} + \frac{M_{Sd,y.net}}{W_{el,y}} + \frac{M_{Sd,z.net}}{W_{el,z}} \tag{11.4}$$

$$\sigma_2 = \frac{N_{Sd}}{A} - \frac{M_{Sd,y.net}}{W_{el,y}} + \frac{M_{Sd,z.net}}{W_{el,z}} \tag{11.5}$$

$$\sigma_3 = \frac{N_{Sd}}{A} - \frac{M_{Sd,y.net}}{W_{el,y}} - \frac{M_{Sd,z.net}}{W_{el,z}} \tag{11.6}$$

$$\sigma_4 = \frac{N_{Sd}}{A} + \frac{M_{Sd,y.net}}{W_{el,y}} - \frac{M_{Sd,z.net}}{W_{el,z}} \tag{11.7}$$

where $W_{el,y} = BL^2/6$ and $W_{el,z} = LB^2/6$. Note the maximum bearing stress is σ_1 and the minimum σ_3.

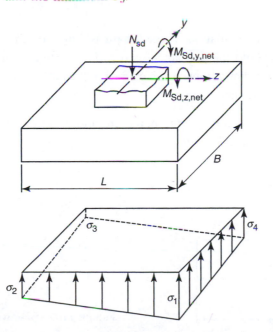

FIGURE 11.3 Base subjected to biaxial bending

The traditional British geotechnical sign convention has been used for bearing pressures, i.e. compression has been taken as positive. Elsewhere, the stress analysis sign convention is used, i.e. tension positive (compression negative).

11.3.3 Offset Columns

With high moments at the base of a column and relatively low axial forces the ratio of the maximum to minimum bearing pressures may be unacceptable. One option that can be considered is to offset the column from the centre line of the base (Fig. 11.4).

The net moment applied to the base is then given by

$$M_{Sd,y,net,1} = M_{Sd,y,net} - N_{Sd}e_{z1} \tag{11.8}$$

where e_{z1} is the displacement between the column and the base centre line.

Note, care must be taken while using this method to equalize bearing stresses, if it is possible for the moments applied to the foundation to reverse their sense of application.

EXAMPLE 11.1   Determination of base dimensions. The characteristic actions applied to a foundation are 300 kN permanent and 350 kN variable together with a net uniaxial moment. If the ground has a design ultimate bearing pressure of 250 kPa and the ratio between maximum and minimum bearing pressures is not to exceed 1,5, determine

(a) the width of foundation, if the length is 3 m with characteristic moments of 40 kNm for both the permanent and variable actions, and
(b) suitable base dimensions, if the characteristic moments are 60 kNm permanent and 80 kNm variable.

In each case, also determine the minimum and maximum design bearing stresses using the final foundation size.

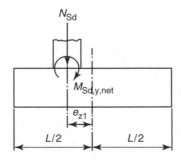

FIGURE 11.4   Base with offset column

(a) Rearranging Eq. (11.2) gives

$$B = \frac{1}{\sigma_1}\left(\frac{N_{Sd}}{L} + \frac{6M_{Sd,y.net}}{L^2}\right) \qquad (11.9)$$

and

$$B = \frac{1}{\sigma_2}\left(\frac{N_{Sd}}{L} - \frac{6M_{Sd,y.net}}{L^2}\right) \qquad (11.10)$$

The only case that needs to be considered is the limit state STR/GEO. In all cases the load set M1 or M2 does not need considering.

Consider each of the Design Approaches:

Design Approach 1: Combination 1
Action set is A1,

so

$$N_{Sd} = 1,35 \times 300 + 1,5 \times 350 = 930\,\text{kN}$$

$$M_{Sd} = 1,35 \times 40 + 1,5 \times 40 = 114\,\text{kNm}.$$

Resistance set is R1, so $\gamma_{R;v} = 1,0$ and $\sigma_{des} = 250/1,0 = 250\,\text{kPa}$.

From Eq. (11.9)

$$B = (930/3 + 6 \times 114/3^2)/250 = 1,544\,\text{m}$$

From Eq. (11.10)

$$B = (930/3 - 6 \times 114/3^2)/(250/1,5) = 1,404\,\text{m}$$

Design Approach 1: Combination 2
Action set is A2,
so

$$N_{Sd} = 1,0 \times 300 + 1,3 \times 350 = 755\,\text{kN}$$

$$M_{Sd} = 1,0 \times 40 + 1,3 \times 40 = 92\,\text{kNm}.$$

Resistance set is R1, so $\gamma_{R;v} = 1,0$ and $\sigma_{des} = 250/1,0 = 250\,\text{kPa}$.

From Eq. (11.9)

$$B = (755/3 + 6 \times 92/3^2)/250 = 1,252\,\text{m}$$

From Eq. (11.10)

$$B = (755/3 - 6 \times 92/3^2)/(250/1,5) = 1,142\,\text{m}$$

For foundations where there are no actions due to the soil, Combination 1 will always be critical.

From Combination 1, set $B = 1,6$ m, this will give $\sigma_1 = 241$ kPa and $\sigma_2 = 146$ kPa.

Design Approach 2:
Action set is A1,
so

$$N_{Sd} = 1,35 \times 300 + 1,5 \times 350 = 930 \text{ kN}$$

$$M_{Sd} = 1,35 \times 40 + 1,5 \times 40 = 114 \text{ kNm}.$$

Resistance set is R2, so $\gamma_{R;v} = 1,4$ and $\sigma_{des} = 250/1,4 = 179$ kPa.

From Eq. (11.9)

$$B = (930/3 + 6 \times 114/3^2)/179 = 2,156 \text{ m}$$

From Eq. (11.10)

$$B = (930/3 - 6 \times 114/3^2)/(179/1,5) = 1,961 \text{ m}$$

This Design Approach would require a base width of 2,2 m giving $\sigma_1 = 175$ kPa and $\sigma_2 = 106$ kPa.

Design Approach 3:
Action set is A1,
so

$$N_{Sd} = 1,35 \times 300 + 1,5 \times 350 = 930 \text{ kN}$$

$$M_{Sd} = 1,35 \times 40 + 1,5 \times 40 = 114 \text{ kNm}.$$

Resistance set is R3, so $\gamma_{R;v} = 1,0$ and $\sigma_{des} = 250/1,0 = 250$ kPa. This gives the same results as Design Approach 1: Combination 1.

It would appear that Design Approach 2 gives conservative results for design of spread foundations.

(b) From Eqs (11.2) and (11.3) the following relationship may be derived

$$L = \frac{6 M_{Sd,y.net}[(\sigma_1/\sigma_2) + 1]}{N_{Sd}[(\sigma_1/\sigma_2) - 1]} \quad (11.11)$$

Consider Design Approach 1: Combination 1
Action set is A1,
so

$$N_{Sd} = 1,35 \times 300 + 1,5 \times 350 = 930 \text{ kN}$$

$$M_{Sd} = 1,35 \times 60 + 1,5 \times 80 = 201 \text{ kNm}.$$

Resistance set is R1, so $\gamma_{R;v} = 1,0$ and $\sigma_{des} = 250/1,0 = 250$ kPa.

From Eq. (11.11)

$$L = 6 \times 201 \times (1{,}5 + 1)/(930 \times (1{,}5 - 1)) = 6{,}48\,\text{m}.$$

This will give a base width of 0,69 m. This clearly is not an economic solution. One approach is to offset the column from the base centre line by 100 mm. From Eq. (11.8), $M_{\text{Sd,y,net,1}} = 201 - 0{,}1 \times 930 = 108$ kNm.

From Eq. (11.11), $L = 3{,}48$ m with a resultant width of 1,28 m.
Design Approach 1: Combination 2 will give

$$N_{\text{Sd}} = 1{,}0 \times 300 + 1{,}3 \times 350 = 755\,\text{kN}$$
$$M_{\text{Sd}} = 1{,}0 \times 60 + 1{,}3 \times 80 = 164\,\text{kNm}.$$

Reduced moment with an offset of 100 mm is $164 - 0{,}1 \times 755 = 88{,}5$ kNm. As these values of M_{Sd} and N_{Sd} are less than Combination 1, they will not be critical.

Adopt $L = 3{,}5$ m and $B = 1{,}3$ m as practical values. These will give $\sigma_1 = 245$ kPa and $\sigma_2 = 164$ kPa.

Design Approach 2 will not be checked for this particular example.

11.4 CALCULATION OF INTERNAL STRESS RESULTANTS IN PAD FOUNDATIONS

11.4.1 Flexure

11.4.1.1 Longitudinal Bending Moment

Although the critical section is that at the face of the column (EN 1991-1-1, cl 5.3.2.2(3)), the effective span should be taken to the centre line of the column as the base should be considered as a continuous cantilever (Fig. 6.13). This will be slightly conservative when the base is thinner than the column size. Thus, for a foundation with either no or uniaxially applied moments (Fig. 11.5), the design moment M_{Sd} per unit width of base at the face of the column is

$$M_{\text{Sd}} = \frac{(L - h_c)^2}{48}\left[5\sigma_1 + \sigma_2 + \frac{h_c}{L}(\sigma_1 - \sigma_2)\right] \tag{11.12}$$

where h_c is the depth of the column.

11.4.1.2 Transverse Bending

The bending moment M_{Sd} per unit length is given by

$$M_{\text{Sd}} = \frac{\sigma_1 + \sigma_2}{8}(B - b_c)^2 \tag{11.13}$$

where b_c is the width of the column.

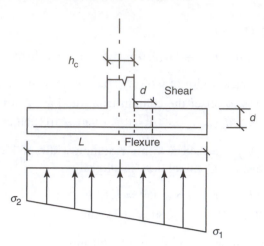

FIGURE 11.5 Critical sections for flexural and shear design

For bases under biaxial bending, analytical expressions may be derived, but it is sufficiently accurate to use the above expression with σ_1 taken as the mean of the two largest adjacent bearing pressures and σ_2 taken as the mean of the remaining two.

11.4.2 Shear

11.4.2.1 Flexural Shear

This should be checked at a distance equal to the effective depth of the base d from the face of the column (cl 6.2.1(8)). The value of the applied shear V_{Sd} is given by

$$V_{Sd} = \frac{d}{4}\left[3\sigma_1 + \sigma_2 + \frac{h_c + 2d}{L}(\sigma_1 - \sigma_2)\right] \tag{11.14}$$

Only $V_{Rd,c}$ and V_{Ed}, the shear capacities of the concrete alone and due to concrete crushing, need be checked as it is not usual to supply shear reinforcement. In addition punching shear may also need checking.

11.4.2.2 Punching Shear

- Column Face

  The design shear V_{Sd} at the column face should not exceed V_{Ed}.

- Successive perimeters

  These are at perimeters of $2d$ out from the column face and the allowable shear stress is determined as for normal shear cases except that the reinforcement ratio, ρ, used is the geometric mean of the reinforcement ratios in the two orthogonal directions. Care should be taken when part of the critical perimeter lies outside the area of the base, while only those parts of the perimeter lying within the base should be considered. Since however the design shear force is the applied axial force less the effect of the upward reaction on the base, this case will not usually be critical except for very highly loaded columns on raft foundations. Where the complete shear perimeter lies within the base, the design shear force V_{Sd} is given by

$$V_{Sd} = N_{Sd} - \frac{\sigma_1 + \sigma_2}{2} A_p \tag{11.15}$$

where A_p is the area contained within the shear perimeter.

For the purpose of determining shear resistance (and shear reinforcement, if necessary) foundations are treated as slabs, i.e. if $V_{Rd,c}$ is exceeded then $\frac{3}{4} V_{Rd,c}$ may be used in determining the link requirement.

Where the ultimate shear strength V_{Ed} is exceeded the base will need thickening (as the effect of increasing the concrete strength is only marginal and the column size is generally predetermined).

If the strength $V_{Rd,c}$ is exceeded then additional tension reinforcement will need to be supplied or the base thickened. In general, shear links are not provided in column bases unless the base is a long thin member supporting a series of column loads in which case the foundation system will act closer to a beam than a slab. Generally column bases should not be thinner than around 600 mm.

11.4.3 Anchorage (cl 9.8.2.2)

It is necessary to check the anchorage of tensile reinforcement to resist a force F_s where F_s is given by

$$F_s = R \frac{z_e}{z_i} \tag{11.16}$$

where R is the resultant force from the ground pressure, z_i is the lever arm between the centre of compression in the concrete and the tensile reinforcement which may be taken as $0,9d$ and z_e is the distance between the vertical load N_{Sd} and the resultant R (Fig. 11.6).

The eccentricity e is determined as M_{Sd}/N_{Sd} subject to a minimum of $015h_c$. For notionally straight bars, the critical section may be taken at a distance of $x = h/2$.

11.5 PILE CAPS

These are needed to distribute the loading from the structure or structural element to the individual piles. Piles are rarely used singly and thus some form of structural

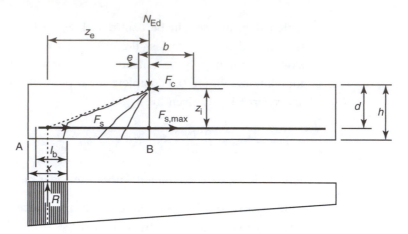

FIGURE 11.6 Model for tensile force with regard to inclined cracks

element such as a pile cap must be used at the head of the piles. Pile caps should be generously dimensioned in plan owing to the likely tolerances in driving or boring the piles. Unless very lightly loaded pile caps are rarely less than a metre deep. Pile caps with only two piles should be connected by ground beams to adjacent pile caps to prevent accidental flexure about the weak axis of the pile system.

To transfer the loading from the structure to the piles the pile cap may be designed either in flexure as a series of simply supported (deep) beams, or using a truss analogy with the tension elements provided by reinforcement and the concrete acting as struts. Truss analogy may only be used when the pile cap is not subject to any moments from the column above the pile cap. The reason for this is that it is impossible to achieve force equilibrium with unequal pile loads.

The pile cap must also be checked for punching shear around the column and the piles themselves. The forces in the piles are determined from the column loads, assuming the pile cap is rigid.

Both methods are illustrated separately in the following examples.

EXAMPLE 11.2   Pile Cap design using beam analogy. A triangular pile cap with a pile at each corner and the centre distance between the piles of 1,5 m carries an axial load at the centroid of 3600 kN and moment of ± 400 kNm about yy axis.

Determine

(a) the forces in the piles and
(b) the resultant bending moment and shear force diagrams.

(a) Pile forces:
   Force in each pile due to the axial load $= 3600/3 = 1200$ kN.

The force in each pile due to the moment is $Mz_i/\Sigma z_i^2$ where z_i is the distance of a pile from the centroid of the pile cap.

For Pile 1 $z_i = 0,866$ m and for Piles 2 and 3, $z_i = 0,433$,

so $\sum z_i^2 = 2 \times 0,433^2 + 0,866^2 = 1,125$ m^2.

Case 1: Moment giving compression in Pile 1
The force in Pile 1 is $400 \times (0,866/1,125) = 308$ kN (compression), and in piles 2 or 3, $400 \times (0,433/1,125) = 154$ kN (tension).
Total load in Pile 1 is $1200 + 308 = 1508$ kN, and
Total load in Piles 2 or 3 is $1200 - 154 = 1046$ kN.

Case 2: Moment giving tension in Pile 1:
Total load in Pile 1 is $1200 - 308 = 892$ kN, and
Total load in Piles 2 or 3 is $1200 + 154 = 1354$ kN.

(b) Determination of SFD and BMD
The loading on the pile cap is taken by two beams, the first of which (Beam A) spans from Pile 1 to the mid-point of Beam B. Beam B spans between Piles 2 and 3.

The loading for both cases on Beam A is given in Fig. 11.7(a) and the resultant SFD and BMD in Figs 11.7(b) and (c).

The reaction at X is now applied to the centre of Beam B to produce the loading, SFD and BMD in Fig. 11.8(a) to (c).

Typical reinforcement details for this type of design are given in Fig. 11.9.

EXAMPLE 11.3 Pile Cap design using truss analogy. A triangular pile cap with a pile at each corner and the centre distance between the piles of s carries an axial load at the centroid of the pile cap of N_{Sd}. Determine the tension force if the effective depth of the reinforcement is d (Fig. 11.10).

The force in each pile is $N_{Sd}/3$.
For convenience refer to pile 1.
The compression force in the strut between the pile and load point is C_{p1}, so resolving vertically

$$C_{p1} \sin \phi = \frac{N_{Sd}}{3} \tag{11.17}$$

or

$$C_{p1} = \frac{N_{Sd}}{3 \sin \phi} \tag{11.18}$$

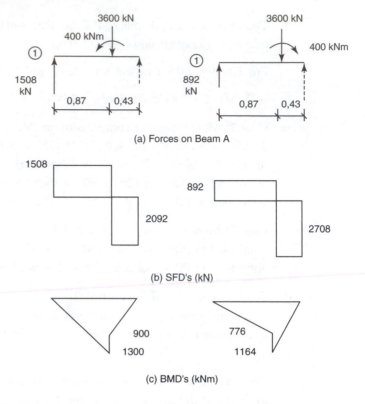

(a) Forces on Beam A

(b) SFD's (kN)

(c) BMD's (kNm)

FIGURE 11.7 Forces and internal stress resultants on Beam 'A'

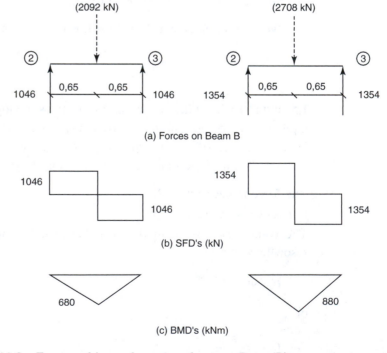

(a) Forces on Beam B

(b) SFD's (kN)

(c) BMD's (kNm)

FIGURE 11.8 Forces and internal stress resultants on Beam 'B'

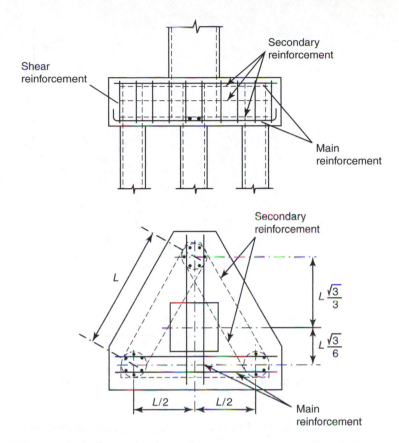

FIGURE 11.9 Detail of pile cap for three piles

The tension forces are T_{12} and T_{13} and are equal. Resolving horizontally

$$T_{12} \cos 30 + T_{13} \cos 30 = C_{p1} \cos \phi \qquad (11.19)$$

or

$$T_{12} = \frac{C_{p1} \cos \phi}{\sqrt{3}} \qquad (11.20)$$

Substitute Eq. (11.18) into Eq. (11.20) to give

$$T_{12} = \frac{N_{Sd}}{3\sqrt{3} \tan \phi} \qquad (11.21)$$

The distance from the centroid to Pile 1 is $s/\sqrt{3}$, thus

$$\tan \phi = \frac{d\sqrt{3}}{s} \qquad (11.22)$$

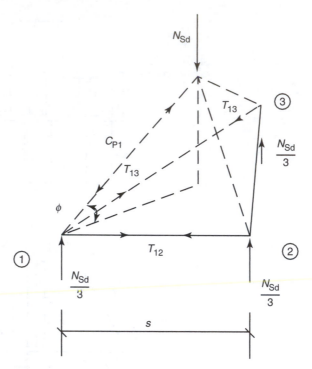

FIGURE 11.10   Schematic diagram for three pile truss analogy

Substitute Eq. (11.22) into Eq. (11.21) to give

$$T_{12} = \frac{N_{\mathrm{Sd}} s}{9 d} \qquad (11.23)$$

The area of reinforcement is then determined using a design stress of $f_{\mathrm{yk}}/\gamma_{\mathrm{s}}$. This reinforcement must be anchored in the horizontal plane to ensure there is no bursting at the top of the pile.

11.6 RETAINING WALLS

Earth Retaining walls are required where either the land take required for a road cutting is less than that available if natural slope stability were relied on or where it would be uneconomic to utilize such stability. Retaining walls can be mass concrete or reinforced concrete (either as plain walls or counterfort walls where the walls are high) (Fig. 11.11). To a certain extent traditional retaining walls which require a substantial working area for their execution have been superseded by either reinforced earth walls, including bridge abutments (Clayton, Mililitsky and Woods, 1993) or diaphragm walling which can be executed within the width of a bentonite slurry filled trench (Henry, 1986). The principles of design for retaining walls also applies to situations such as water retaining systems for reservoirs.

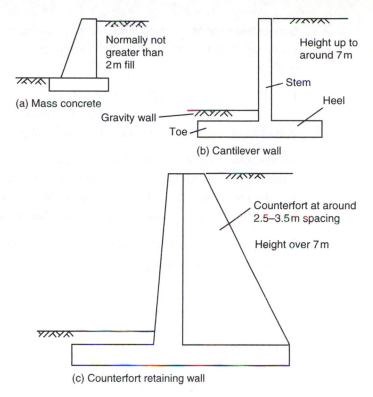

FIGURE 11.11   Types of retaining wall

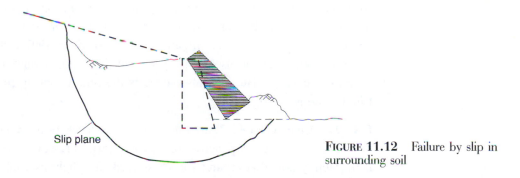

FIGURE 11.12   Failure by slip in surrounding soil

Only the structural aspects of design are covered and not overall geotechnical stability such as in Fig. 11.12 which are covered in any standard geotechnics text book.

11.6.1 Stability of Retaining Walls

Excluding the overall failure mode of Fig. 11.12 a retaining wall needs checking for

(a) Overturning using the limit state of EQU and
(b) Sliding and bearing pressures using the limit states of GEO or STR.

Serviceability limits for total and differential settlement or movement may also need checking. Additionally it is prudent to check cracks widths explicitly for retaining wall stems.

It is important to note that where surcharge is an unfavourable action, then any favourable effect should be ignored by not considering the vertical effect within the zone where the action is favourable (see Examples 11.4 to 11.6).

Where passive pressure is needed to resist actions due to sliding, EN 1997 cl 9.3.2.2 (2) indicates that the ground level should be lowered to an amount Δ_a, where Δ_a should be taken as 10 per cent of the distance between the base and the lowest support if the wall be propped. However, in no case should Δ_a exceed 0,5 m.

To estimate settlement or movement, the following equations (Henry, 1986) may be used. The settlement s is given by

$$s = \frac{2{,}12\,B\left(1 - v^2\right)}{E_v}\,\sigma_{av} \tag{11.24}$$

and the tilt β by

$$\beta = \arctan\left[\frac{5{,}1\left(1 - v^2\right)M_0}{E_v B^2}\right] \tag{11.25}$$

where E_v and v are the Young's Modulus and Poisson's ratio of the underlying soil stratum, B is the width of the base, σ_{av} the mean bearing pressure at serviceability limit state beneath the foundation and M_0 is moment of the resultant thrust at SLS about the centre line of the foundation. Annex H of EN 1997-1 gives some guidance on acceptable levels of displacement and tilt of 50 mm (Annex H cl (4)) and 1/500 (Annex H cl (2)). Clause (2) also suggests the limit on tilt to produce an Ultimate Limit State is 1/150.

EN 1997 Annex C (cl 8.5.2) an indication of the levels of movement whether lateral or rotational to develop active or passive earth pressures. The front of a wall is frequently sloped at around 1 in 25 to mask any slight inevitable tilt. Values of E_v and v are given in Table 11.5 (Martin, Croxton and Purkiss, 1989). Equations (11.24) and (11.25) can only provide estimates, and should more exact information be required, especially where the underlying strata are not able to be categorized by a single type, then recourse must be made to more complex forms of analysis such as finite element techniques.

11.6.2 Mass Concrete Retaining Walls

These are generally restricted for economic reasons to heights of around 2 to 3 m if the section remains of constant thickness (other than any slight front face slope).

TABLE 11.5 Values of Young's Modulus and Poisson's ratio for soils (Martin, Croxton and Purkiss, 1989).

Soil type	E_v (GPa)	v
Dense sand	60–00	0,25
Loose sand	15–50	0,30
Clayey sands/silts	15–30	0,30
Firm and stiff sand clays	2–20	0,35
Firm and stiff clays	2–15	0,40
Very stiff clays	30–55	0,20

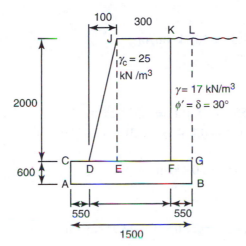

FIGURE 11.13 Design data for a mass concrete retaining wall

It is possible by using stepped walls where the thickness varies with height to exceed 3 m (Henry, 1986). Oliphant (1997) suggests that if the overall height of the wall including the base is H then the width of the base is between $0,4H$ and $0,6H$ with the wall having a vertical front face and sloping back face. The wall should have an equal extension of the base from either face with a width at the top of around 300 mm.

EXAMPLE 11.4 Design of a mass concrete retaining wall. The design data for this example are given in Fig. 11.13.

Overturning is checked using the EQU limit state.

For this case a partial safety factor $\gamma_{\phi'}$ is applied to $\tan \phi'$, thus an effective angle of friction ϕ_{eff} needs determining from

$$\phi_{\text{eff}} = \arctan \left[\frac{\tan \phi'}{\gamma_{\phi'}} \right] \tag{11.26}$$

With $\phi' = 30°$ and $\gamma_{\phi'} = 1,25$, Eq. (11.26) gives a value of $\phi_{\text{eff}} = 24,8°$.

The effect of the active earth pressure is given by $0,5K_a\gamma h^2$, where $K_a = (1 - \sin \phi_{\text{eff}})/(1 + \sin \phi_{\text{eff}}) = 0,409$.

Moment about point A (Fig. 11.13) due to the earth pressure is

$$M_{earth} = \tfrac{1}{3} \times \tfrac{1}{2} \times K_a \times \gamma_{soil} \times H^3 = \tfrac{1}{3} \times \tfrac{1}{2} \times 0{,}409 \times 17 \times 2{,}6^3 = 20{,}4\,\text{kNm/m}.$$

The partial safety factor $\gamma_{G;\,dst} = 1{,}1$, thus factored moment is $1{,}1 \times 20{,}4 = 22{,}4\,\text{kNm/m}$.

Restoring (or favourable) moment about A is due to the concrete in the base and the wall and the retained soil mass:

Concrete base:

$$\tfrac{1}{2} \times 25 \times 0{,}6 \times 1{,}5^2 = 16{,}9\,\text{kNm/m}$$

Concrete wall:

$$\tfrac{1}{2} \times 25 \times 0{,}1(0{,}55 + \tfrac{2}{3} \times 0{,}1) = 1{,}5\,\text{kNm/m}$$
$$25 \times 2 \times 0{,}3 \times (0{,}55 + 0{,}1 + \tfrac{1}{2} \times 0{,}3) = 12{,}0\,\text{kNm/m}$$

Total due to the concrete $= 30{,}4\,\text{kNm/m}$

Retained soil:

$$17 \times 2 \times 0{,}55 \times (1{,}5 - \tfrac{1}{2} \times 0{,}55) = 22{,}9\,\text{kNm/m}$$

Total due to the concrete and earth $= 53{,}3\,\text{kNm/m}$

A factor $\gamma_{G;\,stb}$ of 0,9 must be applied to this, thus the restoring moment is given by $0{,}9 \times 53{,}3 = 48{,}0\,\text{kNm/m}$. This is greater than the factored overturning moment of $22{,}4\,\text{kNm/m}$, and is therefore satisfactory.

The Limit state of STR/GEO is used to determine sliding and bearing pressures.

Design Approach 1: Combination 1 [A1:M1:R1]

As there is no factor applied to $\tan \phi'$, $\phi' = 30°$ and $K_a = \tfrac{1}{3}$.
Moment about centre line of the base due to soil pressure:

$$\tfrac{1}{3} \times \tfrac{1}{2} \times K_a \times \gamma_{soil} \times H^3 = \tfrac{1}{3} \times \tfrac{1}{3} \times \tfrac{1}{2} \times 17 \times 2{,}6^3$$
$$= 16{,}6\,\text{kNm/m [counterclockwise]}$$

Moment about centre line of the base due to concrete and soil

Base $= 0$
Wall: $\tfrac{1}{2} \times 25 \times 0{,}1 \times 2 \times (0{,}75 - 0{,}55 - \tfrac{1}{3} \times 2 \times 0{,}1) = 0{,}33\,\text{kNm/m [counterclockwise]}$
$25 \times 2 \times 0{,}3 \times (0{,}75 - 0{,}55 - 0{,}1 - \tfrac{1}{2} \times 0{,}15) = -0{,}75\,\text{kNm/m [clockwise]}$

Net moment due to the concrete is 0,42 kNm/m [clockwise]. The concrete wall can only have a single partial safety factor attached as it is a single entity, the net effect must be determined before its state of favourability or unfavourability is determined.

Retained soil:

$$17 \times 2 \times 0,55 \times (0,75 - \tfrac{1}{2} \times 0,55) = 8,88 \text{ kNm/m [clockwise]}.$$

Net moment due to the soil and concrete is $0,42 + 8,88 = 9,3$ kNm/m.

As the higher moment is due to the earth pressure this is the unfavourable moment and must be factored by $\gamma_{G;dst}$ $(=1,35)$,
so net moment is $1,35 \times 16,6 - 9,3 = 13,1$ kNm/m

Vertical effects:

Concrete:

> Base: $25 \times 0,6 \times 1,5 = 22,5$ kN/m
> Wall: $\tfrac{1}{2} \times 25 \times 2 \times (0,3 + 0,4) \times 2 = 17,5$ kN/m
> Total $= 40,0$ kN/m

Soil: $17 \times 2 \times 0,55 = 18,7$ kN/m
Overall Total $= 58,7$ kN/m

Bearing pressures:
For the bearing pressures the vertical action is unfavourable and must be factored by $\gamma_{G;dst}$ $(=1,35)$, so vertical load is $1,35 \times 58,7 = 79,3$ kN/m.

$$\sigma_1 = 79,3/1,5 + 6 \times 13,1/1,5^2 = 88 \text{ kPa},$$
$$\sigma_2 = 79,3/1,5 - 6 \times 13,1/1,5^2 = 18 \text{ kPa}.$$

Sliding:
Force due to earth pressure $= \tfrac{1}{2}K_a H^2 = \tfrac{1}{2} \times \tfrac{1}{3} \times 17 \times 2,6^2 = 19,2$ kN/m.
Apply factor of $\gamma_{G;dst}$ $(=1,35)$ to give $1,35 \times 19,2 = 25,9$ kN/m
Total (unfactored) vertical load $= 58,7$ kN/m.
Friction force available is $58,7 \tan \delta = 58,7 \tan 30 = 33,9$ kN/m. This is greater than the unfavourable action, therefore satisfactory.

Design Approach 1: Combination 2 [A2:M2:R1]

A factor of $\gamma_{\phi'}$ is applied to $\tan\phi'$, so $K_a = 0,409$ (as in the EQU limit state).

Moment about centre line of the base due to soil pressure:

> $\tfrac{1}{3} \times \tfrac{1}{2} \times K_a \times \gamma_{soil} \times H^3 = \tfrac{1}{3} \times \tfrac{1}{2} \times 0,409 \times 17 \times 2,6^3 = 20,4$ kNm/m
> [counterclockwise]

Moment about centre line of the base due to concrete and soil:

Base $= 0$

Wall: $1/2 \times 25 \times 0,1 \times 2 \times (0,75 - 0,55 - 2/3 \times 0,1) = 0,33$ kNm/m [counterclockwise]

$25 \times 2 \times 0,3 \times (0,75 - 0,55 - 0,1 - 1/2 \times 0,15) = 0,75$ kNm/m [clockwise]

Net moment due to the concrete is 0,42 kNm/m [clockwise].

Retained soil:

$17 \times 2 \times 0,55 \times (0,75 - 1/2 \times 0,55) = 8,88$ kNm/m [clockwise].

Net moment due to the soil and concrete is $0,42 + 8,88 = 9,3$ kNm/m [clockwise]
The net moment is $20,4 - 9,3 = 11,1$ kNm/m.

Vertical effects:

Concrete:

    Base: $25 \times 0,6 \times 1,5 = 22,5$ kN/m
    Wall: $1/2 \times 25 \times 2 \times (0,3 + 0,4) \times 2 = 17,5$ kN/m
    Total $= 40,0$ kN/m

Soil: $17 \times 2 \times 0,55 = 18,7$ kN/m
Overall Total $= 58,7$ kN/m

Bearing pressures:

$$\sigma_1 = 58,7/1,5 + 6 \times 11,1/1,5^2 = 69 \text{ kPa}$$
$$\sigma_2 = 58,7/1,5 - 6 \times 11,1/1,5^2 = 10 \text{ kPa}.$$

Sliding:
Force due to earth pressure $= 1/2 K_a H^2 = 1/2 \times 0,409 \times 17 \times 2,6^2 = 23,5$ kN/m.
Total (unfactored) vertical load $= 58,7$ kN/m.
Friction force available is $58,7 \tan \delta = 58,7 \tan 30 = 33,9$ kN/m. This is greater than the unfavourable action, therefore satisfactory.

Design Approach 2: [A1:M1:R2]
The action set is the same as Design Approach 1: Combination 1, but the resistance set [R2] for bearing is 1,4 and sliding 1,1. Thus the maximum permissible bearing pressure would need to 1,4 times that of Design Approach 1: Combination 1, i.e. $1,4 \times 88 = 123$ kPa (or if the values in Design Approach 1: Combination were critical, the base would need to be some 20 to 30 per cent larger).

Sliding: Unfavourable action is 25,9 kN/m (as Design Approach 1: Combination 1).

A factor $\gamma_{R;h}$ ($= 1,1$) is applied to the favourable action, i.e. the available force is $33,89/1,1 = 30,8$ kN/m. This is still satisfactory.

Design Approach 3:
As there are no structural actions, the safety factors applied to the geotechnical action is Set $A2$ and the resistance set $R3$ is equivalent to $R1$, this approach reduces to Design Approach 1: Combination 2.

Although the wall has been described as mass concrete it will be prudent to reinforce the front and back faces of the wall and the lower face of the base with mesh in order to control cracking and any thermal movement.

11.6.3 Cantilever Retaining Wall

The stability calculations follow those for unreinforced mass retaining walls. The thickness of the base should be such that no shear reinforcement is necessary. This is also true of the stem of the wall, together with the requirement that the minimum thickness of the wall is twice the cover plus the width of two layers (vertical and horizontal) of reinforcement in both faces and a minimum clearance of 100 mm between each reinforcement layer to allow full compaction of the concrete. This generally gives a minimum width of at least 300 mm, although 400 mm may be preferable. Given the nature of such walls the control of crack widths should be performed using full calculations using limiting values appropriate to liquid retaining structures. An allowance needs making for early thermal cracking in addition to flexural cracking. Oliphant (1997) suggests that for a wall of overall height including the base, then the base width should lie between $0,5H$ to $0,8H$, the front face of the wall set back by $0,25B$ from the front of the base whose thickness should be $0,1H$. For low to moderate heights of wall, the suggested base thickness may be too small as it could give problems with flexural and shear design. It is therefore recommended that the minimum base thickness should be around 600 mm.

Shear keys may be necessary to prevent sliding, but a longer heel may help avoid the problem.

EXAMPLE 11.5   Cantilever retaining wall (Fig. 11.14). For EQU and STR/GEO where a factor $\gamma_{\phi'}$ is applied to $\tan \phi'$, a shearing angle ϕ_{eff} needs to be determined. This is given by

$$\phi_{\text{eff}} = \arctan \frac{\tan \phi'}{1,25} \tag{11.27}$$

With $\phi' = 30°$, Eq. (11.27) gives $\phi_{\text{eff}} = 24,8°$, and

$$K_a = (1 - \sin \phi_{\text{eff}})/(1 + \sin \phi_{\text{eff}}) = 0,409.$$

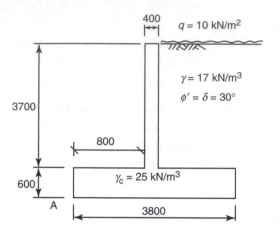

FIGURE 11.14 Design data for a cantilever retaining wall

EQU:

Overturning moment due to soil:

$$\tfrac{1}{3} \times \tfrac{1}{2} \times K_a \times \gamma \times H^3 = \tfrac{1}{3} \times \tfrac{1}{2} \times 0{,}409 \times 17 \times 4{,}3^3 = 92{,}1\,\text{kNm/m}$$

Overturning moment due to surcharge:

$$\tfrac{1}{2} \times q \times H^2 = \tfrac{1}{2} \times 10 \times 4{,}3^2 = 92{,}5\,\text{kNm/m}$$

A factor of $\gamma_{G;dst}$ ($=1{,}1$) is applied to the earth pressure and a factor of $\gamma_{Q;dst}$ ($=1{,}5$) is applied to the surcharge, thus the total factored moment is given by

$$1{,}1 \times 92{,}1 + 1{,}5 \times 92{,}5 = 240{,}6\,\text{kNm/m}.$$

Taking moments about A:

Base: $\tfrac{1}{2} \times 25 \times 0{,}6 \times 3{,}8^2 = 108{,}3\,\text{kNm/m}$
Wall: $25 \times 3{,}7 \times 0{,}4 \times (0{,}8 + \tfrac{1}{2} \times 0{,}4) = 37{,}0\,\text{kNm/m}$
Total $= 145{,}3\,\text{kNm/m}$

Due to the soil:

$$17 \times 3{,}7 \times 2{,}6 \times (3{,}8 - \tfrac{1}{2} \times 2{,}6) = 408{,}9\,\text{kNm/m}$$

Total $= 554{,}2\,\text{kNm/m}$

A factor $\gamma_{G;stb}$ ($=0{,}9$) is applied to the restoring moment, thus the factored restoring moment is given as $0{,}9 \times 554{,}2 = 498{,}8\,\text{kNm/m}$.

This exceeds the factored overturning moment of $240{,}6\,\text{kNm/m}$, and is therefore satisfactory.

The Limit state of STR/GEO is used to determine sliding and bearing pressures.

Design approach 1: Combination 1 [A1: M1: R1]

As there is no factor applied to $\tan\phi'$, $\phi' = 30°$ and $K_a = \frac{1}{3}$.

Moment about centre line of the base:
due to soil pressure:

$$\frac{1}{3} \times \frac{1}{2} \times K_a \times \gamma_{soil} \times H^3 = \frac{1}{3} \times \frac{1}{3} \times \frac{1}{2} \times 17 \times 4{,}3^3 = 75{,}1 \text{ kNm/m}$$
$$[\text{counterclockwise}],$$

due to surcharge:

$$\frac{1}{2} \times q \times H^2 = \frac{1}{2} \times 10 \times 4{,}3^2 = 92{,}5 \text{ kNm/m [counterclockwise]}$$

Moment about centre line of the base due to concrete and soil

Base $= 0$

Wall: $\frac{1}{2} \times 25 \times 0{,}4 \times 3{,}7 \times (\frac{1}{2} \times 3{,}8 - \frac{2}{3} \times 0{,}8 - \frac{2}{3} \times 0{,}4) = 16{,}7 \text{ kNm/m}$ [counter-clockwise]

Retained soil:

$17 \times 3{,}7 \times 2{,}6 \times (\frac{1}{2} \times 3{,}8 - \frac{1}{2} \times 2{,}6) = 98{,}1 \text{ kNm/m [clockwise]}$.

A factor of γ_G of 1,35 is applied to the permanent unfavourable and 1,0 to the permanent favourable and a factor of γ_Q of 1,5 to the unfavourable variable.

Taking the counterclockwise direction as unfavourable, the net factored moment is

$$1{,}5 \times 92{,}5 + 1{,}35 \times 75{,}1 + 1{,}35 \times 16{,}7 - 1{,}0 \times 98{,}1 = 164{,}6 \text{ kNm/m}.$$

Vertical actions:
Concrete:
Base: $25 \times 3{,}8 \times 0{,}6 = 57{,}0 \text{ kN/m}$
Wall: $25 \times 3{,}7 \times 0{,}4 = 37{,}0 \text{ kN/m}$

Total due to the concrete $= 94{,}0 \text{ kN/m}$
Soil: $17 \times 3{,}7 \times 2{,}6 = 163{,}5 \text{ kN/m}$
Total vertical action $= 257{,}5 \text{ kN/m}$

Sliding:
Factored action due to the surcharge:

$$\gamma_Q qH = 1{,}5 \times 10 \times 4{,}3 = 64{,}5 \text{ kN/m}$$

Factored action due to earth pressure:

$$\gamma_G \frac{1}{2} K_a \gamma H^2 = 1{,}35 \times \frac{1}{2} \times \frac{1}{3} \times 17 \times 4{,}3^2 = 70{,}7 \text{ kN/m}$$

Total $= 135,2\,\text{kN/m}$

Available friction force is $257,5\tan30° = 148,7\,\text{kN/m}$. This exceeds the effect of the surcharge and earth pressure and is therefore satisfactory.

Bearing pressures:
The vertical action being unfavourable needs factoring by γ_Q, to give $1,35 \times 257,5 = 347,6\,\text{kN/m}$.

$$\sigma_1 = 347,6/3,8 + 6 \times 164,6/3,8^2 = 160\,\text{kPa},$$
$$\sigma_2 = 347,6/3,8 - 6 \times 164,6/3,8^2 = 23\,\text{kPa}.$$

Design Approach 1: Combination 2 [A2:M2:R1]
As there is a factor of 1,25 is applied to $\tan\phi'$, $\phi_{\text{eff}} = 24,8°$ and $K_a = 0,409$.

Moment about centre line of the base:
due to soil pressure:

$$\tfrac{1}{3} \times \tfrac{1}{2} \times K_a \times \gamma_{\text{soil}} \times H^3 = \tfrac{1}{3} \times \tfrac{1}{2} \times 0,409 \times 17 \times 4,3^3 = 92,1\,\text{kNm/m}$$
[counterclockwise],

due to surcharge:

$$\tfrac{1}{2} \times q \times H^2 = \tfrac{1}{2} \times 10 \times 4,3^2 = 92,5\,\text{kNm/m[counterclockwise]}$$

Moment about centre line of the base due to concrete and soil
Base $= 0$
Wall: $\tfrac{1}{2} \times 25 \times 0,4 \times 3,7 \times (\tfrac{1}{2} \times 3,8 - 0,8 - \tfrac{2}{3} \times 0,4) = 16,7\,\text{kNm/m[counterclockwise]}$

Retained soil:

$$17 \times 3,7 \times 2,6 \times (\tfrac{1}{2} \times 3,8 - \tfrac{1}{2} \times 2,6) = 98,1\,\text{kNm/m [clockwise]}.$$

A factor of γ_G of 1,0 is applied to the permanent unfavourable and favourable actions and a factor of γ_Q of 1,3 to the unfavourable variable.

Taking the counterclockwise direction as unfavourable, the net factored moment is

$$1,3 \times 92,5 + 1,0 \times 92,1 + 1,0 \times 16,7 - 1,0 \times 98,1 = 131\,\text{kNm/m}.$$

Vertical actions:
Concrete:

Base: $25 \times 3,8 \times 0,6 = 57,0\,\text{kN/m}$
Wall: $25 \times 3,7 \times 0,4 = 37,0\,\text{kN/m}$
Total due to the concrete $= 94,0\,\text{kN/m}$

Soil: $17 \times 3,7 \times 2,6 = 163,5 \, \text{kN/m}$
Total vertical action $= 257,5 \, \text{kN/m}$

Sliding:
Factored action due to the surcharge:

$$\gamma_Q qH = 1,3 \times 10 \times 4,3 = 55,9 \, \text{kN/m}$$

Factored action due to earth pressure:

$$\gamma_G \tfrac{1}{2} K_a \gamma H^2 = 1,0 \times \tfrac{1}{2} \times 0,409 \times 17 \times 4,3^2 = 64,3 \, \text{kN/m}$$

Total $= 123,2 \, \text{kN/m}$

Available friction force is $257,5 \tan 30° = 148,7 \, \text{kN/m}$. This exceeds the effect of the surcharge and earth pressure and is therefore satisfactory.

Bearing pressures:
The vertical action being unfavourable needs factoring by γ_Q, to give $1,0 \times 257,5 = 257,5 \, \text{kN/m}$.

$$\sigma_1 = 257,5/3,8 + 6 \times 131/3,8^2 = 122 \, \text{kPa},$$
$$\sigma_2 = 257,5/3,8 - 6 \times 131/3,8^2 = 13 \, \text{kPa}.$$

Design Approach 2:
This essentially follows Design Approach 1: Combination 1, except that

(a) The available friction force of $148,7 \, \text{kN/m}$ needs to be factored by $\gamma_{R;h}$ ($= 1,1$) to give $148,7/1,1 = 135,2 \, \text{kN/m}$. This equals the unfavourable force due to earth pressure and surcharge of $135,2 \, \text{kN/m}$.
(b) The maximum allowable bearing pressure would need to be $1,4$ ($\gamma_{R;v}$) higher.

Design Approach 3:
In this case the factors applied to geotechnical actions are Set $A1$ and to structural actions $A2$. It is assumed the surcharge is a structural action. The soil set is $M2$ and the Resistance set is $R3$ (equivalent to $R1$).
As there is a factor of $1,25$ is applied to $\tan\phi'$, $\phi_{\text{eff}} = 24,8°$ and $K_a = 0,409$.

Moment about centre line of the base:
due to soil pressure:

$$\tfrac{1}{3} \times \tfrac{1}{2} \times K_a \times \gamma_{\text{soil}} \times H^3 = \tfrac{1}{3} \times \tfrac{1}{2} \times 0,409 \times 17 \times 4,3^3 = 92,1 \, \text{kNm/m}$$
$$[\text{counterclockwise}],$$

due to surcharge:

$$\tfrac{1}{2} \times q \times H^2 = \tfrac{1}{2} \times 10 \times 4,3^2 = 92,5 \, \text{kNm/m} \ [\text{counterclockwise}]$$

Moment about centre line of the base due to concrete and soil

Base $= 0$

Wall: $\frac{1}{2} \times 25 \times 0,4 \times 3,7 \times (\frac{1}{2} \times 3,8 - 0,8 - \frac{2}{3} \times 0,4) = 16,7 \, \text{kNm/m}$
[counterclockwise]

Retained soil:

$$17 \times 3,7 \times 2,6 \times (\frac{1}{2} \times 3,8 - \frac{1}{2} \times 2,6) = 98,1 \, \text{kNm/m [clockwise]}.$$

A factor of γ_G of 1,0 is applied to the permanent favourable actions and a factor of γ_G of 1,0 to the permanent unfavourable (geotechnical) and a γ_G of 1,5 to the unfavourable variable.

Taking the counterclockwise direction as unfavourable, the net factored moment is

$$1,5 \times 92,5 + 1,0 \times 92,1 + 1,0 \times 16,7 - 1,0 \times 98,1 = 149,5 \, \text{kNm/m}.$$

Vertical actions:
Concrete:
$$\text{Base: } 25 \times 3,8 \times 0,6 = 57,0 \, \text{kN/m}$$
$$\text{Wall: } 25 \times 3,7 \times 0,4 = 37,0 \, \text{kN/m}$$
Total due to the concrete $= 94,0 \, \text{kN/m}$
Soil: $17 \times 3,7 \times 2,6 = 163,5 \, \text{kN/m}$
Total vertical action $= 257,5 \, \text{kN/m}$

Sliding:
Factored action due to the surcharge:

$$\gamma_Q q H = 1,5 \times 10 \times 4,3 = 64,5 \, \text{kN/m}$$

Factored action due to earth pressure:

$$\gamma_G \tfrac{1}{2} K_a \gamma H^2 = 1,0 \times \tfrac{1}{2} \times 0,409 \times 17 \times 4,3^2 = 64,3 \, \text{kN/m}$$

Total $= 28,8 \, \text{kN/m}$

Available friction force is $257,5 \tan 30° = 148,7 \, \text{kN/m}$. This exceeds the effect of the surcharge and earth pressure and is therefore satisfactory.

Bearing pressures:
The vertical action being unfavourable needs factoring by γ_Q [Set A1], to give $1,35 \times 257,5 = 347,6 \, \text{kN/m}$.

$$\sigma_1 = 347,6/3,8 + 6 \times 149,5/3,8^2 = 154 \, \text{kPa},$$
$$\sigma_2 = 347,6/3,8 - 6 \times 149,5/3,8^2 = 29 \, \text{kPa}.$$

To summarize the results of the various design approaches it will be noted that Design Approach 2 is the most conservative as the sliding calculations only just work and would require a larger base (or higher permissable bearing pressures). There would appear to be little to choose between the other design approaches except that Design Approach 1 requires two combinations to be considered.

It is intended to use Design Approach 1 to determine the stress resultants to carry out the structural design. Only combination 1 will be considered as this produces higher forces.

This is used for the structural design.

With $\phi = 30°$, $K_a = 0{,}333$

For exposure Class XC2 (Foundations) with a concrete strength of C40/50, structure Class is SC3 (as the concrete strength reduces the basic class SC4 by 1), $c_{min,dur}$ is 20 mm. The final design cover c_{nom} is given by $c_{min} + \Delta c_{dev} = 20 + 10 = 30$ mm.

Design of stem:
Effective depth (assuming 25 mm bars):

$d = 400 - 30 - 25/2 = 357{,}5$ mm. Use $d = 355$ mm.

Flexural design:
At the base of the stem take the effective span as the length of the cantilever plus half the base thickness, i.e.

$$3{,}7 + \tfrac{1}{2} \times 0{,}6 = 4\,\text{m},$$
$$M_{Sd} = 1{,}35 \times \tfrac{1}{2} \times \tfrac{1}{3} \times \tfrac{1}{3} \times 17 \times 4^3) + 1{,}5 \times \tfrac{1}{2} \times 10 \times 4^2 = 201{,}6\,\text{kNm/m}$$
$$M_{Sd}/bd^2 f_{ck} = 201{,}6 \times 10^3/(40 \times 355^2) = 0{,}04$$

From Eq. (6.22),

$$\frac{A_s f_{yk}}{bd f_{ck}} = 0{,}652 - \sqrt{0{,}425 - 1{,}5\frac{M_{Sd}}{b\,d^2\,f_{ck}}}$$
$$= 0{,}652 - \sqrt{0{,}425 - 1{,}5 \times 0{,}04} = 0{,}048 \tag{11.28}$$

$A_s = 0{,}048 \times 1000 \times 355 \times 40/500 = 1363\,\text{mm}^2/\text{m}$. Fix D20 at 200 mm spacing $[1572\,\text{mm}^2/\text{m}]$.

The maximum shear may be determined at d above the top of the base, i.e. at $4 - 0{,}355 = 3{,}645$ m.

$$V_{Sd} = 1{,}35 \times \tfrac{1}{2} \times \tfrac{1}{3} \times 17 \times 3{,}645^2 + 1{,}5 \times 10 \times 3{,}645 = 105{,}5\,\text{kN/m}.$$

Shear:

$$V_{Ed} = 0,5vf_{cd}b_wd$$
$$v = 0,6(1 - f_{ck}/250) = 0,6(1 - 40/250) = 0,504$$
$$f_{cd} = f_{ck}/\gamma_c = 40/1,5 = 26,7\,\text{MPa}.$$
$$V_{Ed} = 0,5 \times 0,504 \times 26,7 \times 1000 \times 355) \times 10^{-3} = 2388\,\text{kN/m} > V_{Sd}$$

With no axial force (ignoring self weight of the stem):

$$V_{Rdc} = C_{Rd,c}k(100\rho_1 f_{ck})^{1/3}bd$$
$$\rho_1 = A_s/bd = 1572/(1000 \times 355) = 0,00443 < 0,02$$
$$C_{Rd,c} = 0,18/\gamma_c = 0,18/1,5 = 0,12$$
$$k = 1 + \sqrt{200/d} = 1 + \sqrt{200/355} = 1,75 < 2,0$$

$$V_{Rdc} = 0,12 \times 1,75 \times (100 \times 0,00443 \times 40)^{1/3} \times 355 \times 1000 \times 10^{-3} = 194,4\,\text{kN/m} > V_{Sd},$$
Therefore no shear reinforcement required.

As there is the possibility of water, crack widths need explicit calculation Table 7.1N of EN 1992-1-2 indicates that for exposure class XC2 the limiting crack width is 0,3 mm. Consider the surcharge always at its full value.

$$M_{Sd} = 1/3 \times 1/3 \times 1/2 \times 17 \times 4^3 + 1/2 \times 10 \times 4^2 = 140,4\,\text{kNm/m}$$

Determine critical spacing from (Section 5.2.2.1):

$5(c + \phi/2) = 5(30 + 20/2) = 200$ mm. Actual spacing is 200 mm, therefore treat as close spaced (cl 7.3.4 (3)).

$$s_{r,max} = 3,4c + 0,425k_1k_2\phi/\rho_{p,eff}$$
$$k_1 = 0,8\ \text{(high bond bars)}$$
$$k_2 = 0,5\ \text{(flexure)}$$

Determine neutral axis depth x:

$$E_{cm} = 22((f_{ck} + 8)/10)^{0,3} = 22(40 + 8)/10)^{0,3} = 32\,\text{GPa}.$$
$$E_s = 200\,\text{GPa}; \alpha_e = E_s/E_{cm} = 200/32 = 6,25$$
$$\rho\alpha_e = 0,00443 \times 6,25 = 0,0277.$$

From Eq. (5.25),

$$\frac{x}{d} = \alpha_e\, \rho \left[\sqrt{1 + \frac{2}{\alpha_e\, \rho}} - 1 \right]$$

$$= 0,0277 \left[\sqrt{1 + \frac{2}{0,0277}} - 1 \right] = 0,209 \tag{11.29}$$

$$x = 0,209 \times 355 = 74,2\,\text{mm}.$$

Determine $h_{c,ef}$:

$$2,5(h - d) = 2,5(400 - 355) = 112,5 \, \text{mm}; (h - x)/3 = (400 - 74,2)/3 = 108,6 \, \text{mm};$$
$$h/2 = 400/2 = 200 \, \text{mm}$$

The least of these values is 108,6 mm.

$$A_{c,eff} = 1000 \times 108,6 = 108600 \, \text{mm}^2.$$
$$\rho_{p,eff} = A_s/A_{c,eff} = 1572/108600 = 0,0145$$
$$s_{r,max} = 3,4 \times 30 + 0,425 \times 0,8 \times 0,5 \times 20/0,0145 = 366 \, \text{mm}.$$

Determine the crack width w_k from Eq. (5.31)

$$w_k = s_{r,max}(\varepsilon_{sm} - \varepsilon_{cm})$$

where $\varepsilon_{sm} - \varepsilon_{sm}$ is given by Eq. (5.32),

$$\varepsilon_{sm} - \varepsilon_{cm} = \frac{\sigma_s - k_t(f_{ct,eff}/\rho_{p,eff})(1 + \alpha_e \rho_{p,eff})}{E_s} \geq 0,6 \frac{\sigma_s}{E_s} \tag{11.30}$$

Determine I_{cr} from Eq. (5.24),

$$\frac{I_{cr}}{bd^3} = \frac{1}{3}\left(\frac{x}{d}\right)^3 + \rho \alpha_e \left(1 - \frac{x}{d}\right)^2$$
$$= \frac{1}{3} \times 0,209^3 + 0,0277(1 - 0,209)^2 = 0,204 \tag{11.31}$$
$$I_{cr} = 0,0204 \times 1000 \times 355^3 = 0,913 \times 10^9 \, \text{mm}^4$$
$$\sigma_s = \alpha_e M(d - x)/I_{cr} = 6,25 \times 140,4 \times 10^6 \times 355(1 - 0,209)/0,913 \times 10^9$$
$$= 270 \, \text{MPa}$$

Determine $f_{ct,eff}$:
This may be taken as $f_{ctm} = 0,3(f_{ck})^{2/3} = 0,3(40)^{2/3} = 3,5 \, \text{MPa}$
The loading may be considered long term, thus $k_t = 0,4$.
Thus $\varepsilon_{sm} - \varepsilon_{sm}$ is

$$\varepsilon_{sm} - \varepsilon_{cm} = \frac{270 - 0,4(3,5/0,0145)(1 + 6,25 \times 0,0145)}{200 \times 10^3}$$
$$= 823,5 \times 10^{-6} \tag{11.32}$$

Check limiting value of $\varepsilon_{sm} - \varepsilon_{sm}$:

$$0,6\sigma_s/E_s = 0,6 \times 270/200 \times 10^3 = 810 \times 10^{-6}.$$

Thus $\varepsilon_{sm} - \varepsilon_{sm}$ is $823,5 \times 10^{-6}$, and

$$w_k = s_{r,max}(\varepsilon_{sm} - \varepsilon_{sm}) = 366 \times 823,5 \times 10^{-6} = 0,30 \, \text{mm}.$$

Wall deflection using span-effective depth ratio:

Steel ratio $\rho = 1572/355 \times 1000 = 0{,}00443$

$$\rho_0 = 0{,}001\sqrt{f_{ck}} = 0{,}001 \times \sqrt{40} = 0{,}00632$$

For a cantilever $K = 0{,}4$, as $\rho < \rho_0$, so l/d is given by

$$\frac{l}{d} = K\left[11 + 1{,}5\sqrt{f_{ck}}\frac{\rho_0}{\rho} + 3{,}2\sqrt{f_{ck}}\left(\frac{\rho_0}{\rho} - 1\right)^{3/2}\right]$$

$$= 0{,}4\left[11 + 1{,}5\sqrt{40}\frac{0{,}00636}{0{,}00443} + 3{,}2\sqrt{40}\left(\frac{0{,}00636}{0{,}00443} - 1\right)^{3/2}\right]$$

$$= 12{,}2 \tag{11.33}$$

Actual span/effective depth $= 4000/355 = 11{,}3$. This is less than the allowable, therefore satisfactory.

Longitudinal Reinforcement:

This is needed to prevent early thermal cracking. Use Anchor (1992), as a data source for temperature rise. Anchor indicates that the crack width w_k due to early cracking may be calculated as

$$w_k = 0{,}5\alpha(\theta_1 + \theta_2)s_{r,max} \tag{11.34}$$

where $s_{r,max}$ is the maximum spacing of the cracks, θ is the coefficient of thermal expansion of the concrete ($10\,\mu$strain/°C (cl 3.1.2.5.4)), θ_1 is the temperature generated due the dissipation of the heat of hydration of the cement and θ_2 is temperature dependant upon the time of year in which the concreting took place. The 0,5 factor appears in Eq. (11.34) as an indication that there is in normal practice some restraint to free thermal expansion.

Determination of θ_1 and θ_2:

From Anchor (1992), θ_2 should be taken in the UK as 10°C for winter concreting and 20°C for summer. Take θ_2 as 20°C (conservative).

From Table 5.3 (Anchor, 1992) θ_1 for a wall thickness of 400 mm, cement content of 350 kg/m³ and 18 mm plywood formwork may be taken as 37°C (interpolating).

Use a design crack width of 0,3 mm to determine $s_{r,max}$.

$$s_{r,max} = w_k/0{,}5\alpha(\theta_1 + \theta_2) = 0{,}3/(0{,}5 \times 10 \times 10^{-6} \times (37 + 20)) = 1053\,mm$$

Determination of steel:

Anchor gives the following equation for $s_{r,max}$:

$$s_{r,max} = \frac{f_{ctm}}{f_b}\frac{\phi}{2\rho_c} \tag{11.35}$$

where ϕ is the bar diameter, f_{ctm} is the tensile strength of the concrete, f_{bd} is the bond strength and ρ_c is the reinforcement area. Note that in Eq. (11.35) the partial safety factor has been omitted from both f_{bd} and f_{ctm} as only the ratio between the two values is required.

From cl 8.4.5.2(2), f_{bd} is given by $f_{bd} = 2{,}25\eta_1\eta_2 f_{ctd} = 2{,}25\eta_1\eta_2 f_{ctm}/\gamma_c$.

Thus $f_{ctm}/f_b = 1/(2{,}25\eta_1\eta_2)$

η_1: Conservatively 0,7
η_2: 1,0 (bar size less than 32)

$f_{ctm}/f_b = 1/(2{,}25\eta_1\eta_2) = 1/(2{,}25 \times 0{,}7 \times 1{,}0) = 40/63$. It should be noted that this value closely agrees with the value of $2/3$ for type 2 bars quoted in Table 5.2 of Anchor (1992).

Equation (11.35) becomes

$$s_{r,max} = \frac{40}{63}\frac{\phi}{2\,\rho_c} \tag{11.36}$$

For a thickness less than 500 mm, A_s should be calculated on the basis of an effective concrete thickness of $h/2$ and for thicknesses greater than 500, an effective concrete thickness of 250 mm (Anchor, 1992). In this case the wall is 400 thick at the top and 800 at the bottom. As the majority of the wall is thicker than 500 use 250 mm effective thickness.

If the spacing of the reinforcement is s,

then $A_s = (1000/s)\pi\phi^2/4$,

$\rho_c = A_s/(250 \times 1000) = 4 \times 10^{-6}$ As, therefore
$\rho_c = 4 \times 10^{-6}(1000/s)\pi\phi^2/4 = 0{,}00314\phi^2/s$. Thus $\phi/\rho_c = 318s/\phi$.

From Eq. (11.36),

$$s_{r,max} = \frac{40}{63}\frac{\phi}{2\,\rho_c} = \frac{40}{63} \times \frac{1}{2} \times 318\frac{s}{\phi} = 101\frac{s}{\phi} \tag{11.37}$$

$s_{max} = 1052$ mm (from above), so $(s/\phi)_{req} = 1052/101 = 10{,}4$ mm

Use D10 at 100 ctrs ($A_s = 785$ mm^2/m) each face.

Base:
$\sigma_1 = 154$ kPa (at A) and $\sigma_2 = 29$ kPa (at B)

Design of reinforcement:
Based on an estimate of 32 mm bars, $d = 600 - 30 - 16 = 554$ mm.

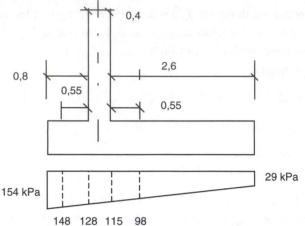

FIGURE 11.15 Ultimate bearing pressure distribution

Use $d = 550$ mm.

Moment at front of wall due to bearing pressure (Fig. 11.15):

$$M_{Sd} = 128 \times 0{,}8(\tfrac{1}{2} \times 0{,}8 + 0{,}2) + \tfrac{1}{2}(154 - 128) \times 0{,}8(\tfrac{2}{3} \times 0{,}8 + 0{,}2) = 69\,\text{kNm/m}$$

Moment at rear of wall due to bearing pressure:

$$M_{Sd} = 29 \times 2{,}6(\tfrac{1}{2} \times 2{,}6 + 0{,}2) + \tfrac{1}{2}(115 - 29) \times 2{,}6(\tfrac{1}{3} \times 2{,}6 + 0{,}2) = 232\,\text{kNm/m}$$
$$M_{Sd}/bd^2 f_{ck} = 232 \times 10^3/(550^2 \times 40) = 0{,}0192.$$

From Eq. (6.22) $A_s f_{yk}/bdf_{ck} = 0{,}0226; A_s = 994\,\text{mm}^2/\text{m}.$
Fix D16 at 200 mm spacing (to line up with vertical steel) [1005 mm^2/m].

Shear at 'd' from either face of the wall.

Front face:

$$V_{sd} = \tfrac{1}{2} \times 0{,}25(154 + 148) = 37{,}8\,\text{kN/m}$$

Back face:

$$V_{sd} = \tfrac{1}{2} \times 2{,}05(98 + 29) = 130{,}2\,\text{kN/m}$$

Check shear capacity V_{Rdc} only.

$$V_{Rdc} = C_{Rd,c}k(100\rho_1 f_{ck})^{1/3}bd$$
$$\rho_1 = A_s/bd = 1005/(1000 \times 550) = 0{,}00183 < 0{,}02$$
$$C_{Rd,c} = 0{,}18/\gamma_c = 0{,}18/1{,}5 = 0{,}12$$
$$k = 1 + \sqrt{200/d} = 1 + \sqrt{200/550} = 1{,}63 < 2{,}0$$

$V_{Rdc} = 0,12 \times 1,63 \times (100 \times 0,00183 \times 40)^{1/3} - 550 \times 1000 \times 10^{-3} = 209\,kN/m > V_{Sd}$,
Therefore no shear reinforcement required.

Anchorage (cl 9.8.2.2)

Front face:
Take $x = h/2 = 250\,mm$.

$$R = \tfrac{1}{2}(154 + 148) \times 0,25 = 37.8\,kN/m$$

$$e = 0,15b = 0,15 \times 400 = 60\,mm$$

$$z_e = 800 - x/2 + e = 800 - \tfrac{1}{2} \times 250 + 60 = 735\,mm$$

$$z_i = 0,9d = 0,9 \times 550 = 495\,mm$$

$$F_s = Rz_e/z_i = 37,8 \times 735/495 = 56,1\,kN/m.$$

There are five bars per metre, therefore force per bar $= 56,1/5 = 11,2\,kN$.
Bond strength $= 2,25 \times 0,7 \times 1,0 \times 3,5/1,5 = 3,68\,MPa$.
Perimeter of bar $= 16\pi = 50\,mm$.
Anchorage length $= 11200/(50 \times 3,67) = 61\,mm$. This is less than x, therefore there are no anchorage problems.

Back face:
Take $x = h/2 = 250\,mm$.

$$R = \tfrac{1}{2}(29 + 37) \times 0,25 = 8,25\,kN/m$$

$$e = 0,15b = 0,15 \times 400 = 60\,mm$$

$$z_e = 800 - x/2 + e = 2600 - \tfrac{1}{2} \times 250 + 60 = 2535\,mm$$

$$z_i = 0,9d = 0,9 \times 550 = 495\,mm$$

$$F_s = Rz_e/z_i = 8,25 \times 2535/495 = 42,3\,kN/m.$$

This is less than the figure for the front face.

Longitudinal steel:
From cl 9.3.1.1(2), the distribution reinforcement should be at least 20 per cent of the main longitudinal reinforcement.
$A_s = 1005\,mm^2m$; $0,2A_s = 201\,mm^2/m$. Minimum steel $= 0,0013bd = 0,0013 \times 550 \times 1000 = 715\,mm^2$. Maximum spacing 450 mm or $3,5h$ ($= 1,75\,m$). Fix D12 at 150 centres (A $= 754\,mm^2/m$).

Settlement and tilt:
Assuming dense sand $E_v = 80\,MPa$ and $v = 0,3$.

Serviceability analysis:
$K_a = \tfrac{1}{3}$.

Moment about centre line of the base:
due to soil pressure:

$$\tfrac{1}{3} \times \tfrac{1}{2} \times K_a \times \gamma_{soil} \times H^3 = \tfrac{1}{3} \times \tfrac{1}{3} \times \tfrac{1}{2} \times 17 \times 4{,}3^3 = 75{,}1\,\text{kNm/m [counterclockwise]},$$

due to surcharge:

$$\tfrac{1}{2} \times q \times H^2 = \tfrac{1}{2} \times 10 \times 4{,}3^2 = 92{,}5\,\text{kNm/m [counterclockwise]}$$

Moment about centre line of the base due to concrete and soil
Base $= 0$

Wall: $\tfrac{1}{2} \times 25 \times 0{,}4 \times 3{,}7 \times (\tfrac{1}{2} \times 3{,}8 - 0{,}8 - \tfrac{2}{3} \times 0{,}4) = 16{,}7\,\text{kNm/m [counterclockwise]}$

Retained soil:

$17 \times 3{,}7 \times 2{,}6 \times (\tfrac{1}{2} \times 3{,}8 - \tfrac{1}{2} \times 2{,}6) = 98{,}1\,\text{kNm/m [clockwise]}.$

Taking the counterclockwise direction as unfavourable, the net moment is

$$92{,}5 + 75{,}1 + 16{,}7 - 98{,}1 = 86{,}2\,\text{kNm/m}.$$

Vertical actions:

Concrete:
$$\text{Base: } 25 \times 3{,}8 \times 0{,}6 = 57{,}0\,\text{kN/m}$$
$$\text{Wall: } 25 \times 3{,}7 \times 0{,}4 = 37{,}0\,\text{kN/m}$$
Total due to the concrete: $= 94{,}0\,\text{kN/m}$
Soil: $17 \times 3{,}7 \times 2{,}6 = 163{,}5\,\text{kN/m}$
Total vertical action $= 257{,}5\,\text{kN/m}$

Bearing pressures:

$$\sigma_1 = 257{,}5/3{,}8 + 6 \times 86{,}2/3{,}8^2 = 104\,\text{kPa},$$
$$\sigma_2 = 257{,}5/3{,}8 - 6 \times 86{,}2/3{,}8^2 = 32\,\text{kPa}.$$

The average bearing pressure is needed at serviceability together with the service out of balance moment. Assuming dense sand $E_v = 80\,\text{GPa}$ (from Table 11.5).

The average bearing pressure $\sigma_{av} = 0{,}5(104 + 32) = 68\,\text{kPa}.$

$B = 3{,}8\,\text{m}$, so from Eq. (11.24),

$$s = 2{,}12 \times 3{,}8 \times (1 - 0{,}3^2) \times 68/80 = 6{,}2\,\text{mm}$$

Out of balance moment $M_0 = 86{,}2\,\text{kNm/m}$, so from Eq. (11.25) the angle of tilt β is given as

$$\beta = \arctan(5{,}1 \times (1 - 0{,}3^2) \times 86{,}2 \times 10^3/(80 \times 3800^2)) = 0{,}0198°.$$

Movement at top of wall $= 400 \times \tan 0{,}0198 = 1{,}4$ mm (mov/h = 1/280)

EN 1997-1 Table C.1 suggests that a rotational movement of the order 0,004 to 0,005H is required for dense sand to mobilize active pressures. In this case the rotational movement is around 0,0036H not including the effect of deflection of the wall. Thus active pressures should be capable of being mobilized.

11.6.4 Counterfort Retaining Walls

These are needed when the height exceeds around 7 m, although an alternative option is to use normal plain stem retaining walls with the addition of ground anchors to resist part of the effects of the lateral thrust. Typical dimensions of a counterfort retaining wall are given in Fig. 11.16. Full joints are usual at around 20 to 30 m. To carry out the initial stability and bearing pressure calculations the specific

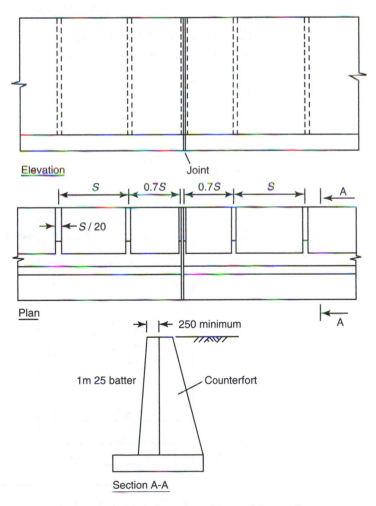

FIGURE 11.16   General arrangement of a counterfort retaining wall

weight of the concrete in the counterforts can be taken as that of the soil in order to simplify the calculations. The internal moments and shears in both the vertical stem and the portion of the base between the counterforts are best determined using the Hillerborg strip method as this allows the designer both to select the manner in which the reinforcement is detailed and enables the reactions on the counterfort to be determined easily. The counterforts are then designed as cantilever 'T' beams resisting these reactions.

EXAMPLE 11.6   Counterfort retaining wall. The basic data for an internal bay are given in Fig. 11.17.

The basic dimensions have sized using the guidelines for normal retaining walls. The overall height is 11 m, thus the base thickness is $0,1H$ which will be slightly rounded down to 1 m. The width of the base B is between $0,5H$ and $0,8H$ giving a range of between 5,5 m and 8,8 m, a width of 8 m will be adopted. The distance between the front of the base and the wall is $0,25B$, i.e. 2 m.

Use EQU to check against overturning:
The effective shearing resistance angle is given by

$$\phi_{eff} = \arctan\left[\frac{\tan 30}{1,25}\right] = 24,8°$$

and K_a by

$$K_a = \frac{1 - \sin \phi_{eff}}{1 + \sin \phi_{eff}} = 0,409$$

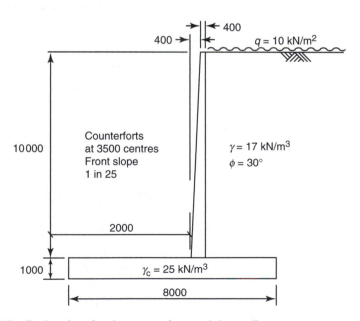

FIGURE 11.17   Design data for the counterfort retaining wall

Moment about A due to earth pressure:

$$^1/_3 \times ^1/_2 \times K_a \times \gamma \times H^3 = ^1/_3 \times ^1/_2 \times 0{,}409 \times 17 \times 11^3 = 1542 \, kNm/m$$

due to surcharge:

$$^1/_2 \times q \times H^2 = ^1/_2 \times 10 \times 11^2 = 605 \, kNm/m$$

Apply a factor $\gamma_{G;dst}$ ($=1{,}1$) to the earth pressure and $\gamma_{Q;dst}$ ($=1{,}5$) to the surcharge to give a factored overturning moment of

$$1{,}1 \times 1542 + 1{,}5 \times 605 = 2604 \, kNm/m.$$

Restoring moment:
Concrete:

Base: $^1/_2 \times 25 \times 1{,}0 \times 8{,}0^2 = 800 \, kNm/m$
Wall: $^1/_2 \times 25 \times 10 \times 0{,}4 \times (2 + ^2/_3 \times 0{,}4) = 113 \, kNm/m$

$$25 \times 10 \times 0{,}4 \times (2 + 0{,}4 + ^1/_2 \times 0{,}4) = 260 \, kNm/m$$

Total $= 1173 \, kNm/m$

Retained earth:

$$17 \times 10 \times 5{,}2(8 - ^1/_2 \times 5{,}2) = 4774 \, kNm/m$$

Total unfactored restoring moment $= 5947 \, kNm/m$

This takes a factor $\gamma_{G;dst}$ ($=0{,}9$) to give a net restoring moment of $0{,}9 \times 5947 = 5352 \, kNm/m$. This is greater than the overturning moment and is therefore satisfactory.

STR/GEO:

For this example only Load Case 1 Combination 1 will be considered. For this case $\tan\phi'$ is unfactored and $K_a = ^1/_3$.

Taking moments about the base centre line.
Earth pressure:

$$^1/_3 \times ^1/_2 \times K_a \times \gamma \times H^3 = ^1/_3 \times ^1/_3 \times ^1/_2 \times 17 \times 11^3 = 1257 \, kNm/m$$

due to surcharge:

$$^1/_2 \times q \times H^2 = ^1/_2 \times 10 \times 11^2 = 605 \, kNm/m$$

Restoring moment:
Concrete:

Base: $=0$
Wall: $^1/_2 \times 25 \times 10 \times 0{,}4 \times (4 - (2 + ^2/_3 \times 0{,}4)) = 87 \, kNm/m$

$$25 \times 10 \times 0{,}4 \times (4 - (2 + 0{,}4 + ^1/_2 \times 0{,}4)) = 227 \, kNm/m$$

Total $= 227 \, kNm/m$ [counterclockwise]

Retained earth:

$17 \times 10 \times 5{,}2 \times (4 - \frac{1}{2} \times 5{,}2) = 1238\,\text{kNm/m}$ [clockwise]
Apply a factor of γ_G of 1,0 to the favourable permanent loads, a factor γ of 1,35 to the permanent unfavourable and γ_Q of 1,5 to unfavourable variable to give a net factored unfavourable moment of $1{,}5 \times 605 + 1{,}35 \times 1257 + 1{,}35 \times 227 - 1{,}0 \times 1238 = 1673\,\text{kNm/m}$.

Vertical:

Concrete:

Base: $25 \times 8 \times 1 = 200\,\text{kN/m}$
Wall: $\frac{1}{2} \times 25 \times (0{,}4 + 0{,}8) \times 10 = 150\,\text{kNm/m}$

Total $= 350\,\text{kN/m}$
Soil: $17 \times 10 \times 5{,}2 = 884\,\text{kN/m}$
Total $= 1234\,\text{kN/m}$

As this is unfavourable, apply a factor γ_G of 1,35 to give a factored vertical load of $1{,}35 \times 1234 = 1666\,\text{kN/m}$.

Earth pressures:

$$\sigma_1 = 1666/8 + 6 \times 1673/8^2 = 365\,\text{kPa},$$
$$\sigma_2 = 1666/8 - 6 \times 1673/8^2 = 51\,\text{kPa}.$$

Sliding:

Disturbing force due to

surcharge: $qh = 10 \times 11 = 110\,\text{kN/m}$
retained earth: $\frac{1}{2}k_a\gamma h^2 = \frac{1}{2} \times \frac{1}{3} \times 17 \times 11^2 = 343\,\text{kN/m}$

Total factored disturbing force $= 1{,}5 \times 110 + 1{,}35 \times 343 = 628\,\text{kN/m}$.

Vertical force (unfactored) is 1234 kN/m, so available friction force is $1234 \tan 30° = 712\,\text{kN/m}$. This is greater than the disturbing force, therefore satisfactory.

Structural Design:

The loading is determined using Design Approach 1: Combination 1.

Wall:

The Hillerborg Strip Method will be used to determine the bending moments in the wall. The distribution of loading is given in Fig. 11.18, where it will be noted the distribution slightly varies from the traditional 45° dispersion for ease of calculation.

(a) Horizontally spanning strips:

The wall is divided into 1 m wide strips and the load calculated at the middle of each strip and assumed to be uniform over the width of the strip. The effective depth is also calculated at the middle of each strip.

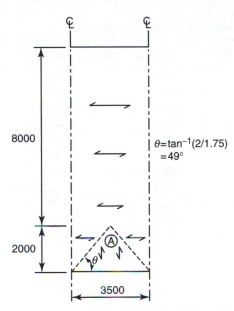

$\theta = \tan^{-1}(2/1.75)$
$= 49°$

FIGURE 11.18   Load dispersion on front wall

For each strip the uniformly distributed load $q_{av,n}$ (kPa) is given by

$$q_{av,n} = 1{,}5q + 1{,}35K_a\gamma\left(n - \frac{1}{2}\right) \qquad (11.38)$$

where q is the surcharge (10 kPa), γ the soil specific weight (17 kN/m³), K_a the active pressure coefficient ($^1/_3$) and n the strip number from the top. Thus Eq. (11.38) becomes

$$q_{av,n} = 15 + 7{,}65\left(n - \frac{1}{2}\right)$$

As there is continuity across the counterforts, the hogging and sagging moments will be taken as equal. Thus for

Strips 1 to 8,

$$M_{Sd,n} = \frac{q_{av,n}\,3{,}5^2}{16}$$

Strip 9,
Determine the load distribution factor K from Eq. (10.141) with $L_1 = 1{,}75$ m and $L_2 = 0{,}875$ m,

$$K = \frac{4}{3}\left(1 - \frac{1}{2 + (L_1/L_2) + (L_2/L_1)}\right)$$

$$= \frac{4}{3}\left(1 - \frac{1}{2 + (1{,}75/0{,}875) + (0{,}875/1{,}75)}\right) = 1{,}04$$

$$M_{Sd,9} = K\,q_{av,9}\frac{(L_1 + L_2)^2}{16}$$

TABLE 11.6   Horizontal flexural design of wall.

Strip	$q_{av,n}$ (kPa)	$M_{Sd,n}$ (kNm/m)	d (mm)	$M_{Sd,n}/bd^2f_{ck}$
1	18,8	14,4	380	0,0025
2	26,5	20,3	420	0,0029
3	34,1	26,1	460	0,0031
4	41,8	32,0	500	0,0032
5	49,4	37,8	540	0,0032
6	57,1	43,7	580	0,0032
7	64,7	49,6	620	0,0032
8	72,4	55,5	660	0,0032
9	80,0	35,8	700	0,0018
10	87,7	5,6	740	0,0003

Strip 10,

The load distribution factor K from Eq. (10.141) with $L_1 = 0,875$ m and $L_2 = 0$ is 4/3, so the design moment is given by

$$M_{Sd,10} = K\,q_{av,10}\frac{(L_1 + L_2)^2}{16}$$

The flexural design is carried out in Table 11.6 with $f_{ck} = 50$ MPa, $f_{yk} = 500$ MPa, and a cover of 30 mm. The effective depth d is calculated assuming a bar size of 16 mm, but the cover plus half bar size rounded to 40 mm.

Minimum reinforcement requirement (cl 9.2.1.1):

$$A_{s,min} = 0,26(f_{ctm}/f_{yk})b_t d \geq 0,0013b_t d.$$

Determine ratio $0,26f_{ctm}/f_{yk}$:

$$f_{ctm} = 0,3f_{ck}^{1/3} = 0,3 \times 40^{1/3} = 3,5\,\text{MPa}.$$

$0,26f_{ctm}/f_{yk} = 0,26 \times 3,5/500 = 0,00182$, thus

$A_{s,min}/b_t d = 0,00182$ or $A_{s,min}f_{yk}/b_t df_{ck} = 0,0228$.

Use Eq. (6.21) to determine the value of M_{Sd}/bd^2f_{ck} corresponding to $a_{s,min}f_{yk}/bdk_{ck}$. This gives a value of M_{sd}/bd^2f_{ck} of 0,019. This is greater than any of the values of M_{Sd}/bd^2f_{ck} in Table 11.6, thus supply minimum horizontal reinforcement all the way up the wall in both faces. For convenience split the wall into two zones using the value of d at the base of the zone:

5–10 m:

$A_{s,min} = 0,0228 \times 1000 \times 740 \times 40/500 = 1350\,\text{mm}^2/\text{m}$. Fix D20 at 200 ctrs [1572 mm^2/m]

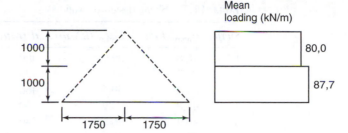

FIGURE 11.19   Loading on area A (of Fig. 11.18)

0–5 m:

$A_{s,min} = 0,0228 \times 1000 \times 540 \times 40/500 = 985 \text{ mm}^2/\text{m}$. Fix D16 at 200 ctrs [1005 mm^2/m]

(b) Vertically spanning strips (Fig. 11.19)

From Strip 10,

$$M_{Sd,10} = 87,7[{}^1\!/_2 \times 1,75 \times 1,0^2 + {}^1\!/_3 \times 1,75 \times 1,0^2] = 52,0 \text{ kNm}$$

From Strip 9,

$$M_{Sd,9} = {}^1\!/_2 \times 80,0 \times 1,0 \times 1,75[1,0 + {}^1\!/_3 \times 1,0] = 93,3 \text{ kNm}$$

Total moment, $M_{Sd} = 145,3 \text{ kNm}$

At the base of the wall, $d = 760$ mm, so

$$M_{Sd}/bd^2 f_{ck} = 145,3 \times 10^6/3500 \times 760^2 \times 40 = 0,0018.$$

This will give minimum reinforcement, therefore fix D20 at 200 centres. This will need to extend the whole height of the back face of wall (and will also be needed in the front face to support the horizontal steel).

Shear:

(a) Horizontally spanning strips (Table 11.7):

The shear force V_{Sd} has been calculated as half the load on each strip. In all cases the shear resistance $V_{Rd,c}$ is greater than the applied shear V_{Sd}.

(b) Vertically spanning strips:

From Fig. 11.19, the total shear force V_{Sd} is given by

$$V_{Sd} = {}^1\!/_2 \times 1,75 \times 1,0 \times 80 + {}^1\!/_2 \times (3,5 + 1,75) \times 1,0 \times 87,7$$
$$= 300,2 \text{ kN (or } 300,2/3,5 = 85,8 \text{ kN/m width of wall).}$$

TABLE 11.7   Shear design of wall.

Strip	$q_{av,n}$ (kPa)	$V_{Sd,n}$ (kN/m)	d (mm)	A_s/bd	$V_{Rd,c}$ (kN/m)
1	18,8	32,9	380	0,0026	171,7
2	26,5	46,4	420	0,0024	181,0
3	34,1	59,7	460	0,0022	189,1
4	41,8	73,2	500	0,0020	195,9
5	49,4	86,5	540	0,0019	204,9
6	57,1	99,9	580	0,0027	244,2
7	64,7	113,2	620	0,0025	251,3
8	72,4	126,7	660	0,0024	261,0
9	80,0	105,0	700	0,0022	266,1
10	87,7	38,4	740	0,0021	274,3

As this figure is below the value of $V_{Rd,c}$ for the 10th strip, then the shear is satisfactory.

(c) Reactions:

The loading from the wall to the counterforts is tensile and must therefore be transmitted through shear links acting at a design strength of f_{yk}/γ (i.e. 500/1,15 = 434,8 MPa).

Divide the wall into two bands 0–5 m and 5–10 m;

0–5 m:

From Table 11.7, the maximum shear is 126,7 kN [Strip 8], therefore the steel area required is $126,7 \times 10^3/434,8 = 291$ mm^2/m. Using a spacing of 200 mm, the area of the leg of the link is

$291/5 = 58$ mm^2, i.e. fix D10 [79 mm^2/ per leg]

5–10 m:

From Table 11.7, the maximum shear is 86,5 kN, therefore the steel area required is $86,5 \times 10^3/434,8 = 199$ mm^2/m. Using a spacing of 200 mm, the area of the leg of the link is $199/5 = 40$ mm^2, i.e. fix D10 [79 mm^2/ per leg].

The shear links will need anchoring around the D40 bars in the sloping faces of the back of the counterforts and additional 2D40 bars in the front sloping face of the wall.

(d) Counterforts:

The loading on each counterfort is TWICE the shear V_{Sd} from Table 11.7. This gives the resultant BMD (Fig. 11.20(a)) and SFD (Fig. 11.20(b)). It has been assumed the load acts at the centre height of any strip.

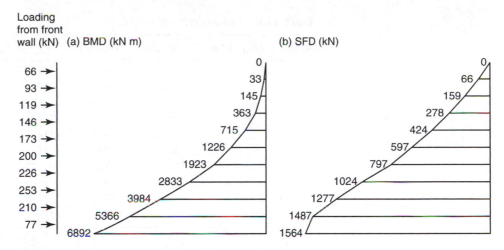

FIGURE 11.20 Bending moment and shear force diagrams for the counterforts

Effective width of slab (cl 5.3.2.1) (Section 6.4):

The value of l_0 is taken as the effective span (to centre line), i.e. 10,5 m.

The width b can be taken typically as $s/20$ ($=175\,$mm)(Fig. 11.11). This will be too narrow to place two layers of reinforcement and compact the concrete. The minimum width is given by twice the cover plus twice link size plus twice flexural bar size place 75 mm spacing between the layers of the reinforcement cage, i.e. $2 \times 30 + 2 \times 10 + 2 \times 40 + 75 = 235$. Use 300 mm.

$$b_{\text{eff},1} = 0{,}2\,b_1 + 0{,}1l_0 = 0{,}2(1750 - 300) + 0{,}1 \times 10500 = 1340\,\text{mm}$$

limiting value $= 0{,}2l_0 = 0{,}2 \times 10500 = 2100\,$mm

Use limiting value of 1340 mm.

Thus $b_{\text{eff}} = 2 \times 1340 + 300 = 2980\,$mm (this is less than the counterfort spacing of 3500 mm.

Assume two layers of 40 mm bars, so effective cover $=$ cover $+$ links $+$ bar diam $+\,{}^{1}\!/_{2}$ spacing $= 30 + 8 + 40 + {}^{1}\!/_{2}40 = 100\,$mm.

Assuming the neutral axis to be in the flange, all the values of $M_{\text{Sd}}/bd^2f_{\text{ck}}$ given in Table 11.8 are below those requiring minimum reinforcement, so using values of d in the 5th and 10th strip.

Note, the minimum reinforcement requirement is based on the width of the web for a 'T' where the flange is in compression.

5–10 m:

$$A_{\text{s,min}} = 0{,}0228 \times 300 \times 7900 \times 40/500 = 4323\,\text{mm}^2.$$

TABLE 11.8   Counterfort design data.

Strip	M_{Sd} (kNm)	h(mm)	d(mm)	$M_{Sd}/bd^2 f_{ck}$	t_f (mm)
1	33	800	700	0,0006	420
2	145	1600	1500	0,0005	460
3	363	2400	2300	0,0006	500
4	715	3200	3100	0,0006	540
5	1225	4000	3900	0,0007	580
6	1923	4800	4700	0,0007	620
7	2833	5600	5500	0,0008	660
8	3984	6400	6300	0,0008	700
9	5366	7200	7100	0,0009	740
10	6892	8000	7900	0,0009	780

Fix 4D40 [5026 mm^2]

0–5 m:

$$A_{s,min} = 0{,}0228 \times 300 \times 3900 \times 40/500 = 2134 \text{ mm}^2.$$

Fix 2D40 [2513 mm^2], i.e. curtail the inner layer of bars above 5 m up the wall. Check position of neutral axis depth using Eq. (6.19)

$$\frac{x}{d} = 1{,}918 \frac{A_s f_{yk}}{b \, d \, f_{ck}}$$

For 0–5 m:

$$A_s = 2513 \text{ mm}^2, A_s f_{yk}/bd f_{ck} = 2513 \times 500/2980d \times 40 = 10{,}54/d$$
$$x/d = 1{,}918 A_s f_{yk}/bd f_{ck} = 1{,}918 \times 10{,}54/d = 20/d \text{ or}$$

$x = 20$ mm. In all cases the flange thickness t_f exceeds the neutral axis depth.

For 5–10 m:

$$A_s = 5026 \text{ mm}^2, A_s f_{yk}/bd f_{ck} = 5026 \times 500/2980d \times 40 = 21{,}1/d$$
$$x/d = 1{,}918 A_s f_{yk}/bd f_{ck} = 1{,}918 \times 21{,}1/d = 40{,}5/d \text{ or}$$

$x = 40{,}5$ mm. In all cases the flange thickness t_f exceeds the neutral axis depth.

Shear:

As in all but the first 2 m down from the top, the concrete shear resistance is inadequate (i.e. $V_{Sd} > V_{Rd,c}$), links must be provided.

Minimum shear links:

$$\rho_{w,min} = 0{,}08\sqrt{f_{ck}/f_{yk}} = 0{,}08\sqrt{40/500} = 1 \times 10^{-3}.$$
$$(A_s/s)_{min} = \rho_{w,min} b_t = 1 \times 10^{-3} \times 300 = 0{,}3.$$

Max spacing is 0,75d.

TABLE 11.9   Shear design of counterforts.

Strip	V_{Sd} (kN)	$d(mm)$	$A_s/b_t d$	$V_{Rd,c}$ (kN)	$V_{Rd.s,min}$ (kN)	$V_{Rd,s}$ (kN)
1	66	700	0,012	142	83	
2	159	1500	0,0055	208	177	
3	278	2300	0,0036	262	270	364
4	424	3100	0,0027	310	364	491
5	597	3900	0,0021	351	457	618
6	797	4700	0,0036	498	551	1446
7	1024	5500	0,0030	541	645	1691
8	1277	6300	0,0027	592	738	1937
9	1487	7100	0,0024	635	832	2183
10	1564	7900	0,0021	671	925	2428

$V_{Rd,s}$ corresponding to minimum link requirement is given by

$$V_{Rd,s} = (A_{sw}/s)zf_{yw}/\gamma.$$

Taking $z=0,9d$ and f_{ywd} as $500/1,15=434,8$ MPa, then $V_{rd,s}$ corresponding to minimum link requirement is given by

$$V_{Rd,s} = 0,3 \times 0,9d \times 434,8 \times 10^{-3} = 0,117d \text{ kN}.$$

Link requirement:
The design resistance of the shear links required is tabulated in the last column of Table 11.9.

0−2 m: Minimum links.

2−5 m: D8 at 250 mm ctrs

$$V_{Rd,s} = (A_{sw}/s)0,9f_{ywd} = (101/250) \times 0,9d \times 434,8 = 0,158d \text{ kN}$$

5−10 m: D10 at 250 mm ctrs

$$V_{Rd,s} = (A_{sw}/s)0,9f_{ywd} = (157/250) \times 0,9d \times 434,8 = 0,307d \text{ kN}$$

Surface reinforcement (cl J.1):
As the bars are D40, then surface reinforcement is required.

Since the neutral axis is in the wall (i.e the flange of the 'T' section), take $d-x$ as being the whole of the web at its deepest, i.e. $800 - 8 - 30 - 40 - 1/2 40 = 702$ mm.

The surface of the concrete external to the links is

$$A_{ct,ext} = 702(30 + 8) + (300 - 2 \times (30 + 8))(32 + 8) = 35180 \text{ mm}^2.$$

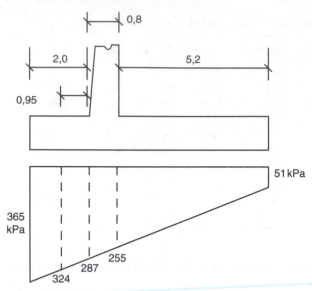

FIGURE 11.21 Ultimate bearing pressure distribution

It is recommended that the area of surface reinforcement $A_{s,surmin}$ is taken as $0,01A_{ct,ext} = 352\,mm^2$ in each direction. Thus additional bars (or mesh) needs to be provided. Note, longitudinal bending reinforcement and shear reinforcement may contribute to $A_{s,surmin}$.

Base:
The bearing pressure distribution is given in Fig. 11.21.
Cover 30 mm, assumed bar size 32 mm, $d = 1000 - 30 - 16 = 954\,mm$.
Use 950 mm.

Toe:

$$M_{Sd} = 287 \times 2(^1/_2 2,0 + 0,4) + ^1/_2 \times (365 - 287) \times 2 \times (^2/_3 \times 2 + 0,4) = 939\,kNm/m.$$
$$M_{Sd}/bd^2f_{ck} = 939 \times 10^3/950^2 \times 40 = 0,0260.$$

From Eq. (6.22) $A_s f_{yk}/bd_{ck} = 0,0307$ and

$$A_s = 0,0307 \times 1000 \times 950 \times 40/500 = 2333\,mm^2.$$

Fix D25 at 200 ctrs [2454 mm²/m]

Shear at $d = 950\,mm$ from face of wall:

$$V_{Sd} = ^1/_2(365 + 324) \times 1,05 = 362\,kN/m.$$
$$\rho = 2454/1000 \times 950 = 0,0026$$
$$V_{Rd,c} = 0,12[1 + \sqrt{(200/950)}](100 \times 0,0026 \times 40)^{^1/_3} - 950 = 363\,kN/m.$$

$V_{Rd,c}$ exceeds V_{Sd} and is therefore satisfactory.

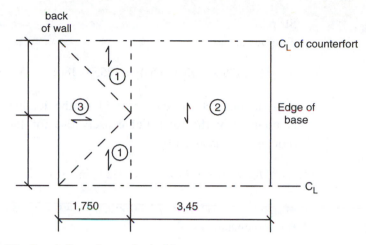

FIGURE 11.22   Load dispersion on heel of base

Heel:

Use the Hillerborg Strip Method with the load dispersal pattern and associated bearing pressure distribution given in Fig. 11.22.

Strip 1:

$L_1 = 1{,}75$ m; $L_2 = 0$, so from (10.141), $K = 4/3$

Mean load on the strip, $q = \frac{1}{2}(255 + 186) = 221$ kPa.

Free bending moment,

$$M = qK(L_1 + L_2)^2/8 = 221 \times (4/3)(1{,}75 + 0)^2/8 = 113 \text{ kNm/m}$$

The design bending moment is half this value as the slab is continuous over the counterfort, so

$$M_{Sd} = 66{,}5 \text{ kNm/m and } M_{Sd}/bd^2f_{ck} = 66{,}5 \times 10^3/950^2 \times 40 = 0{,}0018$$

This will require minimum reinforcement.

Strip 2:

Mean load on the strip, $q = \frac{1}{2}(51 + 186) = 119$ kPa.

Free bending moment,

$$M = qL^2/8 = 119 \times 3{,}5^2/8 = 182 \text{ kNm/m}$$

The design bending moment is half this value as the slab is continuous over the counterfort, so

$$M_{Sd} = 91 \text{ kNm/m and } M_{Sd}/bd^2f_{ck} = 91 \times 10^3/950^2 \times 40 = 0{,}0025$$

This will require minimum reinforcement.

Strip 3:
Total load:

$$\tfrac{1}{2} \times 186 \times 1{,}75 \times 3{,}5 + \times (255 - 186) \times 3{,}5 \times 1{,}75 = 570 + 141 = 711\,\text{kN}.$$

Assume this acts at the centroid of the loaded area (this is true for the first component of the load, but conservative for the second part). Thus the design moment M_{Sd} is given by

$$M_{Sd} = \tfrac{1}{3} \times 1{,}75 \times 711 = 415\,\text{kNm}.$$

$M_{Sd}/bd^2 f_{ck} = 415 \times 10^6/3500 \times 950^2 \times 40 = 0{,}0033$. This will require minimum reinforcement.

Minimum reinforcement per metre width:

$$A_{s,min} = 0{,}0228 \times 1000 \times 950 \times 40/500 = 1733\,\text{mm}^2.$$

Fix D25 at 200 ctrs [2454 mm²/m] in each direction in each face.

REFERENCES AND FURTHER READING

Anchor, R.D. (1992) *Design of Liquid Retaining Concrete Structures* (2nd Edition), Edward Arnold, London.

BS EN 1997-1. Eurocode 7: Geotechnical Design- Part 1: General rules.

Clayton, C.R.I., Milititsky, J. and Woods, R.I. (1993) *Earth Pressure and Earth Retaining Structures* (2nd Edition), Blackie, Academic & Professional, London.

Henry, F.D.C. (1986) *The Design and Construction of Engineering Foundations* (2nd Edition), Chapman Hall, London.

Martin, L.H., Croxton, P.C.L. and Purkiss, J.A. (1989) *Structural Design in Concrete to BS8110*, Edward Arnold, London.

Oliphant, J. (1997) The outline design of earth retaining walls. *Ground Engineering*, **30(8)**, 53–58.

Tomlinson, M.J. (1994) *Pile Design and Construction Practice* (4th Edition), E&F Spon, London.

12.1 BENDING RESISTANCE OF PRESTRESSED CONCRETE MEMBERS AT ULTIMATE LOAD

12.1.1 Introduction

A member where the concrete is prestressed by the longitudinal reinforcing steel (tendons) before being subject to external loading is described as pre-stressed concrete. The tendons are of high strength steel (1000 to 2000 MPa) and the concrete is also at high strength (50 to 90 MPa). There are two methods of manufacturing: (a) pretensioning the tendons before casting the concrete, (b) post-tensioning the tendons after casting the concrete.

A pretensioned member is manufactured by, first, pretensioning the steel tendons over the length of the casting bed, as shown in Fig. 12.1(a), and anchoring the steel to rigid anchor blocks at either end. The tendons are generally in the form of straight high tensile steel wires varying in diameter from 2 to 7 mm. The concrete is then cast around the tendons using shuttering to form the required shape of cross section of the member. Generally within 3 days, using accelerated curing techniques, the concrete has gained sufficient strength to be able to transfer the prestressing force to the concrete by releasing the anchorages. The shuttering is then removed and the casting bed can be used again. This method favours the reuse of shuttering and the manufacture of identical members, but the members have to be transported from the casting yard to the site.

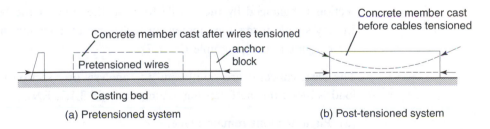

(a) Pretensioned system   (b) Post-tensioned system

FIGURE 12.1   Prestressing systems

A post-tensioned member is manufactured by first preparing the shuttering which supports sheaths or ducts. The ducts contain unstressed tendons (see Fig. 12.1(b)) in the form of straight or curved cables, or bars, which are free to move in the direction of their length but are restrained to prevent movement transversely. The concrete is then poured into the mould and allowed to gain strength. There is less urgency in this method of manufacture and the shuttering is removed after a few days. Later, when the concrete has matured the steel is stressed by jacking against the ends of the beam and anchored using wedges for cables, or nuts on threaded rods for bars. The method is suitable for members of different cross section and for members which are cast in place on the construction sites where road access is difficult.

The advantages of prestressed concrete as compared with ordinary reinforced concrete are:

(a) reduced self weight of members,
(b) no cracks at service load and consequently better durability,
(c) increased shear resistance.

The disadvantages of prestressed concrete as compared with ordinary reinforced concrete are:

(a) increased cost of materials and shuttering,
(b) greater supervision required to ensure correct concrete strength and magnitude of prestress forces,
(c) design calculations are more extensive.

12.1.2 Analysis of a Section in Bending at the Ultimate Limit State (cl 3.1.7, EN)

Analysis entails the determination of the moment of resistance of a section given the size of the section and reinforcement. The behaviour of a prestressed concrete member in bending at the ultimate limit state is similar to that of an ordinary reinforced concrete member. The appearance of a beam element in bending is shown in Fig. 12.2, where tensile cracks, commencing in the extreme tension fibres, extend to the edge of the compression zone. The compressive force at the top of the section is balanced by the tensile force in the steel at the bottom of the section. In very simple terms the moment of resistance of a section in bending is therefore the magnitude of the couple Cz or Tz.

The basic concepts are simple and the analysis of a section in bending at ultimate load is based on the following assumptions (cl 3.1.7, EN).

(a) Plane sections remain plane,
(b) The strain in bonded reinforcement, whether in tension or compression, is the same as that in the surrounding concrete.

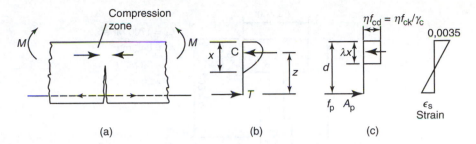

FIGURE 12.2 Bending resistance of a prestressed concrete beam at ultimate load

(c) The tensile strength of concrete is ignored.

(d) The stresses in the reinforcement or prestressing steel are derived from the design curves (Fig. 3.10, EN).

(e) The initial prestrain in the prestressing tendons is taken into account when assessing the stresses in the tendons at the ultimate limit state.

(f) For cross sections subject to pure longitudinal compression, the compressive strain is limited to 0,002 and for bending 0,0035.

(g) The tensile strain in the steel is limited to 0,02.

Applying the assumptions listed above and assuming a rectangular stress block in compression (Fig. 3.5, EN). The design resistance moment of a prestressed beam containing bonded tendons at design strength, all of which are located in the tension zone.

$$M_{Rd} = (f_{pk}/\gamma_s)A_p z = f_{pd}A_p z = f_{pd}A_p(d - \lambda x/2) \tag{12.1}$$

where the factor defining the effective depth of the stress block (Eqs 3.19 and 3.20, EN)

$\lambda = 0,8$ for $f_{ck} \leq 50\,\text{MPa}$

$\lambda = 0,8 - (f_{ck} - 50)/400$ for $50 < f_{ck} \leq 90\,\text{MPa}$

The depth of the compression (x) is obtained by equating the force in the compression zone to the balancing force in the tension zone

$$x/d = \rho f_{pd}/(\lambda \eta f_{cd}) \tag{12.2}$$

where the factor for defining the effective strength (Eqs 3.21 and 3.22, EN)

$\eta = 1,0$ for $f_{ck} \leq 50\,\text{MPa}$

$\eta = 1,0 - (f_{ck} - 50)/200$ for $50 < f_{ck} \leq 90\,\text{MPa}$

These equations assume that the steel and concrete reach their limiting values which must be confirmed.

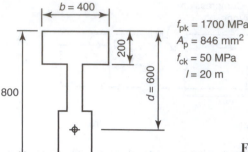

b = 400

200

d = 600

800

f_{pk} = 1700 MPa
A_p = 846 mm^2
f_{ck} = 50 MPa
l = 20 m

FIGURE 12.3   Example: analysis of a section in bending at ultimate load

EXAMPLE 12.1   Analysis of a prestressed beam in bending at ultimate load. Determine the ultimate moment of resistance for the prestressed beam section shown in Fig. 12.3 from basic principles assuming bonded tendons.

From Eq. (12.2) for $f_{ck} = 50$ MPa and $f_{pk} = 1770$ MPa, the ratio

$$x/d = \rho f_{pd}/(\lambda \eta f_{cd})$$

$$= 846/(400 \times 600) \times (1770/1,15)/(0,8 \times 1,0 \times 50/1,5) = 0,203$$

Hence $x = 0,203 \times 600 = 122$ mm which is less than the thickness of the 200 mm top flange (Fig. 12.3), and is therefore satisfactory.

The moment of resistance of the section at the ultimate limit state

$$M_{Rd} = f_{pd} A_p (d - \lambda x/2)$$

$$= (1770/1,15) \times 846 \times (600 - 0,4 \times 0,203 \times 600)/1E6 = 717,8 \text{ kNm}.$$

Check that the steel is at the ultimate design strength. For linear strain distribution over the depth of the cross section the bending strain in the steel

$$\varepsilon_{sb} = 0,0035(d/x - 1) = 0,0035 \times (600/122 - 1) = 0,0137$$

For an initial stress of $0,75 f_{pk}$ in the steel and an estimated loss of prestress at service load of 25 per cent

$$f_{pe}/f_{pk} = 0,75 \times 0,75 = 0,563$$

Total strain in the steel is the sum of the bending and prestress strains $= \varepsilon_{sb} + \varepsilon_{sp} = \varepsilon_{sb} + f_{pe}/E_s = 0,0137 + 0,563 \times 1770/200E3 = 0,0187.$

If the steel is at the maximum design value of f_{pk}/γ_s then from the stress–strain relationship for steel the strain in the steel will be equal to, or greater than

$$f_{pd}/E_s = (1770/1,15)/200E3 = 0,0077.$$

The total strain of $0,0187 > 0,0077$, therefore the assumption that the steel is at the maximum design stress of f_{pk}/γ_s is justified. The total strain is also less than the $\lim \varepsilon_{ud} = 0,02$ (cl 3.3.6, EN).

12.1.3 Design of a Section in Bending at Ultimate Load

The design of a section in bending at ultimate load entails determining the size of section and reinforcement given the magnitude of the bending moment. The design can be carried out from first principles using the equations given in 12.1.2, but this is tedious and can be avoided by using a graph (Annex B2). The design is based on the rectangular concrete stress block in compression and is constructed as follows:

Moment of resistance related to the concrete compression zone

$$M_{Rd} = f_{pd} A_p (d - \lambda x/2)$$

This equation can be rearranged in non-dimensional form for the vertical axis of the graph as

$$M_{Rd}/(bd^2 f_{ck}) = \eta (f_{cd}/f_{ck}) \lambda (x/d)(1 - \lambda x/d)$$

Similarly equating the tensile force to the compression force the horizontal axis of the graph

$$A_p f_{pk}/(bd f_{ck}) = \eta (f_{pk}/f_{pd})(f_{cd}/f_{ck}) \lambda (x/d)$$

A graph can be plotted by iterating values of (x/d) and checking for f_{pd} < yield and $\varepsilon_{ud} > 0,02$ (cl 3.3.6, EN). A typical graph for $f_{ck} = 50$ MPa and $f_{pk} = 1170$ MPa is shown in Annex B2. Others are required for different concrete, steel strength and material factors.

EXAMPLE 12.2 Design of a prestressed beam in bending at ultimate load. Determine the size of a prestressed concrete section to support a uniformly distributed design load of 125 kN at the ultimate limit state over a simply supported span of 15 m. Assuming $f_{pk} = 1770$ MPa, and Grade C50/60 concrete. Many similar beams to be made and access to the sties is good therefore use pretensioned beams.

Design bending moment at ultimate load

$$M_{Ed} = WL/8 = 125 \times 15/8 = 234,4 \text{ kNm}$$

The smallest depth of a simply supported beam, where deflection is not likely to be a problem at service load, is approximately span/30. Try $h = 15E3/30 = 500$ mm. To optimize on the weight an 'I' cross section is generally used in prestressed concrete but members should not be too slender because of instability. Assume breadth $b = 0,5\,h = 250$ mm. Web thickness is generally not less than 100 mm to resist

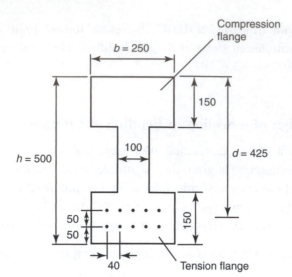

FIGURE 12.4 Example: design of a section in bending at ultimate load

shear and should provide room for wires. Assume $d = 425$ mm (Fig. 12.4). Cover should meet the requirements for durability and fire (Chapter 5).

Having decided on values of h, d and b calculate

$$M_{Ed}/(bd^2 f_{ck}) = 24,4E6/(250 \times 425^2 \times 50) = 0,104$$

Assume $f_{pe}/f_{pk} = 0,5$. From a graph (Annex B2), $A_p f_{pk}/(bd f_{ck}) = 0,13$ and hence

$$A_{ps} = 0,13 \times 250 \times 425 \times 50/1770 = 390\, mm^2, \text{ use } 10-7 \text{ mm wires,}$$
$$A_{ps} = 384, 8\, mm^2.$$

Minimum area of longitudinal steel (cl 9.2.1.1(1), EN)

$$A_{s, min} = 0,26(f_{cm}/f_{yk})b_t d = 0,26 \times (4,1/1770) \times 250 \times 425$$
$$= 63,99 < A_{ps}(384,8)\, mm^2, \text{ satisfactory}$$

From a graph (Annex B2) $x/d = 0,21$ hence

$$x = 0,21 \times 425 = 89,3 \text{ mm. Use top flange thickness of } 150 \text{ mm.}$$

The tension flange area must provide room for the wires with adequate spacing and cover. Try all the wires on one row.

Width required = cover + 9 spaces + cover = $50 + 9 \times 50 + 50 = 550$ mm

This value is greater than the width of the section, therefore use two layers as shown in Fig. 12.4.

Thickness of tension flange = cover + space + cover = 150 mm. Also make the compression flange the same thickness to produce a symmetrical section which makes calculations easier at the service load conditions. The inside faces of the flanges are generally sloped to aid in removing shuttering and to avoid cracks associated with 90° angles.

12.2 BENDING RESISTANCE OF A PRESTRESSED CONCRETE BEAM AT SERVICE LOAD

12.2.1 Introduction

Prestressed concrete members must satisfy two criteria at the serviceability limit state in bending. The tensile stresses in the concrete must not produce cracks, except in special circumstances, and the compressive stresses must not exceed the design values. These two criteria of tensile stress and compressive stress form the basis for analysis and design at the serviceability limit state in bending.

12.2.2 Analysis of a Section in Bending at the Serviceability Limit State

In Section 12.1.3 design formulae are developed with tensile and compressive stress criteria, but because this is complicated it is initially instructive to analyse a section in bending at the serviceability limit state and to identify the critical stresses.

EXAMPLE 12.3   Analysis of a section in bending at the serviceability limit state. Determine the stresses due to prestress and applied service load for a beam of uniform cross section shown in Fig. 12.5(a). The jacking force of 1600 kN is provided by straight high tensile steel wires located as shown. The service load bending moment, which includes the self-weight, is 550 kNm.

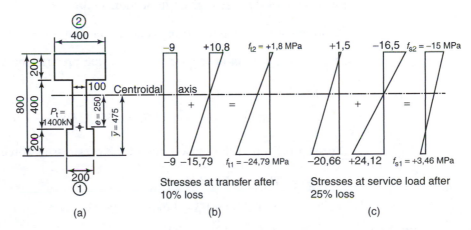

FIGURE 12.5   Example: analysis of a prestressed beam in bending at service load

If the section is not cracked and the compressive stresses are not greater than $\eta f_{ck}/\gamma_c$ then the elastic behaviour may be assumed. The percentage of steel is generally small in relation to the concrete and therefore composite action is ignored. The elastic section properties of the concrete are obtained as follows.

Cross-sectional area of the concrete

$$A_c = (80 + 40 + 40)E3 = 160E3\ \text{mm}^2$$

Centroid of the cross section from moments of areas about the base

$$80E3 \times 700 + 40E3 \times 400 + 40E3 \times 100 = 160E3 \times \bar{y}$$

hence $\bar{y} = 475\ \text{mm}$ from the base.

Second moment of area of the cross section about the centroidal axis

$$I = \Sigma(bh^3/3)$$
$$= (400 \times 325^3 - 300 \times 125^3 + 200 \times 475^3 - 100 \times 275^3)/3 = 10{,}83E9\ \text{mm}^4$$

Elastic section moduli

$$Z_{e1} = I/y_1 = 10{,}83E9/475 = 22{,}80E6\ \text{mm}^3$$

$$Z_{e2} = I/y_2 = 10{,}83E9/325 = 33{,}32E6\ \text{mm}^3$$

The section properties are now used to determine the stresses in the extreme fibres of the cross section. Subscript 1 denotes the bottom of the section and 2 the top of the section. Compressive stresses are negative and tensile stresses are positive.

Critical stresses occur at transfer when the jacking force of 1600 kN is applied. Loss of prestress of approximately 10 per cent occurs immediately because of the elasticity of the concrete which reduces the length of the member. A more accurate assessment of the loss is given in Section 12.3.

Prestressing force applied to the concrete at transfer $P_t = 1400$ kN.

Stress at the bottom of the section due to the prestressing force at transfer

$$f_{t1} = -P_t/A_c - P_t e/Z_{e1}$$
$$= -1400E3/160E3 - 1400E3 \times 250/22{,}80E6 = -9 - 15{,}79 = -24{,}79\ \text{MPa}$$

Stress at the top of the section due to the prestressing force at transfer

$$f_{t2} = -P_t/A_c + P_t e/Z_{e2}$$
$$= -1400E3/160E3 + 1400E3 \times 250/33{,}32E6$$
$$= -9 + 10{,}8 = +1{,}8\ \text{MPa}$$

The axial stresses can be added algebraically to the bending stresses because the principal of superposition is valid for elastic behaviour. Diagrams can now be drawn showing the stresses in the extreme fibres of the beam section at transfer (Fig. 12.5(b)). The extreme fibre stresses can be joined by straight lines because they vary linearly over the depth.

It is important to note that there is a tensile stress of $+1,8$ MPa at the top of the section and a compressive stress of $-24,79$ MPa at the bottom of the section. These stresses are critical and if they are too high in relation to the strength of the concrete at that stage of loading, then cracking, or crushing, occurs. These stresses occur at the supports for a straight member with a constant cross section and straight wires. The stresses at the support for this type of beam are not modified by the self-weight of the beam or the service load.

At midspan however the prestress stresses will be modified by the service load. The critical stresses at the service load occur at midspan after further loss of prestress due to creep in the steel and concrete occurs. The loss of prestress is determined more accurately in Section 12.3, but at this stage it may be assumed that the loss is 25 per cent of the jacking force.

Prestressing force at service load after losses, $P_s = 1200$ kN.

The service load bending moment M_s, which includes self-weight, is 550 kNm. Stress at the bottom of the section due to the prestressing force at service load and the application of the service load bending moment

$$f_{s1} = -P_s/A_c - P_s e/Z_{e1} + M_s/Z_{e1}$$

$$= -1200E3/160E3 - 1200E3 \times 250/22,80E6 + 550E6/22,80E6$$

$$= -7,5 - 13,16 + 24,12 = +3,46 \text{ MPa}$$

Stress at the top of the section due to the prestressing force at service load and the application of the service load bending moment

$$f_{s2} = -P_s/A_c + P_s e/Z_{e2} - M_s/Z_{e2}$$

$$= -1200E3/160E3 + 1200E3 \times 250/33,32E6 - 550E6/33,32E6$$

$$= -7,5 + 9,0 - 16,50 = -15,0 \text{ MPa}.$$

These stresses can also be represented on diagrams as shown in Fig. 12.5(c). For the service load condition it is important to notice that there is a tensile stress of $+3,46$ MPa at the bottom of the section and a compressive stress of $-15,0$ MPa at the top of the section. These stresses are critical and if they are too high in relation to the strength of the concrete at this stage of loading then cracking, or crushing, occurs.

The analysis of a beam is a good method for checking a section, but in practice it is necessary to be able to determine the required size of a section to satisfy the conditions at transfer and the serviceability limit state.

12.2.3 Design of a Section in Bending at the Serviceability Limit State

The critical bending stresses have been identified in the previous section and these may be presented algebraically as follows.

The sign convention for stresses is negative compression and positive tension, with the eccentricity positive below the centroidal axis and negative above. Subscript 1 denotes the bottom of the section and 2 the top of the section.

At transfer of the prestressing force P_t the stresses at the bottom and top of the section

$$f_{t1} = -P_t/A_c - P_t e/Z_{e1} + M_{min}/Z_{e1} \tag{12.3}$$

$$f_{t2} = -P_t/A_c + P_t e/Z_{e2} - M_{min}/Z_{e2} \tag{12.4}$$

where M_{min} is the bending moment that may act immediately the prestress is applied, e.g. the self-weight bending moment.

Eqs (12.3) and (12.4) are applicable to a simply supported beam where the critical section is at midspan e.g. the prestressing force is not parallel to the centroidal axis of the section as shown in Fig. 12.6(a). However if the critical section is at the support e.g. the prestressing force is parallel to the centroidal axis as shown in Fig. 12.6(b), then the term containing M_{min} is ignored.

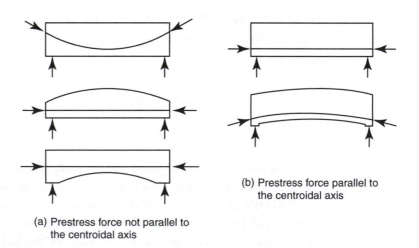

(a) Prestress force not parallel to the centroidal axis

(b) Prestress force parallel to the centroidal axis

FIGURE 12.6   Cable shapes for prestressed concrete beams

After losses at the serviceability limit state the stresses at the bottom and top of the section

$$f_{s1} = -P_s/A_c - P_s e/Z_{e1} + M_s/Z_{e1} \qquad (12.5)$$

$$f_{s2} = -P_s/A_c + P_s e/Z_{e2} - M_s/Z_{e2} \qquad (12.6)$$

where M_s is the service load bending moment and includes the minimum bending moment M_{min}.

If the extreme fibre stresses are now designated as serviceability design stresses such that, $f_{t1} = f_{tc}, f_{t2} = f_{tt}, f_{s1} = f_{st}, f_{s2} = f_{sc}$, then combining Eqs (12.3) to (12.6) and eliminating the eccentricity e, the optimum values of the section moduli can be determined.

Combining Eqs (12.3) and (12.5)

$$Z'_{e1} = (M_s - \eta M_{min})/(f_{st} - \eta f_{tc}) \qquad (12.7)$$

Combining Eqs (12.4) and (12.6)

$$Z'_{e2} = (M_s - \eta M_{min})/(\eta f_{tt} - f_{sc}) \qquad (12.8)$$

where

$$\eta = P_s/P_t = (P_s/P_o)/(P_t/P_o) = \eta_s/\eta_t$$

These values of the section moduli are the minimum theoretical values and must not be confused with the actual values which are written without the accent, i.e. as Z_{e1} and Z_{e2}. In the design process the actual values of the section moduli are greater than the minimum values, and therefore it is impossible to satisfy all the stress constraints simultaneously.

The minimum initial value of the jacking force $P_{o(min)}$ is obtained by satisfying the tensile stress constraints f_{tt} and f_{st}, i.e. combining Eqs (12.4) and (12.5), and dividing by η_t.

$$P_{o(min)} = A_c(f_{st}Z_{e1} + \eta f_{tt}Z_{e2} - M_s + \eta M_{min})/[-\eta_s(Z_{e1} + Z_{e2})] \qquad (12.9)$$

The maximum value of the eccentricity, e_{max} is obtained by satisfying the tensile stress constraints f_{tt} and f_{st}, i.e. combining Eqs (12.4) and (12.5)

$$e_{max} = [Z_{e1}Z_{e2}(f_{st} - \eta f_{tt}) - \eta M_{min}Z_{e1} - M_sZ_{e2}]/$$
$$[A_c(f_{st}Z_{e1} + \eta f_{tt}Z_{e2} - M_s + \eta M_{min})] \qquad (12.10)$$

The maximum initial value of the jacking force $P_{o(max)}$ is obtained by satisfying the compressive stress constraints f_{tc} and f_{sc}, i.e. combining Eqs (12.3) and (12.6) and dividing by η_t

$$P_{o(max)} = A_c(f_{sc}Z_{e2} + \eta f_{tc}Z_{e1} + M_s - \eta M_{min})/[-\eta_s(Z_{e1} + Z_{e2})] \qquad (12.11)$$

The minimum value of the eccentricity e_{min} is obtained by satisfying the compressive stress constraints f_{tc} and f_{sc}, i.e. combining Eqs (12.3) and (12.6)

$$e_{min} = [Z_{e1}Z_{e2}(-f_{sc} + \eta f_{tc}) - \eta M_{min}Z_{e2} - M_sZ_{e1}]/$$
$$[A_c(f_{sc}Z_{e2} + \eta f_{tc}Z_{e1} + M_s - \eta M_{min})] \qquad (12.12)$$

For economy is steel reinforcement it would be ideal to adopt the minimum jacking force and maximum eccentricity. However in design it is often not possible to adopt the minimum value of the jacking force $P_{o(min)}$ because the force is expressed in a discrete number of wires, or cables. Also the associated maximum value of eccentricity e_{max} may not be practical because it lies outside the section. In design calculations therefore a practical magnitude of the practical jacking force, P_{op}, is adopted expressed in number of wires, or cables, fully stressed to the design stress such that

$$P_{o(max)} > P_{op} > P_{o(min)}$$

The eccentricity (e) which is consistent with the practical jacking force P_{op} is then obtained from Eq. (12.13) which is a rearrangement of Eq. (12.5) and satisfies the tensile stress constraint f_{st} at service load.

$$e = (M_s - Z_{e1}f_{st})/(\eta_sP_{op}) - Z_{e1}/A_c \qquad (12.13)$$

A practical value of the eccentricity e_p is then calculated by trial and error such that e_p is approximately equal to e, and also

$$e_{max} > e_p > e_{min}$$

If the critical values $f_{tt}, f_{tc}, f_{st}, f_{sc}$ are related to the characteristic concrete strength then Eqs (12.3) to (12.13) form the basis for a method of design.

In harsh environments where durability is a problem no tensile stress are recommended. For normal use of prestressed concrete tensile stresses are accepted. Cracking is accepted where corrosion or weathering is not a problem, for long span beams where the self-weight is large and must be kept to a minimum, where deflections are not critical, or for temporary structures.

EXAMPLE 12.4 Design of a section in bending at the serviceability limit state. A beam is simply supported over a span of 15 m and supports an imposed service load of 33 kN. The section shown in Fig. 12.4 is satisfactory at the ultimate limit state in bending, check that it is suitable for service load conditions.

Estimate of self weight of member

$$W_{sw} = 10L^3 = 10 \times 15^3/1E3 = 33{,}75 \, \text{kN}$$

Self-weight bending moment

$$M_{sw} = W_{sw}L/8 = 33{,}75 \times 15/8 = 63{,}28 \, \text{kNm}$$

Imposed load bending moment

$$M_i = WL/8 = 33 \times 15/8 = 61{,}88 \, \text{kNm}$$

Total service load bending moment

$$M_s = M_{sw} + M_i = 63{,}28 + 61{,}88 = 125{,}2 \, \text{kNm}$$

The beam has straight wires and a constant cross section. The characteristic strength for steel is 1570 MPa. Assume concrete Grade C30/40 at transfer and C40/50 after 28 days. Assume a jacking stress in a wire of $0{,}9f_{p0{,}1k}$, a loss of prestress of 10 per cent at transfer and 25 per cent at service load. Consequently $\eta_t = 0{,}8$, $\eta_s = 0{,}75$ and $\eta = \eta_s/\eta_t = 0{,}75/0{,}8 = 0{,}938$. Maximum compressive stress $0{,}45f_{ck}$ and maximum tensile stress $0{,}3f_{ck}^{2/3}$.

The design stresses are (tension +ve, compression −ve)

$$f_{tc} = 0{,}45 \times (-30) = -13{,}5 \, \text{MPa}$$

$$f_{tt} = +0{,}3 \times 30^{2/3} = +2{,}90 \, \text{MPa}$$

$$f_{st} = +0{,}3 \times 40^{2/3} = +3{,}51 \, \text{MPa}$$

$$f_{sc} = 0{,}45 \times (-40) = -18{,}0 \, \text{MPa}$$

Theoretical optimum design values of the section moduli from Eqs (12.7) and (12.8)

$$Z'_{e1} = (M_s - \eta M_{min})/(f_{st} - \eta f_{tc})$$

$$= (125{,}16E6 - 0)/[3{,}51 - (0{,}938 \times -13{,}5)] = 7{,}74E6 \, \text{mm}^3$$

$$Z'_{e2} = (M_s - \eta M_{min})/(\eta f_{tt} - f_{sc})$$

$$= (125{,}16E6 - 0)/[0{,}938 \times 2{,}9 - (-18)] = 6{,}04E6 \, \text{mm}^3$$

The value of M_{min} is zero because the prestressing force is parallel to the centroidal axis of the section and consequently the critical section is at the support.

Actual second moment of area for the section (Fig. 12.4)

$$I = \Sigma bh^3/12 = 250 \times 500^3/12 - 150 \times 200^3/12 = 2{,}504E9 \text{ mm}^4$$

Actual section moduli

$$Z_{e1} = Z_{e2} = I/y = 2{,}504E9/250 = 10{,}02E6 \text{ mm}^3$$

The actual values of $Z_e > Z'_e$ which is satisfactory but conservative. The size of the section could be reduced based on these calculations, but the estimated total loss of prestress may be greater than 25 per cent and this would increase the values of Z'_e. The section must also satisfy deflection and ultimate load restraints.

The values of Z'_e also show that the optimum section is not symmetrical about the horizontal centroidal axis and ideally it should be closer to the bottom of the section. The reason for this is that the strength of the concrete at transfer is not as great as at service load conditions. If self-weight of the member is important, or if a very large number of members are to be made, then an iterative process is carried out to reduce the self-weight of the section. However, the more complicated the cross section the more costly the shuttering, but this can be justified if there are large numbers to be manufactured.

Values of the prestressing force and eccentricity at transfer are obtained from Eqs (12.9) to (12.12). Note that in the following calculations actual values of section moduli are used, i.e. $Z_{e1} = Z_{e2} = 10{,}016E6 \text{ mm}^3$.

Cross-sectional area (Fig. 12.4)

$$A_c = 250 \times 500 - 150 \times 200 = 95E3 \text{ mm}^2$$

Maximum value of the eccentricity for minimum jacking force, Eq. (12.10)

$$e_{max} = [Z_{e1}Z_{e2}(f_{st} - \eta f_{tt}) - \eta M_{min}Z_{e1} - M_s Z_{e2}]/[A_c(f_{st}Z_{e1} + \eta f_{tt}Z_{e2} - M_s + \eta M_{min})]$$

$$= [10{,}02E6 \times 10{,}02E6 \times (3{,}51 - 0{,}938 \times 2{,}9) - 0 - 125{,}2E6 \times 10{,}02E6]/$$

$$[95E3 \times (3{,}51 \times 10{,}02E6 + 0{,}938 \times 2{,}9 \times 10{,}02E6 - 125{,}2E6 + 0)]$$

$$= +196{,}7 \text{ mm}$$

Minimum value of the eccentricity for maximum jacking force, Eq. (12.12)

$$e_{min} = \{Z_{e1}Z_{e2}[-f_{sc} + \eta f_{tc}] - \eta M_{min}Z_{e2} - M_s Z_{e1}\}/$$

$$\{A_c[f_{sc}Z_{e2} + \eta f_{tc}Z_{e1} + M_s - \eta M_{min}]\}$$

$$= \{10{,}02E6 \times 10{,}02E6 \times [-18 + 0{,}938 \times (-13{,}5)] - 0 - 125{,}2E6 \times 10{,}02E6\}/$$

$$\{95E3 \times [(-18) \times 10{,}02E6 + 0{,}938 \times (-13{,}5) \times 10{,}02E6 + 125{,}2E6 - 0]\}$$

$$= +41{,}5 \text{ mm}.$$

Maximum value of the prestressing force before losses for minimum eccentricity, Eq. (12.11)

$$P_{o(max)} = A_c(f_{sc}Z_{e2} + \eta f_{tc}Z_{e1} + M_s - \eta M_{min})/[-\eta_s(Z_{e1} + Z_{e2})]$$
$$= 95E3 \times [(-18) \times 10{,}02E6 + 0{,}938 \times (-13{,}5) \times 10{,}02E6 + 125{,}2E6 - 0]/$$
$$[-0{,}75 \times (10{,}02E6 + 10{,}02E6)] = 1150\,kN.$$

Minimum value of the jacking force before losses for maximum eccentricity from Eq. (12.9)

$$P_{o(min)} = A_c(f_{st}Z_{e1} + \eta f_{tt}Z_{e2} - M_s + \eta M_{min})/[-\eta_s(Z_{c1} + Z_{e2})]$$
$$= 95E3 \times [3{,}51 \times 10{,}02E6 + 0{,}938 \times 2{,}9 \times 10{,}02E6 + 125{,}2E6 - 0]/$$
$$[-0{,}75 \times (10{,}02E6 + 10{,}02E6)] = 397\,kN.$$

The different values of the jacking force and eccentricity show that more than one solution is possible. This is because the actual section moduli, $Z_{e1} = Z_{e2} = 10{,}02E6\,mm^3$, are greater than the optimum values $Z'_{e1} = 7{,}74E6\,mm^3$ and $Z'_{e2} = 6{,}04E6\,mm^3$.

A practical value of the prestressing force P_{op} is now chosen such that

$$P_{o(max)}(1150) > P_{op} > P_{o(min)}(397)\,kN.$$

Practically the number (n_w) of 7 mm diameter wires of jacking force

$$P_w = \pi\phi^2/4 \times 0{,}9f_{p0,1k} = \pi \times 7^2/4 \times 0{,}9 \times 1570/1E3 = 54{,}4\,kN.$$

$$P_{o(max)}/P_w > n_w > P_{o(min)}/P_w$$
$$1150/54{,}4 > n_w > 397/54{,}4$$
$$21{,}1 > n_w > 7{,}3$$

The wires are required at ultimate load (Fig. 12.4) and two wires at the top of the section to prevent bending failure if the beam is inadvertently turned upside down. The two top wires are also useful for fixing shear reinforcement.

$$P_{op} = n_wP_w = 12 \times 42{,}3 = 652{,}8\,kN$$

If $P_{op} = 652{,}8\,kN$ then the corresponding eccentricity from Eq. (12.13)

$$e = (M_s - Z_{e1}f_{st})/(\eta_sP_{op}) - Z_{e1}/A_c$$
$$= (125{,}2E6 - 10{,}02E6 \times 3{,}51)/(0{,}75 \times 652{,}8E3) - 10{,}02E6/95E3 = 78{,}4\,mm.$$

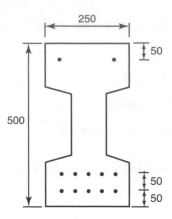

FIGURE 12.7   Example: design of a section in bending at service load

This value of the eccentricity is not the practical value, which is obtained as follows. Cover should meet the requirements for durability and fire (Chapter 5). Spacing should meet the requirements. The distance from the base of the centroid of the 12 wires (Fig. 12.7) is obtained by taking moment of areas about the base.

$$5 \times 50 + 5 \times 100 + 2 \times 450 = 12\bar{y}; \quad \text{hence } \bar{y} = 137,5 \text{ mm}.$$

Practical eccentricity $e_p = h/2 - y = 250 - 137,5 = 112,5 > e$ (78,4) mm.

$e_{max}(196,7) > e_p(112,5) > e_{min}(41,5)$ mm, and e_p is within limits.

$P_{o(max)}(1150) > P_{op}(652,8) > P_{o(min)}(397)$ kN{and P_{op} also within limits.

Using $P_{op} = 652,8$ kN and $e_p = 112,5$ mm check that the fibre stresses at transfer (10 per cent loss) and service load (25 per cent loss), as shown in Example 12.3, are less than the design stresses.

$$f_{t1} = -11,36 < f_{tc} = -13, 50 \text{ MPa}$$

$$f_{t2} = +0,37 < f_{tt} = +2,90 \text{ MPa}$$

$$f_{s1} = +1,84 < f_{st} = +3,51 \text{ MPa}$$

$$f_{s2} = -12,20 < f_{sc} = -18,00 \text{ MPa}$$

Check that, in bending, the cracking of the concrete precedes failure.

$M = (\eta_s f_{tc} + f_{st})Z_{e1} = (0,75 \times 13,5 + 3,51) \times 10,02\text{E6} = 137 < 187,5$ kNm (Example 12.2), therefore satisfactory.

The design continues with checks for loss of prestress, shear resistance and deflection. For this example the design for bending conditions commenced with ultimate load and was modified for service load conditions. The reverse process is also possible and preferred by some designers.

12.2.4 Deflection Limits for Prestressed Concrete Members (cl 7.4, EN)

No definite limits are set for the deflections for prestressed concrete but as with ordinary reinforced concrete the deformations of the structure, or part of it, should not adversely affect its efficiency of appearance. Deflections should be compatible with the degree of movement acceptable by other elements including finishes, services, partitions, glazing and cladding.

The span/effective depth ratios for ordinary reinforced concrete are not applicable to prestressed concrete because generally at service load the concrete is not cracked, and also the prestressing force produces an upward camber of the member. The deflections in prestressed concrete members may be described as short term elastic deflections which occur immediately, and long term creep and shrinkage deflections which occur after the member has been loaded for a long time.

As a guide maximum downward and upward deflections should not exceed span/250. This limit may cause excessive damage to contiguous elements, e.g. partitions, and the limit is reduced to span/500 (cl 7.4.1, EN).

12.2.5 Short Term Elastic Deflections

These occur immediately when the prestressing force is applied at transfer, when the self-weight and dead load act, and when the imposed load is applied. The deflection due to the imposed load becomes zero when the imposed load is removed, but the dead load and prestress deflections remain.

The elastic deformation for a beam of uniform section

$$a_e = (L^2/E_{cm(t)})\Sigma kM \tag{12.14}$$

where

M is the maximum bending moment in the span
k is a factor which depends on the shape of the bending moment diagram as shown in Table 12.1
$E_{cm(t)}$ is the modulus of elasticity (cl 3.1.3, EN) for a particular concrete strength derived from $E_{cm} = 22((f_{ck}+8)/10)^{0.3}$.

12.2.6 Long Term Creep Deflections

These are due to creep in the concrete which is dependent on the stress level and the maturity of the concrete. The creep effect due to transient loads is very small

TABLE 12.1   Deflection factors.

Load	Bending moment diagram	k
w/ unit length L	$wL^2/8$	$\dfrac{5}{48}$
W $L/2$ | $L/2$	$wL/4$	$\dfrac{1}{12}$
e $P \rightarrow \quad \leftarrow P$	$-P_e$	$\dfrac{1}{8}$
P P e Parabolic tendon	$-P_e$	$\dfrac{5}{48}$

and only permanent loads need be considered, i.e. the dead load and permanent imposed load.

Long term deflections can be calculated using an elastic analysis encorporating the secant modulus of elasticity of the concrete ($E_{cm(t)}$) related to the elastic modulus (Eq. 3.5, EN)

$$E_{cm(t)} = (f_{cm(t)}/f_{cm})^{0,3} E_{cm} \tag{12.15}$$

where

$$f_{cm} = f_{ck} + 8\,\text{MPa}$$

EXAMPLE 12.5   Deflection of a prestressed beam.
Calculate the deflection of the simply supported prestressed beam which spans 15 m designed in Example 12.4. Relevant details are $I = 2,504\text{E9 mm}^4$, $f_{ck(3)} = 30$, $f_{ck(28)} = 40\,\text{MPa}$, $P_{op} = 652,8\,\text{kN}$, and $e_p = 112,5\,\text{mm}$.

The calculated deflections which occur during the manufacture and loading of the beam are as follows. Downwards deflections are postive.

Deflections at transfer

Elastic modulus at 3 days (Table 3.1, EN)

$$E_{cm(3)} = 22((f_{ck} + 8)/10)^{0,3} = 22 \times ((30 + 8)/10)^{0,3} = 32,8\,\text{GPa}$$

Elastic modulus at 28 days

$$E_{cm(28)} = 22((f_{ck} + 8)/10)^{0,3} = 22 \times ((40 + 8)/10)^{0,3} = 35,2\,\text{GPa}$$

Elastic deflection (upwards) due to prestress at transfer at three days after 10 per cent loss.

$$a_e = L^2/(E_{cm(3)}I)\Sigma kM = L^2/(E_{cm(3)}I)(1/8)(-\eta_t P_{op}e_p)$$

$$= 15E3^2/(32,8E3 \times 2,504E9) \times (1/8) \times (-0,8 \times 652,8E3 \times 112,5)$$

$$= -20,2\,\text{mm (upwards)}.$$

Elastic deflection (downwards) from self-weight (33,75 kN) at transfer when the beam is lifted from the casting bed

$$a_e = L^2/(E_{cm(3)}I)\Sigma kM = L^2/(E_{cm(3)}I)(5/48)(WL/8)$$

$$= 15E3^2/(32,8E3 \times 2,504E9) \times (5/48) \times (33,75E3 \times 15E3/8)$$

$$= 18,0\,\text{mm (downwards)}$$

The net deflection at transfer (3 days) $= -20,2 + 18,0 = -2,2\,\text{mm}$ (upwards). This value is of interest but generally of little practical importance.

Long term deflections (cl 7.4.3(5), EN) Effective creep modulus of elasticity related to the 3 day concrete strength. National size of section

$$= 2A_c/u = 2 \times (500 \times 250 - 200 \times 150)/[2 \times (500 + 250 + 150)] = 106\,\text{mm}$$

$$E_{c,eff} = E_{cm(3)}/(1 + \phi_{(\alpha,3)}) = 32,8E3/(1 + 3,3) = 7,63\,\text{GPa}$$

Final creep coefficient $\phi_{(\alpha,3)} = 3,3$ is interpolated from Fig. 3.1, EN for humid outside conditions.

Long term creep deflection due to eccentricity of prestress after 25 per cent loss

$$a_c = L^2/(E_{c,eff}I)\Sigma kM = (L^2/E_{c,eff}I)(1/8)(-\eta_s P_{op}e_p)$$

$$= 15E3^2/(7,63E3 \times 2,504E9) \times (1/8) \times (-0,75 \times 652,8E3 \times 112,5)$$

$$= -81,2\,\text{mm (upwards)}$$

Long term creep deflection due to self-weight (33,75 kN) bending

$$a_c = L^2/(E_{c,eff}l)\Sigma kM = L^2/(E_{c,eff}l)(5/48)(WL/8)$$

$$= 15E3^2/(7,63E3 \times 2,504E9) \times (5/48) \times (33,75E3 \times 15E3/8)$$

$$= 77,6 \text{ mm (downwards)}.$$

Short term deflections Elastic deflection (downwards) from imposed load (33 kN) related to the 28 day concrete strength

$$a_e = L^2/(E_{cm(28)}l)\Sigma kM = L^2/(E_{cm(28)}l)(5/48)(WL/8)$$

$$= 15E3^2/(35,2E3 \times 2,504E9) \times (5/48) \times (33E3 \times 15E3/8)$$

$$= 16,4 \text{ mm (downwards)}$$

Net long term creep deflection $= -81,2 + 77,6 = -3,6$ mm (upwards).

Short term elastic deflection from imposed load $= +16,4$ mm (downwards).

As a guide if these deflections are compared with span/250 = 15E3/500 = 60 mm then the previous calculations are within this limit. The limit of span/500 = 15E3/500 = 30 mm, which is to prevent damage to partitions, is also satisfied.

12.3 LOSS OF PRESTRESS (cl 5.10, EN)

12.3.1 Introduction

Loss of prestresses is the difference between the jacking force in the tendon at the prestressing stage and the force after losses have occurred. Some of the losses occur immediately when the force is transferred to the concrete, e.g. elastic shortening of the member. Other losses occur with time, e.g. creep of the concrete. The jacking force does not exceed 90 per cent of the characteristic strength of the tendon and is generally not less than 75 per cent (cl 5.10.3(2), EN). In practice total losses of the jacking force are within the range of 15 to 40 per cent. If large losses of prestress occur then the advantages of prestressed concrete, e.g. no cracks at service load, are lost. It is therefore advantageous to minimize the loss of prestress.

Losses of prestress are related to the jacking stress and expressed as: per cent loss of prestress = 100 (jacking stress − final stress)/(jacking stress)

In general, losses can occur because of shortening of the concrete member, or shortening of the tendon or because of friction acting along the length of the tendon. The following factors affect the loss of prestress:

(a) relaxation of the steel (1−12%),
(b) elastic deformation of the concrete (1−10%),

(c) shrinkage of the concrete (1–6%),

(d) creep of the concrete (5–15%),

(e) draw-in during anchorage (0–5%),

(f) friction in the ducts and at anchorages (3–7%).

The previous list considers each less separately but an alternative is to use Eq. 5.46, EN which calculates time dependent losses in a single formula.

Initially, in design calculations, the loss of prestress is estimated from past experience and later, when the section shape and size has been determined, a more accurate calculation is made. A reasonably accurate estimation of the loss of prestress is required because it affects service load conditions, e.g. the bending moment at which bending cracks occur and also the shear force at which diagonal shear cracks appear.

12.3.2 Relaxation of Steel (cl 5.10.4, EN)

If a steel tendon is stressed to 80 per cent of its characteristic strength and anchored over a fixed length, then after 1000 h the stress will reduce. This phenomena, which is not fully understood, is called relaxation, and may be defined as a loss of stress at constant strain. In prestressed concrete the fixed length is the concrete member, although in reality this does change slightly.

For design calculations the value adopted for the loss should be the 1000 h relaxation value taken from certificates accompanying the consignment.

12.3.3 Elastic Deformation of the Concrete (cl 5.10.4, EN)

If a jacking force from a prestressing tendon is transferred to a concrete member then, because the concrete behaves elastically, the member immediately reduces in length. If the steel and concrete are bonded together, as in pretensioned prestressed concrete, then the reduction in length is the same for the concrete and the steel and is unavoidable. In post-tensioned concrete, because the tendon is not bonded to the concrete during stressing operations, the elastic reduction in length of the concrete can be allowed for when jacking and the loss can be reduced to a very small value.

The simplest method of calculating the loss of prestress due to elastic shortening is as follows. Assuming elastic behaviour for a pretensioned member where the steel is bonded to the concrete before the jacking force is transferred to the concrete. If f_s' is the loss of stress in the steel and f_c is the stress in the concrete adjacent to the steel at transfer strain in the concrete = strain in the steel

$$\sigma_{cp}/E_c = f'_s/E_s$$

Percentage loss of prestress, if f_{pi} is the jacking stress in the steel

$$= 100(f'_s/f_{pi}) = 100(E_s/E_c)\sigma_{cp}/f_{pi} \qquad (12.16)$$

Modulus of elasticity (cl 3.1.3, EN)

$$E_{cm} = 22((f_{ck} + 8)/10)^{0.3} \, \text{GPa}$$

where f_{ck} in MPa is the strength of the concrete at transfer.

The value of the modulus of elasticity for wires and bars is 200 GPa and for strand 190 GPa (cl 3.3.6(3), EN).

Eq. (12.16) can also be applied to a post-tensioned beam but if the elastic shortening is allowed for in the jacking process then the loss will be very small. Losses for elastic shortening can therefore vary between 1 per cent for some post-tensioned beams and 10 per cent for highly prestressed pretensioned members.

12.3.4 Shrinkage of Concrete (cl 3.1.4, EN)

Shrinkage of concrete is a reduction in dimensions of a concrete member and when related to loss of prestress it is the shrinkage in length which is important. The more important factors which influence shrinkage in concrete are:

(a) aggregate used,
(b) original water content,
(c) effective age at transfer,
(d) effective section thickness,
(e) ambient relative humidity.

In the absence of experimental evidence values of shrinkage strain (ε_c) are given in Table 3.2, EN. These values are incorporated in the following theory. Assuming elastic behaviour the loss of prestress in the prestressing steel

$$f_s = E_s \varepsilon_s$$

Assuming equal strains in the steel and concrete

$$\varepsilon_s = \varepsilon_c$$

and the percentage loss of prestress related to the jacking stress

$$= 100(f_s/f_{pi}) = 100 E_s \varepsilon_c/f_{pi} \qquad (12.17)$$

12.3.5 Creep of Concrete (cl 3.1.4, EN)

Creep is the change in strain which occurs very slowly after the immediate elastic strain has taken place. The creep coefficient depends on the following:

(a) original water content,
(b) effective age at transfer,
(c) effective section thickness,
(d) ambient relative humidity,
(e) ambient temperature.

Creep strain in concrete is stress dependent and values are obtained from Fig. 3.1, EN. Expressed as the elastic strain multiplied by the final creep coefficient $\phi > 1$

$$\varepsilon_{cc(\alpha,to)} = (\sigma_c/E_c)\phi_{(\alpha,to)}$$

Assuming elastic behaviour, then loss of prestress in the steel

$$f_s = E_s\varepsilon_s$$

Assuming the strain in the steel equals the creep strain in the concrete

$$\varepsilon_s = \varepsilon_{cc(\alpha,to)} = (\sigma_c/E_c)\phi_{(\alpha,to)}$$

and the percentage loss of prestress related to the jacking stress

$$= 100(f_s/f_{pi}) = 100(E_s/E_c)\phi_{(\alpha,to)}\sigma_c/f_{pi} \tag{12.18}$$

12.3.6 Anchorage Slip (cl 5.10.4, EN)

Anchorage slip reduces the stress in the wires or tendon and results in a loss of prestress.

When the wires for pretensioned prestressed concrete members are being stressed the wires are fixed to a single cross-head and the slip can be allowed for and consequently the loss of prestress is small.

In post-tensioning systems allowance should be made for any slip of the tendon at the anchorage when the prestressing force is transferred from the tensioning equipment to the anchorage. This loss may be large for short members. In practice the slip at the anchorage ∂L is known and the strain $\partial L/L$ can be calculated.

Percentage loss of prestress related to the jacking stress

$$= 100(f_s/f_{pi}) = 100(\partial L/L)E_s/f_{pi} \tag{12.19}$$

12.3.7 Friction Loss (cl 5.10.5.2, EN)

If frictional forces act along the length of a tendon then the force at a point distant from the jack will be less than at the jack.

Generally, for pretensioned members the wires are straight and frictional losses only occur during jacking operations at the anchorage. The magnitude is generally known and allowance can be made.

For post-tensioned members the tendons are contained in straight or curves ducts and during jacking operations the tendons move relative to frictional surfaces.

For a straight duct the force P_x at a distance x from the jack may be calculated from (cl 5.10.5.2, EN)

$$\Delta P_\mu(x) = P_o(1 - e^{-\mu(\theta + kx)}) \tag{12.20}$$

where

P_o is the prestressing force in the tendon at the jacking end
e is the base of the Napierian logarithm
θ is the sum of the angular displacements over a distance x (irrespective of direction of sign)
μ is the coefficient of friction between the tendons and the duct and typical values are

0,17 cold drawn wire
0,19 strand
0,65 deformed bar
0,33 smooth round bar

k generally in the range $0,005 < k < 0,01$

The percentage loss of prestress force in relation to the jacking force (Eq. 5.45, EN)

$$= 100(P_o - P_x)/P_o = 100(1 - e^{-\mu(\theta + kx)}) \tag{12.21}$$

Further information on friction losses is contained in CIRIA Report 74 (1978). In practice the curved ducts are often parabolic but there is little error if a parabolic curve is treated as a circular curve, and from Fig. 12.8 the relation between a chord and a diameter

$$(L/2)^2 = (2r_{ps} - a_t)a_t$$

rearranging

$$r_{ps} = L^2/(8a_t) + a_t/2 \tag{12.22}$$

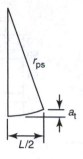

FIGURE 12.8 Approximate shape for a parabolic cable

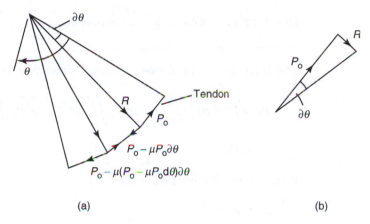

(a)   (b)

FIGURE 12.9 Theory for loss of prestress due to duct friction

A theory to produce the equation for the loss of prestress as expressed in Eq. (12.21) is as follows:

Fig. 12.9(a) represents a tendon with a jacking force P_o subject to successive small changes in direction $\partial\theta$. At each change point there is a reaction R, and at the first change in direction, from geometry (Fig. 12.9(b))

$$R = P_o\partial\theta$$

The frictional force at the first reaction is

$$= \mu R = \mu P_o\partial\theta$$

After the first change in direction the force in the tendon

$$P_x = P_o - \mu P_o\partial\theta$$

After the second change in direction the force in the tendon

$$P_x = P_o - \mu(P_o - P_o\partial\theta)\partial\theta$$

After the third change in direction the force in the tendon

$$P_x = P_o - \mu[P_o - \mu(P_o - P_o\partial\theta)\partial\theta]\partial\theta$$

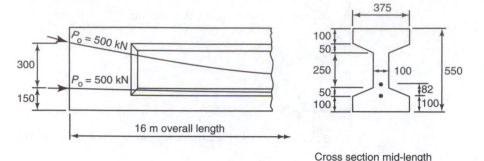

FIGURE 12.10   Example: loss of prestress example

The force in a tendon, which changes direction through an angle θ, is

$$P_x = P_o - \mu P_o \int_0^\theta \partial\theta + \mu^2 P_o \iint_o^\theta \partial\theta\, \partial\theta - \mu^3 P_o \iiint_o^\theta \partial\theta\, \partial\theta\, \partial\theta + \text{etc.}$$

$$= P_o[1 - \mu\theta + (\mu\theta)^2/2 - (\mu\theta)^3/3 + \text{etc.}] = P_o[e - \mu\theta]$$

If the radius of curvature is large $\theta = x/r_{ps}$ and

$$= P_o[e^{-\mu x/rps]}$$

For a straight duct with a wobble $\mu\theta = \mu k x$.

EXAMPLE 12.6   Loss of prestress.
Determine the loss of prestress for the post-tensioned beam (Fig. 12.10) which is subject to a total jacking force $P_o = 1000$ kN. Assume C40/50 Grade concrete, jacking from both ends, and smooth round bars stressed to 70 per cent of the characteristic stress.

Steel stress at jacking

$$f_{pi} = 0{,}7 f_{pk} = 0{,}7 \times 1550 = 1085 \text{ MPa}$$

Assume 4 per cent loss of prestress from relaxation of the steel (cl 5.10.4, EN)

Cross-sectional area of the section

$$A_c = 550 \times 375 - 275 \times 300 = 123{,}75\text{E}3 \text{ mm}^2$$

Second moment of area of the section

$$I = 375 \times 550^3/12 - 275 \times 250^3/12 - 2 \times 275 \times 50^3/36$$

$$- 2 \times 0{,}5 \times 275 \times 50 \times (125 + 50/3)^2$$

$$= 4{,}563\text{E}9 \text{ mm}^4$$

Elastic section modulus of the section

$$Z = I/y = 4{,}563\text{E}9/275 = 16{,}594\text{E}6 \, \text{mm}^3$$

Prestress at the centroid of the prestressing steel

$$\sigma_c = -P_o/A_c - P_o e y/I$$
$$= -1000\text{E}3/123{,}75\text{E}3 - 1000\text{E}3 \times 134^2/4{,}563\text{E}9 = -12{,}0 \, \text{MPa}$$

Elastic modulus of the concrete at 28 days (cl 3.1.3, EN)

$$E_{cm(28)} = 22 \times ((f_{ck(28)} + 8)/10)^{0{,}3} = 22((40 + 8)/10)^{0{,}3} = 35{,}2 \, \text{GPa}$$

Elastic modulus at transfer at three days

$$E_{cm(3)} = 22 \times ((f_{ck(3)} + 8)/10)^{0{,}3} = 22((25 + 8)/10)^{0{,}3} = 31{,}5 \, \text{GPa}$$

Percentage loss of prestress from elastic deformation of the concrete from Eq. (12.16)

$$= 100(E_s/E_{cm(3)})\sigma_{cp}/f_{pi} = 100 \times (200\text{E}3/31{,}5\text{E}3) \times 12/1085 = 7{,}02\%$$

This value can be halved to 3,51 per cent for a post-tensioned beam.

Percentage loss of prestress from drying shrinkage of the concrete (Table 3.2, EN) from Eq. (12.17) for outdoor conditions 80 per cent humidity ($2A_c/u = 111 \, \text{mm}$)

$$= 100E_s\varepsilon_{cd}/f_{pi} = 100 \times 200\text{E}3 \times 310\text{E} - 6/1085 = 5{,}71\%$$

Percentage loss of prestress from creep of the concrete (Section 2.6.2), transfer at 3 days for outdoor exposure (Fig. 3.1, EN). From Eq. (12.18)

$$= 100(E_s/E_{cm(3)})\phi\sigma_c/f_{pi} = 100 \times (200\text{E}3/31{,}5\text{E}3) \times 2{,}5 \times 12/1085 = 17{,}56\%$$

Percentage loss of prestress from slip of 1 mm for half the length of beam during anchorage from Eq. (12.19)

$$= 100(\partial L/L)E_s/f_{pi} = 100 \times (1/8\text{E}3) \times 200\text{E}3/1085 = 2{,}30\%$$

Radius of curvature for bottom cable from Eq. (12.23) assuming the curve approximates to a circle ($a_t = 150 - 100 = 50 \, \text{mm}$)

$$r_{ps} = L^2/(8a_t) + a_t/2 = 16^2/(8 \times 0{,}05) + 0{,}05/2 = 640 \, \text{m}$$

Percentage loss of prestress from friction for the bottom approximately straight duct (cl 5.10.5.2, EN) jacked from both ends from Eq. (12.21)

$$= 100(1 - e^{-\mu(\theta + kx)}) = 100 \times (1 - e^{-0{,}25 \times (8/640 + 0{,}0075 \times 8)}) = 1{,}79\%$$

Radius of curvature for the top curved duct from Eq. (12.22) assuming the curve approximates to a circle ($a_t = 450 - 182 = 268$ mm)

$$r_{ps} = L^2/(8a_t) + a_t/2 = 16^2/(8 \times 0{,}268) + 0{,}268/2 = 119{,}5 \text{ m}$$

Percentage loss of prestress from friction for top cable jacked from both ends from Eq. (12.21)

$$= 100(1 - e^{-\mu(\theta+kx)}) = 100 \times (1 - e^{-0{,}25 \times (8/119{,}5 + 0{,}0075 \times 8)}) = 3{,}12\%$$

Total loss of prestress

$$= \text{relax.} + \text{elastic conc.} + \text{shrink conc.} + \text{creep conc.} + \text{slip} + \text{friction}$$
$$= 4 + 3{,}51 + 5{,}71 + 17{,}56 + 2{,}3 + (1{,}79 + 3{,}12)/2 = 35{,}5\%$$

The total loss can be reduced by rejacking to reduce the concrete elastic loss, or increasing the concrete strength.

The European Code gives an alternative method of calculating the time-dependent losses (cl 5.10.6, EN). These include shrinkage of concrete, relaxation of the steel and creep of concrete. Loss of stress in the tendons

$$\Delta\sigma_{p,c+s+r} = (\varepsilon_{cs}E_p + 0{,}8\Delta\sigma_{pr} + E_p/E_{cm}\phi(t,t_o)\sigma_{c,QP})/$$

$$(1 + E_p/E_{cm}A_p/A_c(1 + A_c/Iz_{cp}^2)(1 + 0{,}8\phi(t,t_o)))$$

$$\Delta\sigma_{p,c+s+r} = (310/1\text{E}6 \times 200\text{E}3 + 0{,}8 \times 0{,}04 \times 1085 + 200\text{E}3/35\text{E}3 \times 2{,}5 \times 12)/$$

$$(1 + 200\text{E}3/35\text{E}3 \times 645{,}3/123{,}75\text{E}3(1 + 123{,}75\text{E}3/4563\text{E}9 \times 134^2)$$

$$\times (1 + 0{,}8 \times 2{,}5)$$

$$= (62 + 34{,}72 + 171{,}4)/(1 + 0{,}0298(1 + 0{,}487) \times 3) = 236{,}7 \text{ MPa.}$$

Percentage loss $= 236{,}7/1085 \times 100 = 21{,}8$

Compare with previous calculation, per cent loss $= 4 + 5{,}71 + 17{,}56 = 27{,}27 > 21{,}8$.

12.4 ANCHORAGE OF TENDONS

12.4.1 Anchorage for Post-tensioned Tendons (cl 8.10.3, EN)

When a force from a post-tensioned cable is transferred to a small area, or zone, at the end of a beam, high localized stresses are produced. These localized stresses do not extend the full length of the member but are confined to the anchor block. The tensile stresses are of particular concern because they can produce cracks in the concrete which need to be controlled by the use of reinforcing steel.

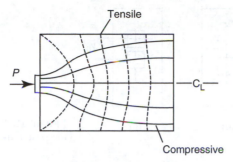

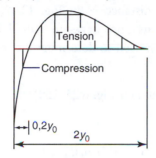

(a) Simplified stress trajectories

(b) Distribution of stress along C_L of block

FIGURE 12.11 Stress distribution in a post-tensioned end block

The cross section of the another blocks is generally rectangular in order to reduce the magnitude of the stresses.

The exact distribution of stresses in an anchor block is complex and is not fully understood theoretically nor has the subject been thoroughly investigated experimentally (CIRIA Guide 1, 1976). The distribution of principal compressive and tensile stresses in a simple rectangular anchor block subject to a single force are shown in Fig. 12.11(a). The distribution of stress along the centre line of the block is shown in Fig. 12.11(b), and this is of particular concern in structural design because reinforcement is required to resist the tensile stresses. These stresses are three-dimensional and consequently tensile stresses are also present at right angles to those shown in Fig. 12.11(b).

In addition to the stresses described previously, where there is more than one jacking force there are stresses close to the loaded face which may cause cracking and spalling of the surface concrete between the anchor forces. These spalling stresses are greatest with unsymmetrical loading of the end block.

When several forces are applied to an end block similar distribution to those described for a single force occur local to each force, and also over the complete end block. It is this fact which is used in design calculations to simplify a complex problem by subdividing the complete end block into simple individual cubic end blocks associated with each anchor force. Examples are shown in Fig. 12.12. The size of each individual cubic end block is based on a controlling dimension y_0

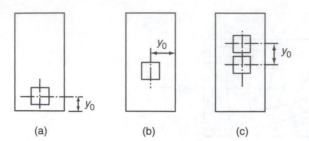

FIGURE 12.12   Critical values of y_0 for post-tensioned end blocks

which may be a vertical edge distance as in Fig. 12.12(a), or a horizontal edge distance as in Fig. 12.12(b), or distance between anchor forces as in Fig. 12.12(c). The minimum dimension is then used to determine the minimum size of cubic end block, $2y_0$.

Tensile force (Eqs 6.58 and 6.59, and Fig. 6.25, EN)

for $b \leq H/2$; $T = 1/4 \times (1 - a/b)F$

for $b > H/2$; $T = 1/4 \times (1 - 0{,}7a/h)F$

Circular bearing plates should be treated as square plates of equivalent area. This bursting force (T) will be distributed in a region extending from $0{,}2y_0$ to $2y_0$ from the loaded face, and should be resisted by reinforcement in the form of spirals or closed links, uniformly distributed throughout this region, and acting at a stress of 300 MPa.

When large blocks contain several anchorages it should be divided into a series of symmetrically loaded prisms and each prism treated in the above manner. However, additional reinforcement will be required around groups of anchorages to ensure overall equilibrium of the end block.

Special attention should also be paid to end blocks having a cross-sectional shape different from that of the general cross section of the beam. Information is given in CIRIA Guide 1 (1976).

The size and spacing of links to resist the bursting force is obtained from equating the bursting force to the resistance of the links stresses to 300 MPa

$$T = (A_{sv}300/s_v)H \tag{12.23}$$

The same reinforcement should be provided in two directions at right angles.

EXAMPLE 12.7   Post-tensioned end block.
Determine the spacing of 8 mm diameter double links to reinforce the end block shown in Fig. 12.13 which is subject to two tendon jacking forces.

Width of the bearing plate $a = 150$ mm.

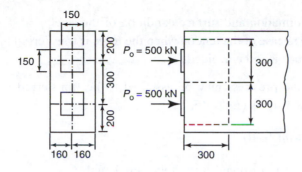

Figure 12.13 Example: post-tensioned end block

From the vertical spacing of the tendon forces the minimum width of a cubic block (Fig. 12.12(c))

$$b = b_{ef} = 300 \text{ mm}.$$

Check the compressive stress beneath the plate. Apply prestressing force (Eq. 6.63, EN) (cl 6.7, EN)

$$F_{Rdu} = A_{co}f_{cd}(A_{c1}/A_{co})^{0.5} = 150^2 \times 25/1{,}5 \times (300^2/150^2)^{0.5} = 750E3 \text{ N}$$
$$3f_{cd}A_{co} = 3 \times 25/1{,}5 \times 150^2 = 1125E3 > 750E3N, \text{ satisfactory.}$$

The tensile force for $b \leq H/2$ (Eq. 6.58, EN)

$$T = 1/4 \times (1 - a/b)F = 1/4 \times (1 - 150/300) \times 500 = 500/8 = 62{,}5 \text{ kN}$$

Rearranging Eq. (12.23) the spacing of double T6 links

$$s_v = (A_{sv}300/T)H = (2 \times 2 \times 28 \times 300/62{,}5E3) \times 300 = 161 \text{ mm}$$

Use double T6-120 mm, in two directions at right angles. For $f_s < = 300$ MPa there is no need to check cracks widths.

12.4.2 Anchorage for Pretensioned Tendons (cl 8.10.2.1, EN)

The prestressing force is zero at the end of a pretensioned member in contrast to a post-tensioned member where it is a maximum. This means that for a pretensioned member the prestressing force is not concentrated at the end of the member and therefore it is not subject to the same localized stresses as a post-tensioned beam.

The distance required for the prestressing force to be transmitted to the concrete for a pretensioned member is called the transmission length. The most important factors affecting the transmission length are:

(a) the degree of compaction of the concrete;
(b) the size and type of tendon;
(c) the strength of the concrete;

(d) the deformation and surface condition of the steel;

(e) sudden release of the tendon when the force is transferred to the concrete which in practice is to be avoided.

At release the prestress may be assumed to be transferred to the concrete by a constant bond stress (Eq. 8.15, EN)

$$f_{bpt} = \eta_{p1}\eta_1\, f_{ctd}(t)$$

It is important to distinguish the following lengths.

(1) The transmission length (l_{pt}) over which the prestressing force (P_o) from a pretensioned tendon is fully transmitted to the concrete, which is related to the type and diameter of tendon and concrete strength (Eq. 8.16, EN)

$$l_{pt} = \alpha_1\alpha_2\phi\sigma_{pmo}/f_{bpt} \tag{12.24}$$

The design value is either $0{,}8l_{pt}$ or $1{,}2l_{pt}$ whichever is less favorable for the effect considered.

(2) The dispersion length (l_{disp}) over which the concrete stresses gradually disperse to a linear distribution across the concrete section. For rectangular cross sections and straight tendons, situated near the bottom of a section, the dispersion length (Eq. 8.19, EN)

$$l_{disp} = (l_{pt}^2 + d^2)^{0{,}5} \tag{12.25}$$

(3) The anchorage length (l_{ba}) over which the ultimate tendon force (F_{pu}) in pretensioned members is fully transmitted to the concrete.

The start of the effective bond should take account of tendons purposely debonded at the ends and a neutralized zone in the case of sudden release. The existence of a transmission length in a pretensioned member means that the full prestressing force is not developed until the end of the transmission length, and consequently, over the transmission length, the bending and shear resistances are reduced.

Example 12.8   Anchorage for pretensioned tendons.

Design the pretensioned prestressed concrete beam shown in Fig. 12.14. The beam is simply supported over a span of 8 m and carries a uniformly distributed design load of 375 kN at the ultimate limit state. The beam has been designed to meet the requirements for bending. At transfer the concrete Grade is C25/30.

The critical section is at the end of the dispersion length.

Basic transmission length (Eq. 8.16, EN)

$$l_{pt} = \alpha_1\alpha_2\phi\sigma_{pmo}/f_{bpt} = 1{,}0 \times 0{,}25 \times 5 \times 1085/1{,}2 = 1130\,\text{mm}$$

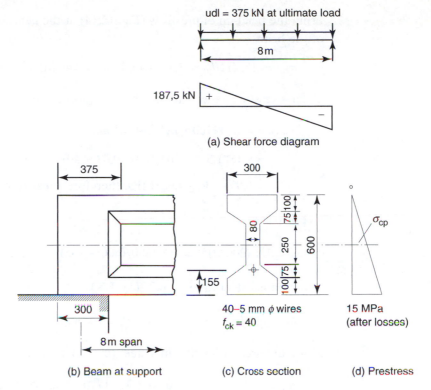

FIGURE 12.14 Example: end block for a pretensioned member

where (Eq. 8.15, EN)

$$f_{bpt} = \eta_{p1}\eta_2 f_{cdt}(t) = 2,7 \times 1,0 \times 0,7 \times 2,57/1,5 = 1,2\,\text{MPa}.$$

Design transmission length (Eq. 8.18, EN)

$$l_{pt} = 1,2 \times 1130 = 1356\,\text{mm}.$$

Dispersion length (Eq. 8.19, EN)

$$l_{disp} = (l_{pt}^2 + d^2)^{0,5} = (1356^2 + 445^2)^{0,5} = 1427\,\text{mm}.$$

Maximum shear force at the supports $V_{max} = W/2 = 375/2 = 187,5\,\text{kN}$

Reduced design shear stress at $d = 445\,\text{mm}$ from end of beam where $b_w = 80\,\text{mm}$ thickness (cl 6.2.1(8), EN)

$$v_{Ed} = (L/2 - d)/(L/2)V_{max}l(b_w d)$$

$$= (8\text{E}3/2 - 445)/(8\text{E}3/2) \times 187,5\text{E}3/(80 \times 445) = 4,68\,\text{MPa}$$

Area of cross section

$$A_c = 600 \times 300 - 220 \times 325 = 108,5\text{E}3\,\text{mm}^2$$

From the build-up of prestress (Fig. 12.14) at the neutral axis 445 mm from the end of the beam

$$\sigma_{cp} = (15/2)d/l_{pt} = 7,5 \times 445/1356 = 2,46 \text{ MPa}$$

Shear stress resistance without shear reinforcement (Eq. 6.2a, EN)

$$v_{Rd,c} = C_{Rd,c}k(100\rho_1 f_{ck})^{1/3} + 0,15\sigma_{cp}$$

$$= 0,18/1,5 \times 1,204(100 \times 0,02 \times 40)^{1/3} + 0,15 \times 2,46$$

$$= 0,991 < V_{Ed}(4,68) \text{ MPa, therefore shear reinforcement required.}$$

Depth factor

$$k = 1 + (200/d)^{0,5} = 1 + (200/(600 - 155))^{0,5} = 1,204 < 2$$

Longitudinal steel ratio (cl 6.2.2(1), EN)

$$\rho_1 = 785/(80 \times 445) = 0,022 > 0,02. \text{ Use } \rho_1 = 0,02.$$

Minimum area of longitudinal steel (cl 9.2.1.1, EN)

$$\rho_{l,min} = 0,26 f_{ctm}/f_{yk} = 0,26 \times 3,5/1770 = 0,000514 < \rho_l(0,02), \text{ acceptable.}$$

Minimum shear resistance Eq. (6.2b, EN)

$$(v_{min} + k_1\sigma_{cp}) = 0,035k^{1,5}f_{ck}^{0,5} + 0,15\sigma_{cp}$$

$$= 0,035 \times 1,204^{1,5} \times 40^{0,5} + 0,15 \times 2,46$$

$$= 0,661 < v_{Rd,c}(0,991) \text{ MPa, acceptable.}$$

Shear reinforcement to resist all of the shear force using T10 single vertical stirrups

$$A_{sw}/s_l = (v_{Ed}b_wd)/(0,9d f_{ywd}) = 4,68 \times 80/(0,9 \times 0,8 \times 500) = 1,04$$

$$s_l = 2 \times 79/1,04 = 151,9 \text{ mm. Use } s_l = 140 \text{ mm.}$$

Maximum shear resistance (Eq. 6.9, EN)

$$v_{Rd,max} = \alpha_{cw}b_w0,9d\, v_1 f_{cd}/[(\cot\theta + \tan\theta)b_wd]$$

$$= 1,09 \times 0,9 \times 0,6 \times (40/1,5)/2 = 7,85 > v_{Ed}(4,68) \text{ MPa, acceptable.}$$

where

$$0 > \sigma_{cp}(2,46) < 0,25 f_{cd}(6,67) \text{ MPa}$$

$$\alpha_{cw} = 1 + \sigma_{cp}/f_{cd} = 1 + 2,46/(40/1,5) = 1,09$$

Shear reinforcement ratio of single T10 vertical stirrups (Eq. 9.4, EN)

$$\rho_w = A_{sw}/(s_1 b_w \sin \alpha) = 2 \times 79/(140 \times 80 \times 1) = 0{,}0141$$

Minimum shear reinforcement (Eq. 9.5, EN)

$$\rho_{w,\min} = 0{,}08 f_{ck}^{0,5}/f_{yk} = 0{,}08 \times 40^{0,5}/500 = 0{,}001 < \rho_w(0{,}0141), \text{ acceptable.}$$

Maximum longitudinal spacing of stirrups (Eq. 9.6, EN)

$$s_{l,\max} = 0{,}75 d(1 + \cot \alpha) = 0{,}75 \times 445 \times 1 = 334 > s_l(140) \text{ mm, acceptable.}$$

Check shear stress failure in web (cl 12.6.3, EN)

$$\tau_{cp} \leq f_{cvd}$$

Check which equation from the European Code is applicable

$$\sigma_{cp} = 2{,}46 \text{ MPa}$$
$$\sigma_{c,\lim} = f_{cd} - 2[f_{ctd}(f_{ctd} + f_{cd})]^{0,5}$$
$$= 40/1{,}5 - 2[(3{,}5/1{,}5)(3{,}5/1{,}5 + 40/1{,}5)]^{0,5} = 10{,}2 \text{ MPa}$$

$\sigma_{cp}(2{,}46) < \sigma_{c,\lim}(10{,}2)$ MPa, therefore use principal tensile stress criterion (Eq. 12.5, EN).

Concrete design stress in shear and compression

$$f_{cvd} = (f_{ctd}^2 + \sigma_{cp}f_{ctd})^{0,5} = ((3{,}5/1{,}5)^2 + 2{,}46 \times (3{,}5/1{,}5))^{0,5} = 3{,}35 \text{ MPa}$$

Shear stress in web of beam (Eq. 12.4, EN)

$$\tau_{cp} = 1{,}5(L/2 - d)/(L/2)V_{\max}/A_c$$

$$= 1{,}5(8E3/2 - 445)/(8E3/2) \times 187{,}5E3/108{,}5E3$$

$$= 2{,}30 < f_{cvd}(3{,}35) \text{ MPa, acceptable.}$$

Compression reinforcement (Eq. 6.18, EN)

$$A_{sl} = 0{,}5 v_{Ed} b_w d(\cot \theta - \cot \alpha)/f_{yd}$$

$$= 0{,}5 \times 4{,}68 \times 80 \times 445/(500/1{,}15) = 191{,}6 \text{ mm}^2. \text{ Use } 2T12 = 226 \text{ mm}^2.$$

12.4.3 Uneven Distribution of Prestress

Cracks sometimes occur in the ends of pretensioned members, which are not rectangular in cross section, as shown in Fig. 12.15. This defect is likely to occur

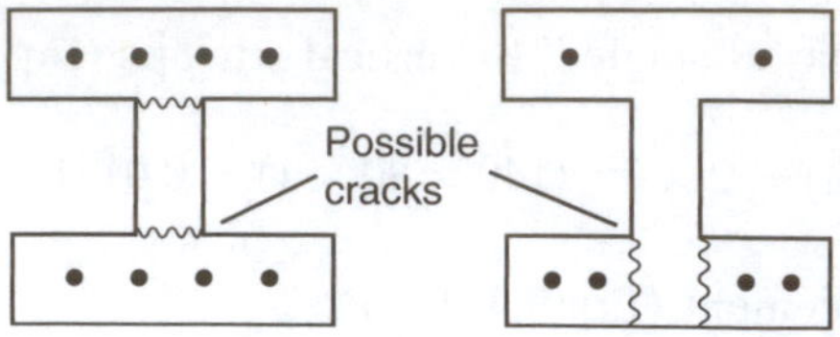

FIGURE 12.15 Cracks in the ends of pretensioned members

if there is an uneven distribution of prestress over the depth and width of a section, and also if there is not rectangular end block. The introduction of horizontal and vertical links ensures that, even if these cracks occur, their development is restricted.

REFERENCES AND FURTHER READING

BS 4486 (1980) *Specification for hot rolled and processed high tensile alloy steel bars for the prestressing of concrete*, BSI.

BS 4757 (1971) *Nineteen-wire steel strand for prestressed concrete*, BSI.

BS 5896 (1980) *Specification for high tensile steel wire strand for the prestressing of concrete*, BSI.

BS 8110 (1985) *Structural use of concrete*, BSI.

CIRIA Guide 1 (1976) *A guide to the design of anchor blocks for post-tensioned prestressed concrete members*. Construction Industry Research and Information Association Publication, 44.

CIRIA Report 74 (1978) *Prestressed concrete, friction losses during stressing*. Construction Industry Research and Information Association Publication, 52.

Neville A.M. (1995) *Properties of Concrete*. Longman.

Cross-sectional area of groups of bars (mm^2).

Bar size (mm)		6	8	10	12	16	20	25	32	40
Number of bars	1	28	50	79	113	201	314	491	804	1256
	2	57	101	157	226	402	628	982	1608	2513
	3	85	151	236	339	603	942	1473	2413	3770
	4	113	201	314	452	804	1257	1964	3217	5026
	5	141	251	393	565	1005	1571	2455	4021	6283
	6	170	302	471	679	1206	1885	2945	4825	7540
	7	198	352	550	792	1407	2199	3436	5630	8796
	8	226	402	628	905	1608	2533	3927	6434	10050
	9	254	452	707	1018	1809	2827	4418	7238	11310
	10	283	503	785	1131	2011	3142	4909	8042	12570
Circumference		18,9	25,1	31,4	37,7	50,3	62,8	78,5	100,5	125,7

Cross-sectional area of bars per metre width (mm^2).

Bar size (mm)		6	8	10	12	16	20	25	32
Pitch of bars	60	471	837	1309	1884	3350	5234	8181	13404
	80	353	628	982	1414	2512	3926	6136	10053
	100	283	503	785	1131	2011	3141	4909	8042
	120	236	419	654	942	1674	2617	4091	6702
	140	202	359	561	808	1435	2244	3506	5745
	150	189	335	524	754	1338	2094	3273	5362
	160	177	314	491	707	1256	1964	3068	5027
	180	157	279	436	627	1115	1746	2727	4468
	200	141	251	393	565	1005	1572	2454	4021
	250	113	201	314	452	804	1257	1964	3217
	300	95	168	262	377	669	1047	1636	2681

Strength and deformation characteristics for concrete.

	Strength classes for concrete														Analytical relation/ Explanation
f_{ck} (MPa)	12	16	20	25	30	35	40	45	50	55	60	70	80	90	
$f_{ck,cube}$ (MPa)	15	20	25	30	37	45	50	55	60	67	75	85	95	105	
f_{cm} (MPa)	20	24	28	33	38	43	48	53	58	63	68	78	88	98	$f_{cm} = f_{ck} + 8$ (MPa)
f_{ctm} (MPa)	1,6	1,9	2,2	2,6	2,9	3,2	3,5	3,8	4,1	4,2	4,4	4,6	4,8	5,0	$f_{ctm} = 0,30 \times f_{ck}^{(2/3)} \leq$ C50/60 $f_{ctm} = 2,12 \cdot \ln(1 + (f_{cm}/10)) >$ C50/60
$f_{ctk,0,05}$ (MPa)	1,1	1,3	1,5	1,8	2,0	2,2	2,5	2,7	2,9	3,0	3,1	3,2	3,4	3,5	$f_{ctk;0,05} = 0,7 \times f_{ctm}$ 5% fractile
$f_{ctk,0.95}$ (MPa)	2,0	2,5	2,9	3,3	3,8	4,2	4,6	4,9	5,3	5,5	5,7	6,0	6,3	6,6	$f_{ctk;0,95} = 1,3 \times f_{ctm}$ 95% fractile
E_{cm} (GPa)	27	29	30	31	33	34	35	36	37	38	39	41	42	44	$E_{cm} = 22[(f_{cm})/10]^{0,3}$ (f_{cm} in MPa)
ε_{c1} (‰)	1,8	1,9	2,0	2,1	2,2	2,25	2,3	2,4	2,45	2,5	2,6	2,7	2,8	2,8	see Figure 3.2 $\varepsilon_{c1}(‰) = 0,7\, f_{cm}^{0,31} < 2,8$
ε_{cu1} (‰)				3,5						3,2	3,0	2,8	2,8	2,8	see Figure 3.2 for $f_{ck} \geq 50$ MPa $\varepsilon_{cu1}(‰) = 2,8 + 27\,[(98 - f_{cm})/100]^4$
ε_{c2} (‰)				2,0						2.2	2,3	2,4	2,5	2,6	see Figure 3.3 for $f_{ck} \geq 50$ MPa $\varepsilon_{c2}(‰) = 2,0 + 0,085\,(f_{ck} - 50)^{0,53}$
ε_{cu2} (‰)				3,5						3,1	2,9	2,7	2,6	2,6	see Figure 3.3 for $f_{ck} \geq 50$ MPa $\varepsilon_{cu2}(‰) = 2,6 + 35\,[(90 - f_{ck})/100]^4$
n				2,0						1,75	1,6	1,45	1,4	1,4	for $f_{ck} \geq 50$ MPa $n = 1,4 + 23,4\,[(90 - f_{ck})/100]^4$
ε_{c3} (‰)				1,75						1,8	1,9	2,0	2,2	2,3	see Figure 3.4 for $f_{ck} \geq 50$ MPa $\varepsilon_{c3}(‰) = 1,75 + 0,55\,[(f_{ck} - 50)/40]$
ε_{cu3} (‰)				3,5						3,1	2,9	2,7	2,6	2,6	see Figure 3.4 for $f_{ck} \geq 50$ MPa $\varepsilon_{cu3}(‰) = 2,6 + 35\,[(90 - f_{ck})/100]^4$

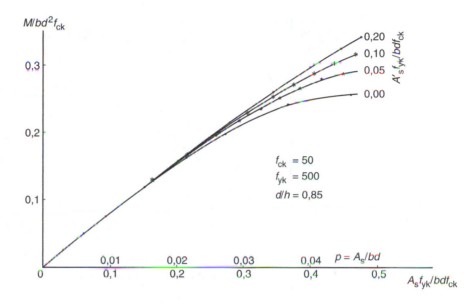

Design graph for doubly reinforced beams

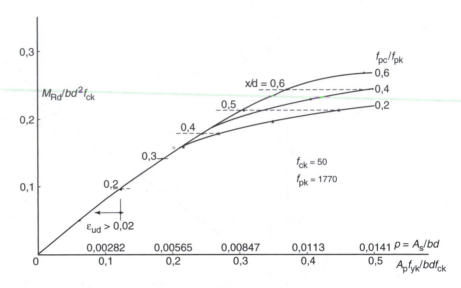

Design graph for prestressed beams (bonded tendons)

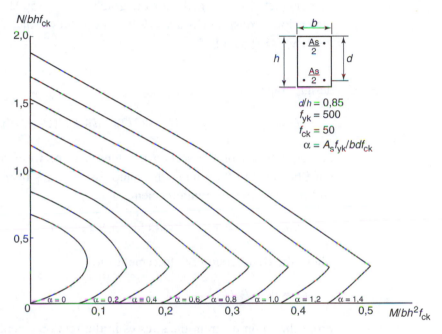

Design graph for a rectangular column

Annex B4 / Derivation of Design Graphs

ANNEX B4-1 DESIGN GRAPH FOR A DOUBLY REINFORCED BEAM

A typical design graph for a doubly reinforced beam is shown in Annex B1. The characteristic strengths for the graph are $f_{ck} = 50$ MPa and $f_{yk} = 500$ MPa. The dimensional constraint is $d'/d = 0{,}15$. The value of the design factors are: $\lambda = 0{,}8$; $\eta = 1$; $\gamma_c = 1{,}5$; $\gamma_s = 1{,}15$.

The ultimate moment of resistance in the dimensionless form required for a design graph

$$M/bd^2f_{ck} = (\eta\lambda/\gamma_c)(x/d)[1 - (\lambda/2)(x/d)] + \alpha'(1 - d'/d)f_s'/f_{yk} \tag{B4-1.1}$$

For equilibrium of the section the tensile force from the tension reinforcement must be equal to the sum of the compressive forces from the concrete stress block and the compression reinforcement.

$$(A_s f_{yk}/bd f_{ck}) = (\eta\lambda/\gamma_c)(x/d)(f_{yk}/f_s) + \alpha'(f_s'/f_s) \tag{B4-1.2}$$

where the compression reinforcement ratio

$$\alpha' = A_s' f_{yk}/bd f_{ck}$$

From the strain diagram the stresses in the reinforcement (compression negative)

$$f_s = \varepsilon_c E_s(1 - x/d)/(x/d) \geq -f_{yk}/\gamma_s \tag{B4-1.3}$$

$$f_s' = \varepsilon_c E_S(x/d - d'/d)/(x/d) \leq f_{yk}/\gamma_s \tag{B4-1.4}$$

Graphs are plotted by evaluating Eqs (B4-1.1) and (B4-1.2) for a series of values of x/d. Particular values of α' are each represented by a separate line, and a separate graph is required for each value of d'/d.

ANNEX B4-2 DESIGN GRAPH FOR A PRESTRESSED BEAM

The equations for a doubly reinforced beam can be modified for prestressed concrete beams with $\alpha' = 0$. The steel stress is increased by the prestress and limited to $f_{pk}/\gamma_c = 1770/1,15$ MPa. The steel strain is also increased by the prestress and limited to 0,02.

ANNEX B4-3 DESIGN GRAPH FOR A RECTANGULAR COLUMN

A typical design graph for a column is shown in Annex B3. The characteristic strength for the column graph are $f_{ck} = 50$ MPa and $f_{yk} = 500$ MPa. The dimensional constraint is $d/h = 0,85$ and the values of the design factors are: $\lambda = 0,8$; $\eta = 1$; $\gamma_c = 1,5$; $\gamma_s = 1,15$.

From equilibrium of axial forces on a column in dimensionless form

$$N/bhf_{ck} = (\eta\lambda/\gamma_c)(x/h) + n\alpha \qquad (B4\text{-}3.1)$$

From equilibrium of bending moments acting about an axis at mid-height of the section

$$M/bh^2f_{ck} = (\eta\lambda/\gamma_c)(x/h)[0,5 - (\lambda/2)x/h] + m\alpha \qquad (B4\text{-}3.2)$$

where
$$n = (f_1 + f_2)/(2f_{yk})$$
$$m = (2d/h - 1)(f_1 - f_2)/(4f_{yk})$$
$$\alpha = A_{sc}f_{yk}/bhf_{ck}$$

Stresses in the steel

$$f_1 = \varepsilon_c E_s(x/h - 1 + d/h)/(x/h) \leq \pm f_{yk}/\gamma_s \qquad (B4\text{-}3.3)$$

$$f_2 = \varepsilon_c E_s(x/h - d/h)/(x/h) \leq \pm f_{yk}/\gamma_s \qquad (B4\text{-}3.4)$$

The case when $\lambda x \geq h$, i.e. when the concrete stress block covers the whole section, must also be considered. In this case f_1 is equal to the design strength in compression and there is no moment from the stress block and hence Eqs (B4-3.1) and (B4-3.2) become

$$N/bhf_{ck} = \eta/\gamma_c + n\alpha \qquad (B4\text{-}3.5)$$

$$M/bh^2f_{ck} = m\alpha \qquad (B4\text{-}3.6)$$

where
$$n = (1 + \gamma_2 f_2/f_{yk})/(2\gamma_s)$$
$$m = (2d/h - 1)(1 - \gamma_s f_2/f_{yk})/(4\gamma_s)$$

This annex details analysis results from simple subframes which may used for preliminary analyses of frame structures. They are now less used for the final analysis owing to the widespread availability of commercial software. Also it should be noted that they take no account of the co-existent horizontal forces that must be applied as part of the load cases involving vertical loads.

The results given here have been obtained using slope-deflection with the following sign convention: clockwise positive for moments and rotations. Settlement of supports has not been considered.

For both ends fixed, the moments at either end M_{AB} and M_{BA} are given by

$$M_{AB} = \frac{2EI}{L_{AB}}(2\theta_A + \theta_B) + M_{12} \qquad (C1)$$

and

$$M_{BA} = \frac{2EI}{L_{AB}}(2\theta_B + \theta_A) + M_{21} \qquad (C2)$$

where EI/L is the stiffness of the member, θ_A and θ_B are the rotations at either end and M_{12} and M_{21} are the fixed end moments at ends A and B, respectively due to the loading applied to the member. The fixed end moments for a partial UDL, a point load and equal end triangular loading are given in Fig. C1.

If one end is pinned (for example, end B, then $M_{BA}=0$), then M_{AB} becomes,

$$M_{AB} = \frac{3EI}{L_{AB}}\theta_A + M_{12} - \frac{M_{12}}{2} \qquad (C3)$$

A similar expression can be derived for end A pinned.

For convenience, the stiffness EI/L has been denoted as k (the full value of k has been used in all the derivations).

Subframe 1: Single beam and column (Fig. C2):

a) End F fixed

$$M_{AB} = -M_{AF} = -\frac{k_{CL}}{k_{B1} + k_{CL}}M_{11} \qquad (C4)$$

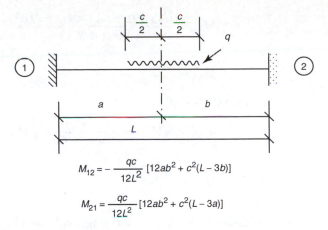

$$M_{12} = -\frac{qc}{12L^2}[12ab^2 + c^2(L - 3b)]$$

$$M_{21} = \frac{qc}{12L^2}[12ab^2 + c^2(L - 3a)]$$

a) Partial UDL

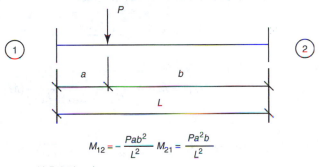

$$M_{12} = -\frac{Pab^2}{L^2} \quad M_{21} = \frac{Pa^2b}{L^2}$$

b) Point Load

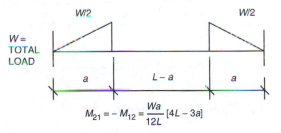

$$M_{21} = -M_{12} = \frac{Wa}{12L}[4L - 3a]$$

c) End triangular loads

FIGURE C1 Fixed end moments

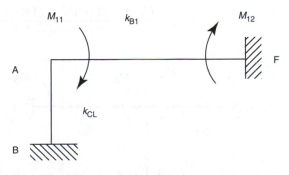

FIGURE C2 Subframe 1 (Single beam, single column)

$$M_{FA} = \frac{2k_{CL}M_{12} + k_{B1}(2M_{12} - M_{11})}{2(k_{B1} + k_{CL})} \tag{C5}$$

$$M_{BA} = \frac{M_{AB}}{2} \tag{C6}$$

b)   End F pinned:

$$M_{AB} = -M_{AF} = -\frac{4k_{CL}}{3k_{B1} + 4k_{CL}}\left(M_{11} - \frac{M_{12}}{2}\right) \tag{C7}$$

$$M_{BA} = \frac{M_{AB}}{2} \tag{C8}$$

Subframe 2: Two beams and single column (Fig. C3):

a)   Ends A and L of beam fixed:

$$M_{FA} = M_{12} - \frac{k_{b1}}{k_{B1} + k_{B2} + k_{CL}}(M_{12} + M_{21}) \tag{C9}$$

$$M_{AF} = M_{11} - \frac{k_{B1}}{2(k_{B1} + k_{B2} + k_{CL})}(M_{12} + M_{21}) \tag{C10}$$

$$M_{FL} = M_{21} - \frac{k_{B2}}{k_{B1} + k_{B2} + k_{CL}}(M_{12} + M_{21}) \tag{C11}$$

$$M_{LF} = M_{22} - \frac{k_{B2}}{2(k_{B1} + k_{B2} + k_{CL})}(M_{12} + M_{21}) \tag{C12}$$

$$M_{FG} = 2M_{GF} = -\frac{k_{CL}}{k_{B1} + k_{B2} + k_{CL}}(M_{12} + M_{21}) \tag{C13}$$

b)   Ends A and L pinned:

$$M_{FL} = \frac{(3k_{B1} + 4k_{CL})[M_{21} - (M_{22}/2)] - 3k_{B2}[M_{12} - (M_{11}/2)]}{3k_{B1} + 3k_{B2} + 4k_{CL}} \tag{C14}$$

$$M_{FA} = \frac{(3k_{B2} + 4k_{CL})[M_{12} - (M_{11}/2)] - 3k_{B1}[M_{21} - (M_{22}/2)]}{3k_{B1} + 3k_{B2} + 4k_{CL}} \tag{C15}$$

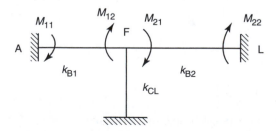

FIGURE C3   Subframe 2 (Two beams, single column)

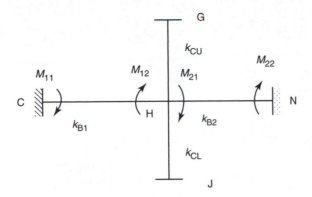

FIGURE C4 Subframe 3 (Two beams, two columns)

$$M_{FG} = 2M_{GF} = \frac{4\,k_{CL}}{3k_{b1} + 3k_{B2} + 4k_{CL}}\left(\frac{M_{22} + M_{11}}{2} - M_{21} - M_{12}\right) \qquad \text{(C16)}$$

Subframe 3: Column above aand below beam either side (Fig. C4):

a) Ends C and N fixed:

$$M_{HC} = -\frac{k_{B1}}{k_{B1} + k_{B2} + k_{CL} + k_{CU}}(M_{12} + M_{21}) + M_{12} \qquad \text{(C17)}$$

$$M_{CH} = -\frac{k_{B1}}{2(k_{B1} + k_{B2} + k_{CL} + k_{CU})}(M_{12} + M_{21}) + M_{11} \qquad \text{(C18)}$$

$$M_{HN} = -\frac{k_{B2}}{k_{B1} + k_{B2} + k_{CL} + k_{CU}}(M_{12} + M_{21}) + M_{21} \qquad \text{(C19)}$$

$$M_{NH} = -\frac{k_{B2}}{2(k_{B1} + k_{B2} + k_{CL} + k_{CU})}(M_{12} + M_{21}) + M_{12} \qquad \text{(C20)}$$

$$M_{HG} = 2M_{GH} = -\frac{k_{CU}}{k_{B1} + k_{B2} + k_{CL} + k_{CU}}(M_{12} + M_{21}) \qquad \text{(C21)}$$

$$M_{HJ} = 2M_{JH} = -\frac{k_{CL}}{k_{B1} + k_{B2} + k_{CL} + k_{CU}}(M_{12} + M_{21}) \qquad \text{(C22)}$$

b) Ends C and N pinned:

$$M_{HN} = \frac{(3k_{B1} + 4k_{CL} + 4k_{CU})[M_{21} - (M_{22}/2)] - 3k_{B2}[M_{12} - (M_{11}/2)]}{3k_{B1} + 3k_{B2} + 4k_{CL} + 4k_{CU}} \qquad \text{(C23)}$$

$$M_{HC} = \frac{(3k_{B2} + 4k_{CL} + 4k_{CU})[M_{12} - (M_{11}/2)] - 3k_{B1}[M_{21} - (M_{22}/2)]}{3k_{B1} + 3k_{B2} + 4k_{CL} + 4k_{CU}} \qquad \text{(C24)}$$

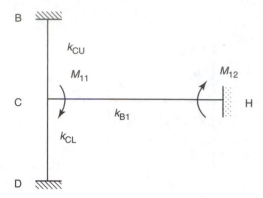

FIGURE C5   Subframe 4 (single beam, two columns)

$$M_{HG} = 2M_{GH} = \frac{4k_{CU}}{3k_{B1} + 3k_{B2} + 4k_{CU} + 4k_{CL}}\left(\frac{M_{22} + M_{11}}{2} - M_{21} - M_{12}\right) \quad \text{(C25)}$$

$$M_{HJ} = 2M_{JH} = \frac{4k_{CL}}{3k_{B1} + 3k_{B2} + 4k_{CU} + 4k_{CL}}\left(\frac{M_{22} + M_{11}}{2} - M_{21} - M_{12}\right) \quad \text{(C26)}$$

Subframe 4: Single beam column above and below (Fig. C5):

a) End H fixed:

$$M_{CH} = \frac{k_{CL} + k_{CU}}{k_{B1} + k_{CL} + k_{CU}}M_{11} \quad \text{(C27)}$$

$$M_{CH} = M_{12} - \frac{k_{B1}}{2(k_{B1} + k_{CL} + k_{CU})}M_{11} \quad \text{(C28)}$$

$$M_{CD} = 2M_{DC} = -\frac{k_{CL}}{k_{B1} + k_{CL} + k_{CU}}M_{11} \quad \text{(C29)}$$

$$M_{CB} = 2M_{BC} = -\frac{k_{CU}}{k_{B1} + k_{CL} + k_{CU}}M_{11} \quad \text{(C30)}$$

b) End H pinned:

$$M_{CH} = \frac{4k_{CL} + 4k_{CU}}{3k_{B1} + 4k_{CL} + 4k_{CU}}\left(M_{11} - \frac{M_{12}}{2}\right) \quad \text{(C31)}$$

$$M_{CD} = 2M_{DC} = -\frac{4k_{CL}}{3k_{B1} + 4k_{CL} + 4k_{CU}}\left(M_{11} - \frac{M_{12}}{2}\right) \quad \text{(C32)}$$

$$M_{CB} = 2M_{BC} = -\frac{4k_{CU}}{3k_{B1} + 4k_{CL} + 4k_{CU}}\left(M_{11} - \frac{M_{12}}{2}\right) \quad \text{(C33)}$$

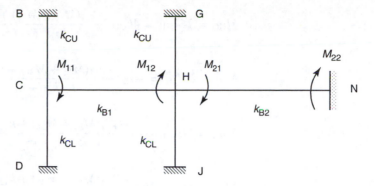

FIGURE C6   Subframe 5 (Two beams, two pairs of columns)

Subframe 5: Two beams, two pairs of columns with the end of one beam not fixed to a column (Fig. C6):

a)   End N fixed:

A complete solution will not be given. The slope deflection equations applied to joints C and H give

$$\begin{pmatrix} k_{11} & k_{12} \\ k_{12} & k_{22} \end{pmatrix} \begin{pmatrix} \theta_H \\ \theta_c \end{pmatrix} = \begin{pmatrix} -M_{11} \\ (-M_{12} - M_{21}) \end{pmatrix} \tag{C34}$$

where

$$k_{11} = 4(k_{CL} + k_{CU} + k_{B1}) \tag{C35}$$

$$k_{22} = 4(k_{CL} + k_{CU} + k_{B1} + k_{B2}) \tag{C36}$$

$$k_{12} = 2k_{B1} \tag{C37}$$

Solving Eq. (C34) in terms of k_{11}, k_{12} and k_{22} gives

$$\theta_C = \frac{k_{12}(M_{12} + M_{21}) - k_{22} M_{11}}{k_{22} k_{11} - k_{12}^2} \tag{C38}$$

and

$$\theta_H = \frac{-k_{11}(M_{12} + M_{21}) + k_{12} M_{11}}{k_{22} k_{11} - k_{12}^2} \tag{C39}$$

$$M_{CB} = 2M_{BC} = 4k_{CU} \frac{k_{12}(M_{12} + M_{21}) - k_{22} M_{11}}{k_{22} k_{11} - k_{12}^2} \tag{C40}$$

or

$$M_{CD} = 2 M_{DC} = 4k_{CL} \frac{k_{12}(M_{12} + M_{21}) - k_{22}M_{11}}{k_{22}k_{11} - k_{12}^2} \tag{C41}$$

$$M_{HG} = 2M_{GH} = 4k_{CU} \frac{-k_{11}(M_{12} + M_{21}) + k_{12}M_{11}}{k_{22}k_{11} - k_{12}^2} \tag{C42}$$

$$M_{HJ} = 2M_{JH} = 4k_{CL} \frac{-k_{11}(M_{12} + M_{21}) + k_{12}M_{11}}{k_{22}k_{11} - k_{12}^2} \tag{C43}$$

$$M_{HN} = M_{21} + 4k_{B2} \frac{-k_{11}(M_{12} + M_{21}) + k_{12}M_{11}}{k_{22}k_{11} - k_{12}^2} \tag{C44}$$

$$M_{NH} = M_{22} + 2k_{B2} \frac{-k_{11}(M_{12} + M_{21}) + k_{12}M_{11}}{k_{22}k_{11} - k_{12}^2} \tag{C45}$$

$$M_{CH} = M_{11} + 2k_{B1} \frac{M_{11}(k_{12} - 2k_{22}) + (M_{12} + M_{21})(2k_{12} - k_{11})}{k_{22}k_{11} - k_{12}^2} \tag{C46}$$

$$M_{HC} = M_{12} + 2k_{B1} \frac{M_{11}(2k_{12} - k_{22}) + (M_{12} + M_{21})(k_{12} - 2k_{11})}{k_{22}k_{11} - k_{12}^2} \tag{C47}$$

b) End N pinned:

$$\begin{pmatrix} k_{11} & k_{12} \\ k_{12} & K_{22} \end{pmatrix} \begin{pmatrix} \theta_C \\ \theta_H \end{pmatrix} = \begin{pmatrix} -M_{11} \\ (-M_{12} - M_{21} + (M_{22}/2)) \end{pmatrix} \tag{C48}$$

where

$$k_{11} = 4(k_{CL} + k_{CU} + k_{B1}) \tag{C49}$$

$$k_{22} = 4k_{CL} + 4k_{CU} + 4k_{B1} + 3k_{B2} \tag{C50}$$

$$k_{12} = 2k_{B1} \tag{C51}$$

Solving Eq. (C48) gives

$$\theta_C = \frac{k_{12}(M_{12} + M_{21} - (M_{22}/2)) - k_{22}M_{11}}{k_{22}k_{11} - k_{12}^2} \tag{C52}$$

and

$$\theta_H = \frac{-k_{11}(M_{12} + M_{21} - (M_{22}/2)) + k_{12}M_{11}}{k_{22}k_{11} - k_{12}^2} \tag{C53}$$

or

$$M_{CB} = 2M_{BC} = 4k_{CU} \frac{k_{12}(M_{12} + M_{21} - (M_{22}/2)) - k_{22}M_{11}}{k_{22}k_{11} - k_{12}^2} \tag{C54}$$

$$M_{CD} = 2M_{DC} = 4k_{CL} \frac{k_{12}(M_{12} + M_{21} - (M_{22}/2)) - k_{22}M_{11}}{k_{22}k_{11} - k_{12}^2} \tag{C55}$$

$$M_{HG} = 2M_{GH} = 4k_{CU} \frac{-k_{11}(M_{12} + M_{21} - (M_{22}/2)) + k_{12}M_{11}}{k_{22}k_{11} - k_{12}^2} \tag{C56}$$

$$M_{HJ} = 2M_{JH} = 4k_{CL} \frac{-k_{11}(M_{12} + M_{21} - (M_{22}/2)) + k_{12}M_{11}}{k_{22}k_{11} - k_{12}^2} \tag{C57}$$

$$M_{HN} = M_{21} - \frac{M_{22}}{2} + 4k_{B2} \frac{-k_{11}(M_{12} + M_{21} - (M_{22}/2)) + k_{12}M_{11}}{k_{22}k_{11} - k_{12}^2} \tag{C58}$$

$$M_{CH} = M_{11} + 2k_{B1} \frac{M_{11}(k_{12} - 2k_{22}) + (M_{12} + M_{21} - (M_{22}/2)) + (2k_{12} - k_{11})}{k_{22}k_{11} - k_{12}^2} \tag{C59}$$

$$M_{HC} = M_{12} + k_{B1} \frac{M_{11}(2k_{12} - k_{22}) + (M_{12} + M_{21} - (M_{22}/2)) + (k_{12} - 2k_{11})}{k_{22}k_{11} - k_{12}^2} \tag{C60}$$

Subframe 6: Four columns, three beams, symmetric about centre line.

This is the same as case (a) Subframe 5 with the stiffness of the centre beam halved to allow for symmetry (FEM's must still be calculated on full span).

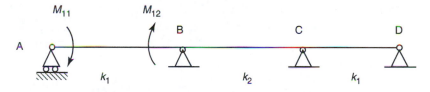

a) Span AB Loaded

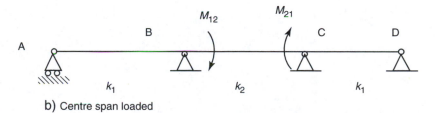

b) Centre span loaded

FIGURE C7   Continuous beam

Subframe 7: Continuous beam.

The only case which will be considered here is a three span continuous beam with the end spans having stiffness k_1 and the centre span having stiffness k_2.

Case 1: Span AB loaded (Fig. C7 (a)):

$$M_{CB} = -M_{CD} = \frac{6k_1k_2}{(4k_2 + 3k_1)^2 - 4k_2^2}[(M_{11}/2) - M_{12}] \tag{C61}$$

$$M_{BC} = -M_{BA} = \frac{12k_2(k_1 + k_2)}{(4k_2 + 3k_1)^2 - 4k_2^2}[(M_{11}/2) - M_{12}] \tag{C62}$$

Case 2: Centre span loaded (Fig. C7 (b)):

$$M_{BA} = M_{CB} = -M_{BC} = -M_{CD} = \frac{3k_1}{(4k_2 + 3k_1)^2 - 4k_2^2}[2M_{22}k_2 - M_{21}(4k_2 + 3k_1)] \tag{C63}$$

Index